Information and Experimental Knowledge

Information and Experimental Knowledge

JAMES MATTINGLY

The University of Chicago Press
Chicago and London

The University of Chicago Press, Chicago 60637
The University of Chicago Press, Ltd., London
© 2021 by The University of Chicago
Published 2021
Printed in the United States of America

30 29 28 27 26 25 24 23 22 21   1 2 3 4 5

ISBN-13: 978-0-226-80464-4 (cloth)
ISBN-13: 978-0-226-80481-1 (paper)
ISBN-13: 978-0-226-80478-1 (e-book)
DOI: https://doi.org/10.7208/chicago/9780226804781.001.0001

Library of Congress Cataloging-in-Publication Data

Names: Mattingly, James, author.
Title: Information and experimental knowledge / James Mattingly.
Description: Chicago ; London : The University of Chicago Press, 2021. |
    Includes bibliographical references and index.
Identifiers: LCCN 2021016878 | ISBN 9780226804644 (cloth) | ISBN 9780226804811
    (paperback) | ISBN 9780226804781 (ebook)
Subjects: LCSH: Science—Experiments—Philosophy. | Knowledge, Theory of. |
    Information theory.
Classification: LCC Q175.32.K45 M324 2021 | DDC 507.2/4—dc23
LC record available at https://lccn.loc.gov/2021016878

♾ This paper meets the requirements of ANSI/NISO Z39.48-1992
(Permanence of Paper).

This book is dedicated to my very dear Natalia, truly sine qua non.

Contents

Introduction

This is a book about experimental knowledge. My main concern is a conceptual one: what is experimental knowledge? I believe that it is not possible to answer in general the question "what is a scientific experiment?" in any other way than by saying "it's what gives experimental knowledge"; so I think in order to understand what experiments are we need first to begin with experimental knowledge. It is easy to define scientific experiments, as Wootton does, by saying that an experiment is "an artificial test designed to answer a question" (Wootton 2015, 348). The problem with this definition is in the words "artificial" and "designed" and the fact that we don't really know what "test" means in any robust way that will discriminate between experimentation and observation. But discriminating between those two concepts is crucial. It would be a peculiar concept of experiment, for example, that would make counting the coins in my pocket to find out how much money I have, or glancing down to see what is in my hand, turn out to be experiments, but this definition seems to do just that. I do not mean to say that Wootton is wrong about experiments being tests, but not every test is an experiment.

That definition and others like it centralize the idea that experimenting is a kind of doing, but they do so by emphasizing the aims and desires of human agents. Experimenting is a kind of doing, but a doing that is centrally a method for coming to know. So I focus on the capacity of various activities to generate knowledge, rather than on the aims and desires of experimenters. I propose that the key distinction between experiment and mere observation is that the former generates knowledge about things *not* observed by way of the observation of other things. Thus I define experimental knowledge as the knowledge generated about something by observing something else. As simple as it may seem, this definition will support a great deal of unpacking.

This is a quite expansive definition of experiment and experimental knowledge, but not inappropriately expansive. I believe, and will argue throughout this book, that it exactly captures the key conceptual features of experimentation. It will include in the category of experiments many things that are not normally counted as such, or are only counted begrudgingly. I embrace their inclusion. The rest of the book, like so many other philosophy books, is mainly directed toward elaborating and defending my definition. (And as in other philosophy books, there will also be plenty of attention paid to knocking others' views.)

Part I details what is wrong with our current conception of how experiments give us knowledge. In this part I direct my attention to three aspects of experimental practice: calibration, intervention, and replication. Each of these activities is in its own way central both to the experimental enterprise and to the way we have misunderstood how that enterprise functions. Over the course of the discussion of those aspects of experimentation and my negative account of current views of experimental knowledge, I will also be developing the outline of my own positive story of the role of information flow in generating experimental knowledge.

In part II I will fill in that outline by explicitly presenting an epistemology appropriate for experimentation and applying that epistemology to an account of experimentation. This will all be done in terms of information flow.

Finally, in part III I will use my account of experimentation to analyze a number of types of thing that, I claim, are as robustly experimental as the standard laboratory experiment discussed in introductions to the scientific method. My account unifies all of these practices under the concept of experimentation, shows that they all provide robustly experimental knowledge, and identifies the peculiarities of each one, especially the information pinchpoints that are characteristic of that type. I conclude that all experimentation is some breed or other of analogical experimentation, so the fact that some practice is analogical cannot as such be grounds for suspicion about its knowledge claims—such suspicion must always be grounded in questions about whether or not information can and did flow in the experiment.

Isn't This All Obvious?

Understanding experimental knowledge may seem easy, but I am going to talk about it because I think it is misunderstood. We will see that experimental knowledge is easy in the way other forms of knowledge are easy: given the right account, they *are* easy to understand. I do not say that it is easy to *get* experimental knowledge; just that it is easy to understand. Even so, I do not think anyone to date has said quite what it is. The trouble in many cases comes from not distinguishing it clearly enough from *other* related things, such as *knowing whether* something is experimental knowledge or knowing *how to get* experimental knowledge. These things are arguably every bit as important as experimental knowledge itself, and I do not have very much to say about them. But as important as they are, it is necessary first to distinguish them from experimental knowledge, else we will not understand any of them. I will distinguish them by clearly explaining what experimental knowledge is. Of course, even with that as a background conceptual framework the other issues will not automatically be fully explained; but it will be clear, against that background, how they are related to experimental knowledge, and what needs to be done in order to explain them more fully—a task for a different book.

What can experimentation tell us about the world? On its own, no experiment can tell us very much. Walk into a lab, perform some manipulations, inscribe some figures on a sheet of paper (or make a graph, or store some data in a computer), and those actions will not constitute experimental knowledge (though they may well describe the entire course of an experiment). Latour and Woolgar in *Laboratory Life* were right about this much: the activities one does in the lab do not by themselves add up to experimental knowledge. And yet it does appear that very often scientists go into the laboratory and then emerge with more knowledge than they had going in. Something that happens in there allows them to generate experimental knowledge.

Experimentation turns out to be a lot like other kinds of knowledge acquisition. (How could it not?) The main role for experimentation in the context of producing scientific knowledge is to provide general claims generated from finite data that can then support myriad other general (and specific) claims. Some important examples: calibrating a device by checking that it is stably generating claims that accord with our background knowledge of the system to which these claims pertain, which can then be used to support the inference that the device is functioning properly; providing subpopulation data that can be used to support inferences about the total population; providing population-level data that can be used to support inferences about other, related populations; providing statistical information about a system that can be used to make predictions about that (or a related) system in the future, or for making retrodictions about it in the past.

I am going to begin my overall argument with two obvious points: (1) we have experimental knowledge; and (2) that knowledge is ineluctably inductive. These points really are obvious. But the consequences are not so easy to see. Here is one direct consequence to which I will return repeatedly:

experimental knowledge cannot be justified, for induction (its foundation) cannot be justified. We must, and should, use induction, but we cannot do so responsibly if we continue to think that knowledge involves justification.

History seems to have shown us that the best way to find out about the world, to gain generalizable, counterfactual supporting knowledge of nature, is through experiment. But how do we get experimental knowledge? What is it, and what kinds of activity count as generating it? This book is an extended answer to these questions, one that follows from trying to make sense of my conviction that more counts as experimentation than just what goes on in the laboratory, or what activities are undertaken involving direct interventions into nature. I am convinced that thought experiments, observational studies of various sorts, simulations, and natural experiments all count as robustly and legitimately productive of experimental knowledge. To experiment is to make trial, which, I will argue, is to urge information of just the right sort to flow to us from the world. This information is about systems in the world generally, and yet it is captured by attending to specific, isolated systems.

Experimental knowledge is not peculiar to the sciences; it is part and parcel of our everyday engagement with the world. We are in constant interaction with various worldly systems, and by that interaction we come to know many properties of these systems. When we then exploit the information borne by those properties about yet other systems to come to know their properties in turn, I say we have experimental knowledge of the latter. What separates the scientific experiment from other means of generating knowledge is the array of techniques that are brought self-consciously (or institutionally?) to bear both on the gathering of observational information about proximal systems and on the use of that information to extract further information about distal systems (by treating it as a source of signals bearing information about those distal systems). The application of these methods serves to give us assurance that in particular cases we really do have information flow from the distal systems that are our concern. It does not, however, distinguish conceptually between the two knowledge-generating enterprises. Experimental knowledge in the sciences is not *better* knowledge than that in everyday life, since it is in fact the same stuff; the techniques of the sciences are simply better at *getting* it than our everyday techniques. This is a small point, but an important one that ramifies throughout the study of experimental knowledge.

I am going to explore some of the consequences for our understanding of scientific experiments if we think of them as tools to control the flow of information from distal systems of interest to our minds by way of the proximal systems that make up the experimental apparatus. This is not intended to be a comprehensive account of all aspects of scientific experimentation.

But this new understanding should reorient us in how we think of important features of experimental practice (intervention, replication, calibration) and what count as instances of full-fledged scientific experiments (natural, analogue, thought, simulated, etc.). We will see that items in the first class provide useful resources for the experimenter but are not themselves of central conceptual importance. And we will see that those in the second class all fall under the same concept of experiment and differ only in having different characteristic bottlenecks to the flow of information—bottlenecks that are, however, present in all experimental systems.

One caution is appropriate here: I do not think that scientific experimentation can be detached from scientific theory, for then it would be unable to give knowledge at all. We do not need to return to the view of the logical empiricists to appreciate that without a knowing subject possessed of a great deal of theoretical knowledge experimental practices are no different from randomly poking the world with a stick. What transforms the practice of experimenters into experimentation is the way these practices transform their (and by extension our) states of knowledge—not only of how to do things, but of the sort relevant to making claims about the nature of the world. Knowledge that is relevant to the sciences is useless without an integration into theory. For whatever we can do with an experiment, it does not become scientific knowledge until it is generalized, and no amount of knowhow can, without theory, produce generalized claims from localized practices. Do these theories need to be formal languages of the sort envisioned by Carnap and company? Of course not. Those accounts of theory are, however, no more what theory itself amounts to than any other accounts. But the attempt to say what theories are and do arises from the correct insight that science is the generation of knowledge by attending to nature in the right way, and an ineliminable part of that attention is integrating the information we get from our experiments into a broader, more comprehensive, less particularized theoretical structure.

Without the tight connection between experimentation and theory we cannot understand how *doing this here* and *observing that now* can give us knowledge of how things are throughout the rest of the world. But experimentation does just that. I therefore propose that for an activity to count as experimental it must be so structured as to generate information about distal systems by way of registering the properties of proximal systems. The rest is details, and I will have a lot to say about them as we proceed.

In these first chapters I will spiral my way into the subject. I begin with some very general and uncontroversial remarks about the nature of experimentation. I then try to make contact with epistemology, because experimentation, as I conceive it, is a knowledge-generating activity and so epistemology is relevant. But I will not concern myself with standard debates about the fundamental nature of the knowledge concepts. Instead I will simply explain how I understand the connection between knowledge and experiment. The epistemology I will adopt is externalist, and that seems necessary for really understanding what is going on in scientific experimentation. I will argue in favor of my grounding principle that having knowledge has nothing much to do with justification, and is instead determined by how knowing subjects are situated with respect to the systems about which they have knowledge. This is the only way there can properly be any knowledge of systems distal to the experiment itself.

To recap, though with different emphasis: In part I, I will turn my attention to three key aspects of experimental practice. The focus will be on intervention, replication, and calibration. My goal there will be first to show that as useful as it is for the generation of experimental knowledge, intervention has very little to do with the fundamental character of experimentation. I argue that it offers choices about which experiments to do, but no more than that. Replication, however, is clearly part of the way we go about securing confidence in various experimental claims. I argue, though, that the way it does so is not what is normally thought. In fact I will show that it amounts in the end to a species of calibration, the significance of which in *understanding* experimental practice has not been widely discussed. I show that much of the so-called crisis of replication in the social sciences boils down to a lack of attention to basic features of calibration, and that calibration failure is at the heart of why so many animal models of disease shed virtually no light on those diseases in humans. Part I concludes with the controversial (but I think correct) claim that it is possible to replicate in the absence of novel data.

This sets the stage for the introduction of an analysis of information and information flow. My account will first make clear the relevance of information for generating empirical knowledge. After general considerations about the connection between information, experiment, and knowledge I will return to epistemology. At this point I will plump for my preferred account, a version of Dretske's information theoretic conception. After introducing the account and addressing some standard concerns with some standard replies, I will consider a friendly amendment to this account: the Barwise and Seligman theory of information flow. This theory broadens the Dretske picture to accommodate much more complicated systems than those considered by Dretske, and clarifies the way such systems and their regularities are relevant to how and whether information flows within them. Finally, I appropriate the Barwise–Seligman account to give my final analysis of experimentation as a method for controlling the flow of information between researchers and the things they are researching through a special type of distributed system: experimental systems.

To conclude part II, I will discuss the nature of experiment, and consider some possible objections and highlight some oddities of the account.

Part III will comprise overviews of various types of experimental practice along with an argument that they count equally in how they produce experimental knowledge. In particular I will argue for the claim that all experimentation is analogical, and that thought experimentation is as robustly experimental as laboratory experimentation.

Why Information?

To be informed is the only way to acquire knowledge. This must be true for experimental knowledge especially. So there is a reason, if we want to understand experimental knowledge and how it is produced, to develop a theory of such knowledge geared to an epistemology based on information flow.

But why think we do not already understand experimental knowledge? My answer is that we make two mistakes we would not make if we really understood it, and one flows from the other. First, we draw a distinction between direct and indirect experiments, or between experiments and other things that are mere analogues of experiment (which amounts to the same thing). Second, we do not take seriously activities other than but similar to normal laboratory experiments as legitimate ways of generating experimental knowledge. I believe that in principle (as well as often in practice), as a conceptual matter, we do get experimental knowledge from models, simulations, "analogical" experiments, and thought experiments. These are in fact mistakes, and they suggest that we have long been confused about the nature of experimental knowledge.

The nature of this confusion suggests that the right way to proceed is to clarify its source, diagnose it, and treat it. The diagnosis will contain the key to the treatment, and along the way I will also be setting up the transition to the information theory of experimental knowledge. The typical understanding of how to do an experiment (simplified in form, but in ways that I believe are appropriate to this discussion) goes like this: calibrate an instrument, intervene in a material system, observe the results of that intervention and draw general conclusions on that basis, and replicate the results. Experimental knowledge comes when that experiment has been successfully replicated. Here then is a third way to chart the course of the book:

Part I is the diagnosis phase, and its focus will be on the calibrating, the intervening, and the replicating.

Part II is the treatment phase, in which I propose the alternative of an information flow–based account of experimental knowledge that properly connects the activities of calibration, intervention, and replication to the observation-based inferences that experimental knowledge comprises.

Part III displays the healthy understanding of experimental knowledge embodied in a better account of the kinds of thing that count as experiments, the things that generate experimental knowledge.

On the Proximal–Distal Distinction

In what follows I will be speaking a lot about proximal and distal systems. While I do not think this distinction is doing a great deal of conceptual work, it is pervasive, and so I want to give a sense here for how I am using it.

In discussions of experiment it is common to distinguish between *target* and *object* systems. For my purposes, however, this is a little misleading. It centralizes too much the desires and interests of the experimenter, rather than keeping the focus where it belongs: on the way information flows through these systems and makes knowledge about them possible. I intend to keep the focus on the distinction between systems with which we have to deal (more or less) directly, and the systems they are bearing information about. I will refer to the former (the systems that are the sources of observational data) as *proximal* systems; they need not be near to the user of the data either in time or space, but are merely the systems about which it makes sense to say that they are being observed. I will call *distal* the systems about which the proximal systems are transmitting information, the systems whose properties can be experimentally known by observing features of the proximal systems. Similarly, these systems need not be far away in either time or space. A proximal system is observationally accessible, a distal system is experimentally accessible. The distinction then is based not on what the agent *cares* about, but rather on the position of agents and what they are observing in a knowledge-producing situation.

"Direct" and "indirect" are clearly terms of art here. Our organs of sight and hearing are never in direct contact with the objects of sight and hearing; at the very least, they are responding to electromagnetic influences arising from the things we sense. Put more prosaically, when I observe a thing in the world, a painting perhaps, it makes sense to say that I am aware of the color of, for example, a curtain in the scene on the canvas, or even the brush marks on that curtain. But from these "direct" observations I can also learn *indirectly* about the brush, the artist, the original scene, etc. Here the proximal system is the painting, while the brush, the artist, and the original scene are various distal systems. The former we observe directly, the latter indirectly.

The Lay of the Land

Before turning to my own story of experimental knowledge, I should say a few words about how this project is situated with respect to the main features of the conceptual space it occupies. The three parts of the book, as outlined above, are: a general analysis of the knowledge-generating features of experiment; a specific grounding of these features in an extension of work on information, knowledge, and distributed systems to the case of experiment, and a development of a general theory of experimental knowledge; and an application of this theory to the various types of experimental activity, showing at once the adequacy of the theory to address these various types and its fruitfulness as a unifying account of experimental activity. These three parts have very different structures, so a guide to the terrain is in order.

With respect to parts I and II, I am attempting here to perform two separate preliminary tasks. The first is to gain clarity on what it is about experiments that makes them suited to give us knowledge of nature. The answer is simple: they allow information to flow to us from systems that, for whatever reason, are beyond our direct inspection, through systems that are not. To understand this in a very general way I have reexamined with philosophical attention three different aspects of experiments that, in my view, have not been clearly understood in terms of their role in generating knowledge from experiments: calibration, replication, and intervention. Their usual ordering is exactly the reverse of the order in which I just listed them. What apparently everyone knows is that to experiment, we intervene in nature, and when that intervening has repeatable results we can claim experimental knowledge. If one takes a closer look, one can see that there are all kinds of technical tricks for getting apparatuses to respond properly; these are the tools of calibration. In my analysis of experimental knowledge, though, I have found that things are quite the reverse.

First, intervention is a mere technical trick for getting to the systems that are useful to us, and has nothing conceptual to do with experimental knowledge. The role of replication, then, is twofold. On the one hand, its role in the persuasive discourse that is part and parcel of how the scientific enterprise connects to society at large is to give justification. I will argue that justification has nothing to do with knowledge generally, and that it plays no role in the generation of scientific knowledge. On the other hand, it operates in service of calibration, in ways I detail below. Then comes calibration itself, which in my view is the key feature of scientific experimental design and execution. Calibration is what makes our experiments apt for transmitting information, their best and highest purpose.

The history of philosophical attention to replication and calibration has largely been confined to their technical aspects. Here I break with much of that literature, focusing instead on their conceptual features, which have received much less attention. I do circle back at the end of my discussion of intervention to make contact with current work in philosophy of science, where I show that the so-called replication crisis in science has everything to do with various disciplines lacking the means to properly calibrate their experimental apparatus.

The second preliminary task, in part II, is to provide a technical account of information, a notion that was used only intuitively in part I. In addition, it is to turn that account to in the purpose of devising the promised account of experimental knowledge. Thus sections on information theory, Dretske's account of knowledge, and the Barwise–Seligman extension are included as background. Barwise and Seligman gave some clues about how to think about distributed systems, which I have modified slightly, and which I have then applied to the particular kind of distributed system that is experimentation. But nothing other than mode of presentation is novel here. The important contribution comes in the rest of part II, in which I show how to understand what that kind of distributed system is. Indeed, it is in understanding the right way to partition intervention, calibration, and replication and the way they facilitate information flow that my analysis gets their structure right for the first time.

Finally, part III takes these preliminary results and, in a series of case studies of various types of experimental settings, displays how clear and straightforward it is to generate solutions to various problems that have for some time been vexing theorists about experimental knowledge. But to do so I have had to delve back rather far into some discussions of these matters and to take an entirely new path before rejoining contemporary debates.

Aspects of Experimentation

1

Introduction to Part I

1.1. About Induction

I cast some aspersions above on the idea that knowledge has anything interesting to do with justification. I will have a lot more to say about that in part II when I rehearse and defend Dretske's basic story of knowledge as information-caused (or -sustained) belief. Now, though, I want to clarify why justification cannot be a legitimate aim in experimentation, or in scientific inquiry generally. I believe that an important source of our preoccupation with justification when thinking about experimental knowledge is Boyle and his use of the matter of fact as the foundation of the experimental method.[1] For Boyle, the matter of fact was the cornerstone on which all further experimental claims would be built. To work as a foundation of that sort, the matter of fact would have to be beyond question and would thus need to be generated through a process that, when done properly, could not fail. Of course there is no such process. Any activity, even when done properly, can sometimes fail, but the rhetoric supporting the method had a heavy burden to bear.[2]

This rhetoric is needed to get the experimental method up and running as a sustainable enterprise, and it contains a very strong commitment to assurance, security, and stability—in short, to justification as a constitutive feature of our experimental knowledge. What goes wrong then—wrong from the perspective of our attempts to *understand* experiment, but not from the

---

1. For useful analyses of Boyle's method, see Shapin and Shaffer (1985), Sargent (1989, 1995), and Shapin (1988).

2. The early experimentalists had to overcome both ridicule and challenges directly to gaining knowledge from experience. See especially in this context Dear (2014), Ragland (2017) (for the connection to the renaissance medicine context), Daston (2011), and Sprat (1667).

perspective of what was needed to get the process of modern science up and running—in the transition beginning in the Renaissance to the philosophy of nature that leads to the Royal Society and from there to our current conception of the ultimate foundation of scientific knowledge, is the conflation of experiment itself with a motley panoply of justificatory techniques. It is understandable that when one of your main tasks with respect to your novel experience-based conception of natural philosophy is to promote and to defend that conception, the result will be strongly oriented toward an epistemology founded on assurance. An important part of the business of the mechanical philosophers was to convince the consumers (and producers) of the new natural philosophy that the method could be trusted to give real, secure knowledge of nature. So it is no wonder that security of results was built into the very conception of what it was to be doing things correctly. It seems that even now the main currents of philosophy of science equate experiment with the epistemological force of experiment, or at least make a very fine distinction between them. Of course this view is not explicitly articulated very often, and by no means is it universally held. However, given the lack of attention to experiment generally in the philosophy of science during the period from, say, the 1920s until the 1990s, there has also been little explicit probing of this implicit view.

We no longer think that any of our experimental claims are absolutely certain, nor do we think that such a lack of certainty is any threat to our scientific knowledge. Indeed, science is often understood to be essentially a self-correcting enterprise that upholds the revisability of all claims as a central tenet. Yet we still tend to think of the category of "scientific fact" as containing only truths on the one hand and constituted by something like Boyle's method on the other. And we do not notice that Boyle's category of fact covers all claims made in accordance with the demand of proper performance of a trial witnessed by disinterested audiences of good character.[3] In addition to that muddle, much of the contemporary literature on the foundations of scientific experiment is concerned with the security of experimental claims and other justificatory notions. These notions, some of which are mentioned above, include: assurance, security, stability, trustworthiness, etc. In short, they are notions that indicate not whether the claims being made are true, nor

---

3. I'm being expansive here and not restricting the category to claims witnessed by gentlemen, as Boyle originally conceived it. The idea that the livelihood of those pursuing scientific truth could be linked to whether or not they generated articles claiming to have found such truth, and that the entire scientific enterprise is run that way, would, I suppose, have been incomprehensible to Boyle.

whether the beliefs they report are properly acquired, but rather what resources a knower has available to defend these claims and to persuade others of their truth. But in my view the latter (justificatory features) need to be kept clearly separate from the former (broadly informational features). I will in due course defend an information theoretic account of knowledge that makes no appeal at all to the things a knower needs to possess or be able to do in order to defend her claims. Naturally there are some who continue to use the language of justification even in the context of externalist accounts of knowledge. However, in my view that simply leads to confusion. Consider: justification, at least on traditional accounts of knowledge, leads to trouble in part because any justification invites the further question of its own justification, and so on. This leads to the well-known trilemma posed by knowledge skeptics: infinite regress, dogmatism, circular reasoning. Part of the motivation (as we will see later) for information as the fundamental principle is as *replacement* for the traditional notion of justification. One of its key virtues is that it does not lead to the trilemma because it is not about persuasion, or defense, or even knowing that one knows. An account of this sort that generates knowledge without justification is naturally immune to skeptical worries, as I argue elsewhere and make clear in the remainder of this book.

Eschewing justification and its concomitant notions, my account is at odds with much of what one sees in epistemological accounts of scientific experimentation. For when philosophers and foundations-oriented scientists speak about the knowledge claims resulting from experiments, they are attempting to provide analyses of how various standards of evidence can be brought to bear to evaluate those knowledge claims, and they are not principally concerned with the nature of experimental knowledge as such. I am going to argue in this chapter that much of the philosophy of science involving experimental knowledge over the last few decades is directed toward an analysis of various classes of justification and not toward an answer to the question of what conceptually is going on when we generate experimental knowledge.

My preoccupation in the bulk of this book is with exactly that question, and so I mostly do not engage directly with justificatory questions. That said, this chapter is an attempt to situate the concerns of the book against the broader background of recent philosophy of science, especially as its concerns relate to experiment. There are several important issues in the philosophy of science in the background of my treatment here, and my approach to these strongly colors my approach to experiment more generally. Perhaps surprisingly, my view of theories (as a hybrid of their formulations and the mathematical models they pick out) provides important structure, as does my view of the nature of scientific explanation. For the latter, I have adopted what I take to be the best reading of Hume's version of scientific explanation—that explanations are cognitively salient for us only when they are couched in terms of local causal interactions. On the other hand, I have no view at all on the perennial realism/antirealism debate in the philosophy of science, other than that it's a red herring distracting us from our real quarry, an account of scientific knowledge. Issues in the confirmation of scientific theories, and the role of experiment in that confirmation process, are downplayed here, and would have to be recast in order to fit them properly into my own account of experimental (and more broadly, scientific) knowledge. I will begin, though, with a polemical discussion of the contexts of justification and of discovery.

1.2. Contexts of Research

The first shocking claim I wish to make is that there is no *scientific* context of justification for empirical knowledge generation. The second is that even if there were, all experimentation would operate not in that context but instead in the context of discovery. My reason for the first claim is simply that there is no such thing as empirical justification at all, and particularly not in the epistemology of the sciences. It was and is a mistake to treat the business of science as *securing* rather than producing knowledge. The former is properly the domain of policy, while the latter is the real aim of the sciences. My reason for the second claim is is elaborated in detail in the rest of this book, but in short, some experiments are exploratory, obviously for discovery, and the rest are for letting information about the world flow through them to scientific researchers. There is simply no place for justification in this enterprise.

The claim that there is no empirical justification shouldn't be shocking, but it is. So a good place to start my discussion of contemporary understandings of experiment might be with Karl Popper, who told us that we could not have any knowledge of nature that went beyond bare token instances.[4] His reason for this outlandish claim is that there are basically two kinds of things we could conceivably know about nature: (1) that something has some property, and (2) that generally things of one type are also of some other type. More concretely, I can know that the object in front of me is rigid, blue mixed with clear, sphondiloidal at one end, etc. I could conceivably also know that dogs are mammals, domesticated, carnivorous, etc. That's about it. But the

---

4. Popper I think was wrong about this, but only because he had a wrong view of the nature of knowledge. He thought knowledge itself required justification. If it did, we wouldn't have any knowledge of natural laws. Fortunately, it doesn't.

latter relies on inductive logic, and there can be no *justification* for that. If what we mean by knowledge is something involving justification, then we have very little of it.

The falsificationist fiasco does not come from a mistake about justification, and its impossibility when induction is involved. Popper is quite right in his view that no inductive claim can ever have enough evidence in its favor to be beyond doubt—for if it did it would be deductive. He is also right that Hume's analysis of induction generally, the necessary step in any inductive argument, shows without question that there can be no evidence for induction that doesn't itself appeal to induction. Hume drives this point home, and yet it is still often underappreciated. Induction can never be justified—at all. There has never been and will never be a single shred of evidence in favor of induction. We may, and do, succeed with its use; we may well even be correct in the conclusions we draw when we use it; but its justification would need to rely on inductive reasoning itself, and indeed on a justified use of that reasoning. Where would that come from? Only induction can provide justification for induction, and it's the kind of thing that, if you need it, you can't have it. Here Popper is on solid ground. His real mistake, and the one we continue to make even after rejecting his model of scientific development, is to think that justification pertains to knowledge, rather than to persuading others of our views or to legitimizing the publication of those views. In laying out the problem of induction, he tells us that it can also be formulated like this ([1959] 2005, 9):

> For many people believe that the truth of these universal statements is "known by experience"; yet it is clear that an account of an experience—of an observation or the result of an experiment—can in the first place be only a singular statement and not a universal one. Accordingly, people who say of a universal statement that we know its truth from experience usually mean that the truth of this universal statement can somehow be reduced to the truth of singular ones, and that these singular ones are known by experience to be true; which amounts to saying that the universal statement is based on inductive inference.

Popper and others thought that without justification there could be no knowledge, and those others' disagreement with Popper about whether we really do have scientific knowledge set the stage for a fruitless attempt to provide justifications for inductive claims. But as we will see below, there is available a much more congenial account of scientific knowledge, one that does not appeal to justification. In contemporary science, we are often making universal claims about nature—claims that support counterfactuals across all of space and time. There is always the possibility that our conclusions are wrong. We keep looking, and attempting to get more data, and higher-order data as well (that is, data about the nature of our original data). But it is not right to say that we do not know *because* the data are not all in. Rather, we may well know, and indeed we may well know based on a very small amount of data. In point of fact, any amount of data we collect will be a small amount, since generally we're making claims with infinitely many possible token instances, and *all* finite quantities are small compared to infinity. In addition to the circularity problem above, there is then an entirely separate issue with the so-called inductive sciences that comes down to just this point. If the game is to get enough data to justify our claims, then we are out of luck, for from that perspective our data are finite and thus have no significant justificatory weight. For we are making universal claims, claims that in principle apply to an infinity of situations, and all finite quantities are exactly as good as zero when stacked up against infinity.

Yet we do know things about nature. So if we wish to analyze empirical knowledge, knowledge which is acquired only inductively in the end, we must immediately reject any account of knowledge founded on justification. Accepting that we do have empirical knowledge, and accepting the trivial logical point above, entails that we cannot accept any model of scientific knowledge of which justification is a necessary component. We must, when it comes time to share the results of our scientific inquiries, make reference to standards of acceptance, to the strength of the data, and so forth. But those are only, and can only be, pragmatic norms of acceptance. They can never be part of what we mean when we say that we have scientific knowledge.

Rejecting the idea that scientific knowledge has any connection to justification dispenses with other concerns that have troubled philosophers. For example, one issue that has dogged the foundations of science for a long time is whether there can be crucial experiments in the sciences, and more generally how we are to respond to the problem posed by Duhem and revived by Quine: that I can maintain any self-consistent hypothesis I wish no matter what data I find that are supposed to bear on that hypothesis, provided I am willing to adopt sufficient auxiliary hypotheses. However, the worry about crucial experiments—crucial in the sense of decisive—makes sense only against a justificationist backdrop. To be sure, we do want guidance on which directions of inquiry are most likely to be fruitful going forward. So it might seem that we could use experiment to settle definitively the question of which theory is correct (if either). We can! But not because the *data* stop me from holding on to my theory, come what may. The data don't stop me. Instead my commitment to understanding what information about nature those data are really carrying stops me. Strategies of justification fail in the

face of Duhem–Quine problems, but the right thing to do is to stop thinking of knowledge claims as having to do with justification. Philosophers are
accustomed to thinking that knowledge and justification go together. They
don't, they can't, and happily they don't need to.

I'm going to begin the book proper with what I take to be a few obvious
points. Then I'm going to use those obvious points to motivate a view of experimental knowledge that also seems reasonably obvious, or at least (once it
is made clear) should seem to be a plausible account of what we already think
is going on in experiments. But then I will use that account to do two things:
(1) to undermine some standard stories about the nature of experimentation
and how it connects to our knowledge of nature, replacing those stories with
one about the standards of communication of knowledge; and (2) to argue that
much more counts as experimental knowledge than is commonly allowed,
and to show that that fact is an easy consequence of my view.

The most obvious point I want to make is that we do have experimental knowledge. That is to say, some of the things that we do really count as
knowledge-generating experimental investigations. Thus, perhaps, we know
that dioxins cause liver cancer in rats; that saccharin causes bladder cancer
in humans; that cathode rays are streams of charged particles; that certain
arrangements of very cold particles support sound waves obeying the Gross–
Pitaewski equation, thus giving knowledge about the cosmos at large; that
agents in perfect knowledge game situations tend to maximize their utili-
ties; that similar agents tend to value objects more when they own them than
when they don't; etc. On the other hand, we might well *not* know some of
these things: it might be that some of those claims are false, while others are
true; it might be that some are true but we *still* do not know them. In any
case, though, I do think that at least sometimes we know things on the basis
of our experimental practices. I also think I might know how to argue with
someone—a skeptic, for example—who thinks that we do not, but I am not
going to try. That's a different project.

Even though I won't give a concerted argument against skepticism, I will
say something on this point. Let's return to induction, because induction is at
the heart of all scientific knowledge. Does it work? Well, it has worked in the
past, and I believe it will continue to work in the future. Probably I am using
induction to generate that belief, and if induction does work then that's OK.
Probably we all believe that induction works. Even if, like the skeptics, we are
not willing to affirm explicitly that it works, still every action we take that is
guided by our past experience betrays an implicit trust in induction: bread
nourishes, it doesn't poison, so we eat it; things fall down, not up, so we avoid
walking underneath construction cranes; mustard is delicious and ketchup is

disgusting, so we put the one on hotdogs and the other in the trash[5]; etc. In addition to our individual beliefs that induction works, let's suppose that it does indeed work. But even assuming that we believe it, and that it works, do we really know that it does?

A standard response is "no," and a pretty standard argument seems to go something like this: Whatever separates those who know from those who are merely making lucky guesses is missing in the case of believers in induction. Why? Because the evidence we are taking to support induction—the fact that it has worked in the past—would be just the same even were induction to fail beginning now. There is simply no inference from the past to the future without the use of induction, and any defense of induction is making an inference from the past to the future. Since question begging does not provide justification, we are not justified in our belief that induction will continue to function.

Given a conception of knowledge that equates having knowledge with having justified true belief, or perhaps justified true belief plus something extra, the above argument is sound.

But so what? Isn't that all just a philosopher's quibble and irrelevant to the true question of how scientific experimentation works? "Sure," one might say, "the justification of induction is a special case. But can't we just forget about it and move on with the real business of science, which is using induction to get scientific knowledge?"

Let me recast this discussion this way: There is at the heart of the experimental enterprise a mystery, an apparent magic, that is the cause of much of the suspicion of empiricism that has been directed toward it since Socrates articulated his worries about coming to know in the *Theaetetus*, if not longer. This is the mystery of induction. Whether we are worried about how we know that the future will be like the past, when we only have the past as a guide, or how it is possible to perform inductions about new instances on the basis of entirely different instances, or which, out of all the infinite possible patterns that could follow the initial series we have observed, is the actual pattern, the difficulty is to infer somehow the distal from the proximal. The problem is the same whether we focus on law expressions or on the properties that kinds of entity have. Worries about induction were suppressed in the later twentieth and early twenty-first centuries, rather than resolved. By suppressing these worries, however, we have lost some clarity about just where and how induction takes place, and that is unfortunate for an operation that is, after all, at the heart of the scientific enterprise. I do not mean to say that we haven't learned a great deal about induction over the last century or so.

5. I am aware that not everyone understands this simple truth—more's the pity.

But I believe that by taking induction to be a general feature of the scientific enterprise, noting that science works, and moving on, we have made less clear where in the generation of experimental knowledge these inductions take place. This lack of clarity has consequences for how the scientific enterprise is understood, both by those involved in it and by those observing it from outside.

The precise character of the worry in the case of specifically experimental knowledge is that we observe only particulars but gain knowledge of universals, and then, by that means, gain knowledge of different particulars. Put another way, we transform knowledge about proximal tokens into into knowledge about distal tokens via knowledge of the distal tokens' type that is drawn from knowledge of the proximal tokens' type. That is, we observe token instances, we discover general truths about the types of which they are tokens, we use that to learn about different types, and then we apply that knowledge to their tokens.[6] How could such an operation fail to give rise to suspicion? We interact with something at this location in space-time (say we break a glass), and instantly we know something about other things arbitrarily far away in space-time (e.g., that all glasses are fragile). Ultimately there is no special worry here (other than the worry about the legitimacy of induction, which can never be satisfactorily resolved); but by showing in terms of information flow *why* there is no special worry, I can also show that in a wide range of types of experimentation our prospects are in principle as good as they are in any experiment. In fact I will claim more: that in the case of paradigmatic experimental knowledge, the logic of the situation is just the logic of analogical experimental knowledge. Here then is the ultimate thesis of the book: subject matter identity is irrelevant to the security of experimental knowledge itself; it is merely one means of establishing the permissibility of token → type (→ type . . . type) → token inferential transitions underwriting knowledge of the properties of distal tokens with the knowledge of proximal tokens.

Notice that in speaking about induction I have said something else that I think is obvious: the knowledge we have from experiment is generated from a finite amount of data, and that knowledge is very often of a universal character covering a potential infinity of cases. If that is true, there must be something about the nature of experimental activity that makes it possible for these finite data to give us this universal knowledge. It seems obvious that that something is that these data must bear the information we use to get this knowledge.

6. In the simplest case the distal tokens are of the same type as the proximal.

Here then is where I begin: We (sometimes) have experimental knowledge that is universal in scope, and we get it by consuming information borne
to us by a finite amount of data. Keeping this basic conviction in mind, I will
now try to understand a little better some of the key features of experimental
practice: intervention, replication, and calibration.

I will be working with a model of experimental knowledge that is decep-
tively simple: experimental knowledge is knowledge about distal systems arising from the observation of proximal systems. It is a kind of observation at a
distance, whether that distance be spatial, temporal, or typic. In part II of this
book I will develop in detail my model of experimental knowledge, and provide a conceptual analysis of the information-bearing structures that make
such knowledge possible. In part III, I will put my model to work. In this first
part I will anticipate that analysis and show that there is a need for it by clarifying the important and, in some sense, central-to-experimental-knowledge
notions of intervention and replication. Typically one sees these treated as
necessary constitutive elements of experimental practice and the knowledge
it generates; they are thought to be elements that single out experimentation
from other kinds of scientific (and indeed nonscientific) activities. I think
that is a mistake, and I will use my analysis of experimental knowledge to
show why the things one typically wants from intervention and replication
are available without it. But I will also clarify the real and significant role that
the activities of intervening and replicating play in our experimental practice.

To that end I will start with two basic questions in the philosophy of scientific experiment: When do experiments begin? And when do they end?
Only the latter has received attention explicitly in that form, but the former
comprises a set of perennial concerns about what constitutes the difference
between experimenting and observing, and what activities are properly experimental. Those issues and the question of when experiments begin naturally find a home in discussions of intervention and natural experimentation.
The question of how and when experiments end is more familiar in that form
and has been consistently associated with issues in experimental replication,
and also (though less prevalently) with the calibration of experimental devices and systems. Three central conclusions of this part set up the discussion
in part III: (1) intervention is a red herring that has less to do with the foundations of experimental knowledge than with our ability to choose the things
we want to get that knowledge about; (2) replication has a curious feature
that has not been noticed before—it can be done without new interventions,
or even new data; and (3) both of these will have more to do with calibration
than is normally thought, and it is calibration that plays the central role in the
generation of experimental knowledge.

Here's the program for this part of the book.

1.3. Calibration as Preparation

In chapter 2 I will be concerned with the notion of calibration. Before we can generate experimental knowledge, we must have an apparatus that is suited to give us the information out of which we make that knowledge. Preparing such apparatuses is the work of calibration. So before I begin my discussion of what are commonly thought of as the more central features of experiment,[7] I will give an account of the essential features of calibration and the role it plays in the generation of experimental knowledge.

1.4. Beginning an Experiment

Once all the apparatus has been laid out and properly prepared, it seems for the next step is to strike out into the world to change the course of nature and generate those data that will be the result of our experiment. This may, however, be one of those instances where seeming is not being.

I am going to argue that a central preoccupation of theorists about experimentation has been its interventional character, and that even when not in the foreground this character is a constant background feature of discussions of the foundation of experimentation. I argue that this is a mistake, and I show that the only issue for the optimal functioning of the opening phases of experimentation is whether the observations that result from our attention to the proximal system are appropriately related to the distal systems of interest. Such observations can and do arise without any intervention. Thus the purpose of intervention is not what is conventionally thought. I show that its sole purpose is to make available for such observation a much wider range of proximal systems than natural experimentation would.

I then apply my arguments to the case of natural experimentation, to make clear that there is no conceptual difference between that kind of experiment and any paradigmatic laboratory experiment. The results of this part of the book are interesting in their own right. But they are also useful for establishing the more provocative claims I make in part III: that there is no conceptual difference between laboratory experimentation and thought, analogical, simulated, or model-based experimentation.

The central importance of double-blind, randomized control trials in the generation of experimental knowledge has been both taken for granted and

---

7. I will be attempting to change this common thought throughout part I of this book.

strongly propounded over the last 200 years.[8] Indeed this experimental method is commonly taught as in many respects identical to proper experimentation. More generally, the idea that experimenting is somehow coextensive with intervening has become popular in the wake of work by Franklin, Hacking, Woodward, and others. The work of these authors suggests that there are two separate kinds of activity that make up the scientific enterprise, characterized roughly as doing versus thinking, experimenting versus theorizing, or representing versus intervening. There is a lot to be said for this distinction, especially when we recognize that the purpose of drawing it was to refocus the attention of philosophers and other students of the sciences on experiments after a long period when their major focus was on theories. That purpose was certainly accomplished. Beyond these arguments and discussions is the simple idea that, as objects of contemplation, theories and experiments simply appear to be very different things. If they are anything, theories are abstract objects: collections of various distinct ways of picking out models, and the models they pick out. Experiments are, if anything, concrete objects: the activities of some spatially extended things involving and monitored by actual people, or something of that sort.

But insofar as they are things we do in order to get knowledge of nature, experimenting and theorizing are not so clearly distinguishable. Because both experimenting and theorizing are directed at the generation of empirical knowledge, and because they involve much of the same content, they will overlap significantly. My position here is that from the point of view of experimental knowledge there is a very tight link between scientific experimentation and scientific theorizing. While there is little doubt that it can be fruitful to consider the natures of theory and experiment separately, I hope to show the fruitfulness of considering them as linked by the activity of experimental knowledge generation.

These chapters then will explore the role intervention into nature plays in the generation of scientific knowledge and evaluate just how centrally important it is that people do intervene as part of scientific experimentation. Put simply, the role of intervention is to secure and display some epistemic goods that are not easy to find without it. As we will see, however, they can be and sometimes are found without intervention. So intervention will not count as a conceptually necessary feature of scientific experimentation, though the impediments to getting information to flow without intervening help to explain why intervention has often been thought necessary for genuine experimentation.

8. See, for example, Bhatt (2010) and Stolberg (2006).

1.5. Ending an Experiment

Another centrally important activity in the generation of experimental knowledge appears to be replicating experiments. A supposed hallmark of a well-done experiment and its attendant report is that others are able to reproduce the experiment to show whether its results do or do not hold. Whether or not one actually does replicate an experiment, showing that such replication is possible and how one might do it is a standard part of the scientific enterprise. And yet the exact meaning of replication is not at all clear. Any given experiment is a unique space-time object, a particular activity occupying some region of space with a beginning and end, however vague and subject to negotiation the demarcation of those end points may be. One cannot replicate the entire space-time object that is the experiment in question, to make a new one exactly like the original. Nor would one want to. For if there were errors or confounding features of the original experiment, those would be present in any exact replication, and replication would therefore be useless.

So after I explore the nature and role of intervention into nature for generating experimental knowledge, I will consider the nature and role of replication. What counts as replication, and why is it important? Again, put simply the answer is that replication secures some epistemic goods that are not so easy to find without it. But the analysis of the nature of replication will make clear that there is no constitutive conceptual connection between replication and experimental knowledge. Rather, replication is pragmatically useful.

In the 1980s philosophy of science enjoyed a fruitful disintegration and redevelopment in part as a result of its incorporation of the methods of analysis used by sociology. Much of this, I suppose, was prompted by Kuhn's historiography of science and the reactions to and interpretations (and misinterpretations) of it. My own view is that, on calm reflection, most of this redevelopment was already part and parcel of the philosophy of science, but had slipped (so to speak) out of our collective self-conception.

There were, of course, a number of missteps along the way. An important one is to mistake the fact that our knowledge is a human, social production for the *non*fact that knowledge, or even truth, is somehow nonobjective. One place where this mistake is apparent is in the philosophical, sociological, and historiographical discussions of replication. Replication as a topic may seem unlikely to excite much controversy, except perhaps over the narrowly technical question of whether any given experiment is, in fact, a replication of some other. Yet by seeing how these controversies go, we can see more clearly than before why anyone would be interested in those narrowly technical questions.

Here I weigh in on the controversy over whether replication is an essentially contested category, as Collins (1985) and, following him, Shapin and Schaffer (1985) have it. It isn't. Indeed its original use in the scientific revolution seems to have more to do with advertising and recruiting than with any interesting epistemic purpose. What I show is that replication *conceptually* has less to do with generating experimental knowledge than with evaluating prior experimental protocols and apparatuses. This may seem strange, given how central and damaging recent failures to replicate are to the epistemic status of certain of our scientific disciplines. I try to make this seem less strange by locating those failures as independent of, and merely uncovered by, attempted replications that have revealed a fundamental weakness in the calibration standards of those disciplines.

1.6. Centralizing Calibration

In fact what I think the discussion of calibration, intervention, and replication makes clear is that of the three, calibration is by far the most central conceptually to the generation of experimental knowledge. The other two either have pragmatic benefits or are conceptually important insofar as they are themselves instances of calibration. I conclude part I by anticipating this centrality of calibration to my general analysis of experimental knowledge in part II. That analysis makes information flow the key to experimental knowledge, and information can flow only in properly calibrated apparatuses.

2

Calibration

There are many steps involved in producing experimental knowledge, and it would be a mistake to treat this book as an instruction manual for experimentation in the special sciences. Instead it is meant to help us as theorists understand what is going on in experiments when they go well, and also when they do not. This part comprises chapters that highlight three important aspects of the production of experimental knowledge, and that ground the machinery of the next part within scientific practice and with attention to the concerns of philosophers. I begin with a review of calibration, the crucial feature of apparatuses that allows them to bear information to us about the world.

Franklin (1997, 31) puts it simply: "Calibration [is] the use of a surrogate signal to standardize an instrument." I think he puts it too simply. First, in the case of scientific instruments for experimentation, the standardization is itself a proxy for something else. That some instrument gives the same output as some other using the same input—that is, that it be standard—is of no real importance. Instead we think that the standard instrument is apt for measuring what we are interested in, so when our device is brought into line with that standard, we are satisfied. That is what we really want: for the device we use to be apt for measuring what we want measured. For certain types of instruments, evaluating their results when they are used to measure known signals that stand as surrogates for the signals they will receive in actual use may well be the best way to ensure that they are apt to receive those signals. For some instruments, however, that may not be best way—and indeed, we may not even *have* such signals available in some circumstances. In those circumstances we would still want to make our devices apt for what we want to measure.

The second way this formulation is too simple is that we may want to say, of a certain instrument, what it is capable of measuring (if anything), even should we be unable to intervene on it. We might properly use "calibrate" to indicate the process of finding that out. Broadly, we have two senses of calibration: (1) to make a device apt for measuring something; and (2) to check what, if anything, a device is apt for measuring (to align its use with its capacities). Finally, we would not want to restrict ourselves to the calibration of instruments alone—unless, that is, we adopt an expansive view of what constitutes an instrument. For example, any signals carried by some extended experimental apparatus from a distal through a proximal system, no matter how complex and distributed, may bear information about that distal system. The most natural way to characterize things in that case would be to say that the whole system, understood as a conduit for experimental information, is an instrument calibrated so that measurements on the proximal system bear information about the distal system.

Experimental knowledge comes from the information-bearing signals that make their way through our experimental apparatuses. In part II of this book I will make explicit how such information-bearing signals require special channels connecting their source to a receiver. Later in the book I will be saying a lot about those connections, and they are what (following Barwise and Seligman 1997) I will be calling "infomorphisms." For now we can think of infomorphisms in just this way: they are the connections between systems that let us know about one by examining the other. Later we will worry about their technical features. But before getting into the nitty-gritty of the logical structure of all that, I will examine carefully but less formally the conditions that must be in place in order to secure such connections between the source and the receiver of information. Generally though, infomorphisms (connections between the parts of distributed systems) are implemented in *devices* broadly construed to include entire channels of information flow. The main *conceptual* requirement is that these devices be calibrated in sense (1) above, and the main method of *seeing that they are* is calibration in sense (2). But these channels are useless without signals transiting them. So if we want that knowledge we must find or create such signals. For experiment to *begin* we need apparatuses that can transmit information-bearing signals, and we need such signals to be transmitted.

We will see toward the end of this part of the book that worries about the potential lack of what Franklin calls "surrogate signals" are not idle ones. A major problem with animal models of disease, for example, is that the notion of surrogate signal there doesn't even make sense. Or rather, the endurance of the surrogacy relation is at best questionable: insofar as there are such signals, perhaps we can test whether some group of rats' susceptibility to cancer from dioxins is the same as the susceptibility of rats used in prior studies that showed that humans are also susceptible to cancer from dioxins. But we have no good reason to think that having the same response *in this case* suits this strain of rats for any further experiment on *human* metabolism. For example, their susceptibility to cancer from saccharin does not bear on human susceptibility. If surrogate signals were the only way to calibrate animal models as apt for measuring features of human biology, then these models would be of no use at all.[1] We know that for the vast majority of animal models, when we find that some effect in humans is detectable by seeing the analogous effect in animals, seeing even very similar effects in those animals simply does not qualify as a detection of the analogous similar effect in humans. Moreover, in the case of animal models especially, but also as a general matter about instruments, the observations we make to discover whether they are calibrated are not generally going to be of the same sort as the observations we make during the course of the experiment.

It is not entirely fair to leave Franklin's account at the surrogate signal stage. His discussion does prompt him to extend his understanding of calibration to include investigations of "sources of background that may mask or simulate the desired signal" (1997, 75) So while he doesn't quite say it, Franklin does seem receptive to the idea that calibration involves evaluating the entire channel for its information-bearing capability. And he at least implicitly recognizes that calibration is more than the tweaking of the device, but also includes evaluation, in context, of the suitability of using observations of its state to learn about the states of another system. Even though Franklin's account of calibration is too simple to give a good understanding of what is going on, still his general conclusions bear repeating. Calibration can work. We have produced, fixed, or found instruments that can be used successfully to generate experimental knowledge. That process is not always straightforward, and skeptical doubts can certainly arise. But experimental knowledge is possible. Moreover, the fact that sometimes our calibration procedures produce instruments that are *not* capable of giving us experimental knowledge does not negate this fact, nor should it undermine our confidence in the procedures in general. Rather, like all epistemic defeaters, the way such procedures function to defeat our claims should be incorporated into any future procedures. And so on.

---

1. As we will see, there is good reason to think that animal models of disease are of very little value for understanding human diseases; that quite generally their calibration does not ramify beyond any surrogate signal; that they cannot be calibrated to produce observations yielding new experimental knowledge. But the point stands that we want a broader notion of what is involved in calibrating a device.

I don't know whether Franklin would endorse the following remarks, but I do think they are in the general spirit of his account of how we get experimental knowledge from our instruments: Skepticism is false. Still, our scientific procedures themselves cannot *show* that it is false. Instead our scientific procedures are constructed around the presupposition that knowledge of nature is possible, and sometimes even actual. Given that skepticism is false, and that inductive methods give us scientific knowledge, our business is to find out how they do so, to exploit that capacity as well as we can, and to avoid things that appear to give knowledge but really do not.

Franklin is quite right to emphasize the importance of calibration; its importance follows almost immediately from my account of experimental knowledge (i.e., knowledge that arises from the inferring of a property of a distal systems on the basis of a property of a proximal system). Indeed, asserting the necessity for well-calibrated channels for the flow of information, and a fortiori the devices that occupy the nodes of those channels, is just another way of stating that experimentation is manipulating the flow of information in such channels.

Readers of Franklin's account are liable to be a little puzzled regarding issues of certainty. For as he says, and as I alluded to just now, sometimes even channels that have been calibrated don't give us knowledge. "Calibration," he tells us, "does not guarantee a correct result; but its successful performance does argue for the validity of the result" (1997, 76). Franklin's remarks are ambiguous: Is he using "calibration" itself as a success term, in the sense that a successful calibration produces a standard instrument (and moreover, a standard instrument capable of generating signals that bear information about their target)? Or is he suggesting that carrying out calibration procedures to the best of our ability will not always result in such an instrument? The difficulty is that only the former reading really supports his conclusion, whereas the sense of the passage seems to support the latter reading. In any case, however, the fact is that channels that are capable of passing information-bearing signals need not be flawless. It is simply that when signals in those channels really do bear information, then an observer can, by receiving those signals, come to have knowledge.

That condition needs to be kept sharply distinct from another, possibly similar-sounding, but false condition: *For a channel that can transmit information-bearing signals, this signal in that channel does bear information.* It is of course true that when the signal does bear the information the channel is functioning well, but just having a well-functioning channel does not tell us that it is functioning perfectly for all signals. Calibration is to be understood as making the channel apt for information flow (or finding out that it is so)—not that every signal in that channel will be information-bearing.

Here, then, is how we are to understand calibration: to calibrate a channel is to make it apt for information transmission or to uncover its aptness as a conduit for information flow. Whatever else we do in experimentation is to no purpose unless the observations we make of the proximal systems of our attention are relevant to the conclusions we draw about our distal systems of interest. The type–token relations of the one, which we gain via token observations, must give information about the type–token relations of the other, where our experimental knowledge gives type information and allows us to infer features of new token instances.

We will now look at a couple of other aspects of experimental practice— intervention and replication—that are useful and important, but do not have this same essential character. Intervention gives us a lot of power and freedom over our choices of experimental system, and replication gives us good reason to trust that others' experiments have been properly done. Neither of these is essential, however, for the generation of experimental knowledge. Calibration is.

3

Intervention's Role

3.1. Introduction

I have asked many people, from scientists to philosophers to engineers to folks who have no professional connection to anything in the neighborhood of scientific knowledge, what constitutes a scientific experiment. They have many different views, but they are nearly uniform in asserting that experimentation requires deliberate intervention into material systems by human (or other knowing) subjects.[1] My view, by contrast, is that experimentation can be legitimate without intervention, because they are conceptually unconnected notions. Nor indeed are material systems a necessary part of our experimental apparatus. In this chapter I will argue for the former claim; the latter will have to wait until almost the end of the book.

A standard view seems to be that experiments begin when a scientist, having constructed and calibrated some apparatus, or having devised some experimental protocol, intervenes in nature to produce some data. Indeed, many seem to believe that taking the data *is* the experiment, and that the rest is merely analysis. But that idea ignores the grounding of the concept of experiment in making trial. Making a trial of what? Typically, making a trial of some claim about an entire class of objects (that some biological species is susceptible to some disease, that economic agents behave rationally in the market, that cathode rays are charged, etc.) based on the features of some other class.

1. It is probably not accurate to say that all theorists of scientific experimentation are committed to the need for intervention. However, examples abound: from Brendel (2004), to Dear (2014), to Hacking (1983), to Parker (2009), to Steinle (2016), to Woodward (2003), etc. Rarely is the view defended, or even noticed I suppose, but it is essentially never denied. Morgan (2003) is an apparent exception to the materiality part, but not the intervention part.

The trial of that claim is not performed by some human perturbing the latter class, nor by taking the data, but rather by the sifting of those data. In any case there is no experimental knowledge until that point.

Most, however, do not focus on sifting the data. Instead, the focus is on its generation. Consider the fabled blinded, randomized control trial (RCT). How does it work? We use randomizing procedures to develop multiple test systems from a single undifferentiated system (a hundred mice randomly assigned to ten different groups, say), and we perform on these systems various classes of intervention, beginning with no intervention and covering the entire range of suspected causal agency, and we see what data result from such interventions (9 groups of mice given drugs and one group left alone; a placebo and nine doses of a drug administered; the standard treatment and nine other treatments administered; etc.). In contemporary trials we hide from the receivers of the intervention, as well as those administering it and those recording any effects, whether or not it is a genuine intervention using the test treatment. The data so generated are subjected by turns to two classes of induction: (1) that the response of each member of a group is typical in the counterfactual sense that the probability of responding as it did is the same for each member,[2] and (2) that the response of the grouping as a whole is typical in the counterfactual sense that our results would not have changed were we to have permuted ad lib the intervention protocol among the groups.

At the end of this process we have experimental knowledge: the experiment generates the data, and the inductions on those data generate the knowledge. But what is really unique about experimentation as a source of knowledge different from any other? and does intervention have anything to do with that uniqueness? The uniqueness of experimentation is primarily in the connection between the systems that are monitored and those that are learned about. Intervention has nothing to do with that. This will be an important theme when I outline my theory of experiment in part II. My theory begins with the distinction between observation and experiment. They are related in the way the systems being observed are connected to the systems being investigated. In observation we go no further that the system being observed and we keep our attention focused on that. In experimentation we use those observations to gain information about further systems. That is the key insight. The details specify how that information flows from those further distal systems to the original proximal system.

2. We cannot really do this, but we assume a uniform distribution across the members of the group of a range of possible responses and their probabilities.

Experimental knowledge is generated when the observational features of some proximal system bear information about the features of a distal system, and correct beliefs about that distal system are caused by that information. In order for the proximal system to bear that information it must be connected to the distal system by a sufficiently rich channel. Channels of that sort are represented in my theory by mappings between the proximal system and the distal system. Nothing here requires any intervention at all.

My own basic notion of what separates experiment from observation can be expressed like this: the former uses the latter to make a test of a statement about a different system than the one observed (although that different system may well be the same one observed at a different time). We may even say that an experiment gathers by observation the information that would be needed to make that test, even were no such test to be carried out. This is in stark contrast to the interventionist conception, but it is easy to see why one might be tempted to opt for the latter: it is the requirement that there be sufficient information to evaluate a distal system on the basis of observations of a proximal system that raises the possibility that intervention might be a constitutive feature of experiments, and hence of experimental knowledge. One could well imagine that what allows for inferences connecting the proximal to the distal system—for example, from the effects of a drug on the members of a sample of a population to its effects on the entire population—is *predicated* on the exercise of control over the way these proximal systems display their observational features, control that derives from choosing to intervene or not on a given element at a given time. And this view is enshrined in the standard story of what really counts as an experiment: random sampling of some population; random assignment of the sample to various groups; random selection of intervention targets. This standard view gives lower status *as experiments* to other kinds of activity. It does so for what seems like a very good reason: these other activities seem unsuited to gain the kinds of epistemic good that intervening can give. In particular, they seem unable to prevent possibly confounding variables from entering the picture and throwing off our experimental conclusions.

Whether or not there is some official view that intervention is necessary for experiment (and how could there be for any public, historical, contested notion like this?), it does appear to be the way people usually think about what does and does not count as experimentation.

I will be arguing that while intervention is a powerful way of generating infomorphisms of certain types, and of generating and getting access to signals transmitted within informationally connected channels, it has essentially nothing to do with the concept of experiment.

3.2. What Is an Intervention?

As with so many other things in this book, "intervention" can be said in more than one way. Two main options are: (1) the initial perturbing of some system *in order to* get it to reveal certain of its features; or (2) interrupting the signals that carry those revelations. The latter is merely what is normally associated with observation—it's how our senses and our sensors work generally, by getting in the way of signals. My concern in this chapter is with a kind of standard equivocation. The former understanding of "intervention" carries the intention of the one making the intervention as part of its conception. It seems to me that people often take that intention seriously as somehow bound up in a constitutive way with the nature of experimental intervention. The idea is that what makes experiments special is that agents want something and then do something to get it: they want data, and they perturb systems to get it. But nothing about agents, neither their intentions nor their actual interventions, is relevant to the generation of experimental knowledge. Nothing about agents' desires tells us how the features and behaviors of one system carry information about another.

3.3. Why "Intervention"?

Why "intervention" (and "interference" and "manipulation," etc.)? In his charming book on mathematical methods for physics, Robert Geroch (1985, 183, for example) tells us that to find out how a system behaves, and more generally to discover its rule of behavior, we should poke it with a stick and see what happens. There is no doubt that this is often a good way to find out such things. In his own book on experimentation, Ian Hacking (1983, 149) reports the apocryphal story of Bacon's injunction that on occasion we must twist the lion's tail, and he adopts this as a model of experimental practice. This blend of poking and twisting has come to be seen as just what we mean by experimenting. The exemplar model of experimental knowledge production considered above is of precisely this sort. By looking at the differences in behavior or properties or whatnot between and among those systems on which we intervene and those we leave alone, we can discern the difference that intervening made. From there we can move on to counterfactual supporting regularities, and then on to causal structures, and finally to laws of nature. The key element in all of this, the element without which we would have had no experiment and thus no experimental knowledge, is the intentional, deliberate intervention by the experimentalist. Without that we would be left with the much weaker knowledge afforded by mere observation. The epistemic power of the experimental method is, in this way of looking at things, carried by the bringing to bear on some physical system the experimenter's *desire* to know as *translated* through the poking stick or the twisting hand.

This is all by way of contrast with the view we have inherited of Aristotle's conception. On that conception, intervening into nature is a guide only to the behavior of systems in artificial states and cannot reliably inform us about their natural states. At this late date the experimental tradition is well established and our confidence in its power is well placed. We should not be concerned that intervention is a guide only to systems *unlike* those we are really interested in. Instead we should agree that it is a good thing that allows us to gather useful information about the systems we are interested in. But we may well wonder "what *is* the good of intervening?" What precisely does intervening contribute to our experimental knowledge?

The language of intervention typically adopted in discussions of experimental science is certainly evocative. Twisting the lion's tail, torturing or putting nature itself on trial, and making it answer as before a judge all support and reinforce the metaphor of nature as hidden in its features, and indeed as actively hiding those features. This kind of metaphorical view of the role of experimentation can make it seem that to find out about nature we must twist, torture, and interrogate it. We must pose it questions that it does not want to answer. This way of speaking, however, is just metaphor, and is appropriate only metaphorically. Making trial is simply not the same as putting on trial, and despite its source in the common origin of those expressions, "experiment" is not fundamentally interrogatory in that way. This talk of interrogation does connect up with a basic attitude of intentionality toward the process of experimentation; however, intentionality has no place in an account of the foundations of experimental knowledge. I think many would acknowledge this without the need for argument. But some continue to use the language of intentionality, and it is at least potentially confusing, and possibly the source of actual error, to continue to use such language while officially repudiating appeal to intentional features.

What do I mean by "intentionality" in this context? Simply that the desires of human agents, or at least the contributions of human agency are taken to be requisite parts of any true experiment. If what the experimenter wants to find out is a fundamental part of a proposed analysis of the concept of experiment, then that analysis structures the concept intentionally. In my own analysis, the state of knowledge of the experimenter *is* a fundamental part of the concept of experiment, but the aims, desires, purposes, and so forth of the experimenter are not.

One thought that might make intervention seem crucial to experiment is that experiments are tests. As I have suggested, the origin of our use of experimentation in science is in the making of trial, what the late Renaissance and Early Modern natural philosophers called "periculum facere." And the evolution of experimentation seems to have preserved the sense that making trial, performing tests, is at the heart of that practice.

The historian David Wootton misses his opportunity to clarify exactly what separates experimentation from other kinds of activity. His thought is that linguistic usage simply lags behind the advance of science, and getting experiment and its cognates across Europe to line up correctly is beside the point. According to him, the question of what constitutes an experiment poses no difficulty. "The answer is simple: an experiment is an artificial test designed to answer a question. The Latin term for this . . . is *periculum facere*, to make a test or trial of something. Such a test usually involves controlled conditions, and often requires special equipment" (Wootton 2016, §5 in chapter 8). But this, as I have said, focuses on all the wrong things. Most off point is his twin focus on artificiality and design. And talk of controlled conditions and special equipment, even if only sometimes relevant, tells us very little. To be fair, Wootton is not at that point in his discussion of experiment focusing on the question of what *constitutes* them; he is instead analyzing the significance of the fact that, while the cultures of ancient Greece and Rome had experiments and scientific communities, it was not until the Toricellian barometer that scientific "consensus developed around what the English call an 'experiment'" (2016, §5 in chapter 8). I'll return to his discussion of this episode when I turn to replication later in this book.

Wootton is, of course, correct about making trial as the key to experimentation; but when, precisely, does the trial at the core of the experimental finding out take place? I think the naive answer most of us would give is that the trial is the poking with the stick and the twisting of the tail. But that is not right. At least in the original sense that both "experience" and "experiment" have in common, the trial takes place during the sifting through or picking out of items that have already been generated. In her "Empire of Observation," Daston points to Bacon (1620, II.xxxvi) as originator and Hooke as continuer ("An attempt to prove the motion of the earth by observations," treating parallax as a crucial experiment) of the view that crucial experimenting arises "in the context of a sifting and comparison of observations" (Daston 2011, 89). This is apparently the sense that *periculum facere* has in its technical use in Renaissance medicine (see Ragland 2017). So scientific experiment appears concerned less with intervention than with evaluation. Clearly etymology is of no use in settling questions about the conceptual underpinnings of a method of investigation as wide-ranging and deeply entrenched in our scientific practice as experimentation. But it does provide a reminder that, from the beginning, the point of the practice was learning general things on the basis of particularized data: when those data suffice to perform the test, then we have the essence of experimentation.

This is also the legal sense of trial that we have today. No evidence is *generated* at trial, typically. Evidence is either admitted or not, the admitted evidence is sifted, and the weight of that evidence in settling the question before the court is evaluated during the trial. To be sure, it is hard to sift through data that do not exist. So we need to get these data, and as we will see, intervening does serve a data-generating purpose; it is not, however, usefully thought of as an epistemological purpose. Making trial is not done with a stick.

Again, while we need the data in the first place, it is the sifting of the extant data that generates the knowledge. Bacon's apocryphal metaphor of twisting the lion's tail is arresting; so is Kant's analogy between law courts and laboratories. But both of these images miss something fundamental. What will allow us to convict someone of a crime committed elsewhere and elsewhen are the connections between those data. Who or what produced them is not relevant.[3] Of course, in contemporary systems of jurisprudence data may be ruled out of court if produced in violation of various norms of justice. In the case of law courts, our interrogations of the prisoner are constrained by the right to privacy of defendants, and even more by the rights of those who are not yet defendants. Such ruling out, however, does not prevent investigators in general from coming to know the truth—it only prevents that truth from becoming a matter of law. In the case of nature itself, though, there are no privacy concerns. We can circumspectly observe as much as we like, and anything that nature reveals is admissible into our scientific canon. Similarly, while twisting nature's tail is sometimes necessary, there is no general necessity that we do it. If we can find and observe situations where another lion, or a tiger, or a bear is twisting the lion's tail, so much the better. Nature is vast and complicated, and that vastness offers many opportunities for nature itself to reveal its secrets to those who are carefully observing.

3.4. Intervention's *Bona Fides*

By the middle of the seventeenth century the division of scientific experience into observation (passive witnessing) and experimentation (active interven-

---

3. As opposed to the situation *about which* they are data: presumably, whether or not it involved the defendant may well be relevant.

ing) had become pretty solidly entrenched.[4] This division, though always contested to some extent, has survived and is now firmly codified in many accounts of the scientific method. It has survived not only as an account of our
various ways of encountering nature, but also as an epistemological division:
while observations do no more than provide collections of individual, particularized experiential facts, experiments—*because* they put nature to the
test, twist its tail, torture it, etc.—are capable of producing generalized knowledge of regularities and laws of nature from the collections they provide. The
division and its significance are clear, starting with Bacon's emphasis on artificial experiments as noted by Daston and proceeding through, for example,
Herschel's *Preliminary Discourse on the Study of Natural Philosophy*, where
the division of experience into "observations" and "experiments" is made explicit (1831, paragraph 67). In the next section we will see that this extends to
Hacking's (1983) *Representing and Intervening*, where experiment and various
modes of intervention are essentially equated with each other, and to Woodward's (2003) manipulationist/interventionist account of causation.

I will try to divorce the concept of *experimentation* from that of intervention. In the meantime, I will try to show that their marriage was, in any case,
only one of convenience: experiment and intervention are joined not by true
epistemological affinity, but instead to serve the defensive strategy of the early
proponents of experimental science.

Observation and experimentation are typically contrasted in the way seeing and doing[5] are. The notion is that experiments are active while observations are passive. However, to account for the difference between observing and experimenting in this way is to misunderstand both things. In the
first place, I may have to actively prod a system in order even to observe the
features I am interested in. Observing the structure of the DNA molecule, for
example, requires bombarding it with X-rays. Similarly, observing the quarks
at the heart of nucleons requires bombarding those nucleons with tremendously energetic particles. More prosaically, to observe the behavior of angry
hornets I may have to poke their nest. On the other hand, we may perfectly
well experiment on systems without doing any prodding at all.[6] The real distinction between these two modes of knowledge acquisition is in the structural

4. See again Daston (2011) for an account of the development of these twin notions, and for
their development as epistemological categories.

5. Or hearing and doing, or smelling and doing, etc.

6. It is perhaps worth mentioning explicitly that neutrino experiments involve no intervention in the construction of the signal-bearing neutrinos, but only in the observing of their
properties.

relation between what we come to know and the systems from which we come to know it. In short, the distinction is that in observation the data I acquire are all signals bearing information about proximal systems about which I am gaining knowledge, while in experimentation those data are signals bearing information about distal systems about which I am gaining information. In each case the signals' source is the proximal system: in observation what is revealed by the experience is a feature of the source; in experiment what is revealed by the experience is a feature of a system distinct from the source.

This contrast between experiment and observation, the former being active and the latter passive, might seem like a view with a lot of precedent. Theorists considering the foundations of experimental science at least seem to have been making this contrast for a long time. Consider Herschel, a terrific scientist and theorist of science, and a favorite of historians of science. He seems to make a sharp distinction between observation and experimentation. In setting up experience as the "only ultimate source of our knowledge of nature and its laws" (1831, 76), he does draw a clear conceptual distinction between observation and experiment (1831, 76):

> Experience may be acquired in two ways: either, first, by noticing facts as they occur, without any attempt to influence the frequency of their occurrence, or to vary the circumstances under which they occur; this is OBSERVATION: or, secondly, by putting in action causes and agents over which we have control, and purposely varying their combinations, and noticing what effects take place; this is EXPERIMENT. To these two sources we must look as the fountains of all natural science.

On this basis one sometimes hears Herschel discussed as an interventionist about experiment, someone who finds the essence of the thing in our liberty to take control over a physical situation. But he is an interesting case, and things are not so clear. I would say he is an interventionist in only a mild sense, for he finishes his remark this way (1831, 76–77):

> It is not intended, however, by thus distinguishing observation from experiment, to place them in any kind of contrast. Essentially they are much alike, and differ rather in degree than in kind; so that, perhaps, the terms *passive* and *active observation* might better express their distinction.

In his judgment that experience and observation are the sole sources of natural science, I think Herschel is quite right. To explain why it is that those sciences in which we cannot make interventions move more slowly than others, he says that it is "highly important to mark the different states of mind" when using one versus the other of these two methods (1831, 77). The former is like listening to a story where we're apt to doze off, while the latter is like a cross-examination.

Herschel at least does not draw a distinction between observation and experiment that bears any epistemological weight. Rather, the distinction for him is between passive and active versions of the same activity. I do think the active/passive distinction still does not get quite right what separates observation from experimentation. In my view the right distinction is between local and nonlocal observation, where experiment is, as I have said, observation at a distance. Notice that in each case the information we actually acquire is by means of local signals originating from the proximal system with which we're operating. And we can be as active as we like in building devices and measuring instruments to receive these signals. The real difference is whether we can, on the basis of those signals, draw inferences about further, distal systems. That is, the distinction centers on whether the system we are observing can stand in for other systems, allowing us to observe them through the properties that are more directly manifested. Herschel correctly points out that we may find ourselves with less data than we like in domains of interest when we cannot intervene in those domains; but our passivity in such cases is not the problem. The problem is simply getting a signal that bears the appropriate information. The distinction between experiment and observation is not aptly drawn at the more/less data junction nor at the active/passive junction, and we can have huge amounts of data in cases where we cannot (or for ethical reasons will not) intervene and very small amounts where we do (in the Large Hadron Collider, for instance, where we are clearly intervening to produce the particles we're working with and we find only very few interactions of interest). Instead the distinction must be drawn at the point where information produced in one system allows further information to flow from some other system.

The historical waters are muddied even more when we note that as early as Hume's *Treatise of Human Nature* there seems to be a bias in favor of the interventionist model of experimentation, a bias that Hume is at some pains to overcome. Hume saw clearly that the model of experiment as intervention had already gripped the imagination of the scientific community, and that that model was not appropriate. An obvious problem interventionism posed for Hume is that on such a model no program of psychological experiments could easily be envisioned, because intervening on subjects generally changes precisely what it is we are trying to study. Hume understood, however, that the key to experimental knowledge was the counterfactual supporting data that interventionist experiments provide; so patient, judicious *non*interventionist experiments could provide data that were just as informative and just as secure. This is not the place for a defense of Hume's philosophy, but it is lamentable that his clear insight into the proper role of intervention in experimentation has not been appreciated.

Here is how Hume (1738, 7) puts it in the introduction to the *Treatise*:

> Moral philosophy has, indeed, this peculiar disadvantage, which is not found in natural, that in collecting its experiments, it cannot make them purposely, with premeditation, and after such a manner as to satisfy itself concerning every particular difficulty which may be. When I am at a loss to know the effects of one body upon another in any situation, I need only put them in that situation, and observe what results from it. But should I endeavour to clear up after the same manner any doubt in moral philosophy, by placing myself in the same case with that which I consider, 'tis evident this reflection and premeditation would so disturb the operation of my natural principles, as must render it impossible to form any just conclusion from the phenomenon.

So far it sounds as though Hume is on the side of the interventionists. But as he goes on, he makes, in abbreviated form, the argument of this chapter (1738, 7):

> We must therefore glean up our experiments in this science from a cautious observation of human life, and take them as they appear in the common course of the world, by men's behaviour in company, in affairs, and in their pleasures. Where experiments of this kind are judiciously collected and compared, we may hope to establish on them a science which will not be inferior in certainty, and will be much superior in utility to any other of human comprehension.

One might think it a little unfair to focus here on the word "experiment," since Hume's English doesn't yet make a sharp distinction between that word and the word "experience." But it is possible to go through the *Treatise* and see that he is using experiment in precisely the way I intend, as the method of learning about *those* types of systems by observing *these* token systems. Put this way, it makes clear again my reasons for rejecting justification as connected to knowledge: there simply cannot be any justification for induction, but we seem to know a lot by its means. The resistance one sometimes sees to taking seriously Hume's use of the word "experiment" is driven more by a conviction that Hume doesn't understand experimental science than by a careful examination of his use of the word. That conviction is misplaced.

Despite the unfortunate fact that Hume has had relatively little influence as a theorist about experimentation, he was very well trained in the Scottish schools in what was then the cutting edge of the foundations of science. The important point he makes about the nature of experimental knowledge at the beginning of the *Treatise* is solidly informed by the best Newtonianism of his day, and the point is this: experiment does not require intervention into material systems. The reason is that the way some system is put into a configuration where its development will give knowledge of other systems (including its future states) is not directly relevant to whether it will give that knowledge. There is, of course, a practical matter here. It is much easier to *find* systems that reveal their nature and that of other systems if we *do* intervene, and so we have more choices when we intervene; but it is a mistake to include intervention as relevant to the concept of experimental knowledge.

3.5. Two Prominent Interventionists

It turns out to be difficult to identify precisely the view about intervention accepted by those who study the conceptual foundations of experiment. Rarely do proponents of a particular view of experimentation come right out and say that for something to count as an experiment it must involve an experimenter making an intervention into nature, forcing it to deviate from its accustomed path. Still, when we hear about the difficulties of Aristotelian science cast, in part, as due to Aristotle's refusal to take as legitimately representative of nature those systems that have been forced from their natural states by interventions, it is hard to discount the import that intervention has for students of experimentation. Similarly, but from the other direction, the veneration for a simplified version of empirical philosophy that makes it out as founded on twisting nature's tail, or on an interrogation of nature, points toward a conviction that there is some power that intervention gives us, a power that separates the experiment from its weaker, less informative cousins: the observational study, the simulation, the natural experiment, etc. Indeed, those cousins are typically treated as not merely weaker, but qualitatively other, being simply unsuited to the role of generating experimental knowledge.

Perhaps not everyone is committed to the idea that replication is *necessary* for experiment, given the existence of prominent examples of natural experiments. However, even these examples of experiment without intervention will tend to be seen as aberrations rather than full members of the class of experiments. Tiles (1993), for example, is a strong advocate for the significance of intervention as part and parcel of our contemporary experimental knowledge. While he does a good job tracing the changing views of that significance over the last few centuries, I don't find much in the way of argument for the claim itself. As far as I can make out, the idea is that if we don't intervene we don't understand. But it sounds like dogma rather than argument. For example, he claims (rather obscurely) that counting noninterventionist activities as experiments relies on illegitimately distinguishing between data and the objects of study. He then says this (1993, 471):

> It is open to the interventionist to challenge the basis on which that line is drawn. Is the electromagnetic radiation arriving on earth not an integral part of an astronomical phenomenon which extends in both space and time far beyond the region of the heavens where for the sake of convenience it is assigned that location in an astronomer's catalogue? To be sure we cannot reach such regions and bring the necessary forces to bear which would alter their general course. To that extent we cannot be said fully to understand what is happening in those regions.

I don't see what to call this other than question begging. If the claim is that without intervention our understanding is limited, more is required than simply asserting it. I am, myself, engaged in what I hope is a better sort of argument for the contrary view.

Tiles does, however, make a clear case that intervention is part and parcel of the contemporary understanding of experiment. Let us see how clearly advocacy of this view manifests in two influential contemporary examples. Ian Hacking and James Woodward in their different ways seem to endorse an interventionist conception of the foundations of experiment. Indeed, many of us, at least in philosophy, are strongly influenced by Woodward and by Hacking—so much so that even where we disagree about their specific claims we will still adopt as background their framings, assumptions, and orientations. Those are interventionist. So even if nobody is actively campaigning for an interventionist view of experimentation, I think it is worthwhile first to identify significant threads of argument that seem to support that view, and then to explicitly argue against them.

Intuitively, of course, the idea that finding out about a material system (especially one about which I don't have much prior knowledge) involves diverting it from a state it would otherwise occupy into a different state is compelling. It is something we do every day when we want to get knowledge of the world from its subsystems. Want to know what someone is doing? Ask! That is, intervene with a vocal signal structured to evoke another vocal signal (or hand wave or head gesture or . . .) that indicates the state of mind of that someone. "Where are you going?" elicits an "I'm going to the store" (or a hand might be waved toward the store, or a head gestured in its direction, or . . .). And from such a signal we can come to know what this person is doing. We must be cautious, though. Taking for granted, as we should, that intervening does in many cases produce subsequent behavior that reveals the nature of

the system intervened on can give us the idea that there is something epistemologically fundamental about the intervention itself. We can come to take
probing as necessary for becoming experimentally informed. However, this
is a mistake. There is nothing fundamentally epistemically interesting about
intervening. Instead, as I am trying to make clear, intervention is fundamentally about *choice*—about having the flexibility to find out about whatever
system interests us—and has nothing interesting to tell us about the epistemology of experiment as such. To continue the image I introduced above: if I
wait and do nothing, I may still be lucky, and somebody may volunteer, "I'm
going to the store." In that case the status of my knowledge that that person is
going to the store is, as Hume told us, no less certain than before.

3.5.1. HACKING

Despite the relative lack of explicit endorsements of the view, it is common to
treat intervening in material systems as fundamental to experimental practice. Hacking for example explicitly contrasts the mode of intervening with
that of representing and suggests that these categories are both exhaustive
and exclusive. He famously tells us to "count as real what we can use to intervene in the world to affect something else, or what the world can use to
affect us" (1983, 146). Why is this? Because "reality has to do with causation
and our notions of reality are formed from our abilities to change the world"
(1983, 146). It is clear from these remarks, and from the rest of *Representing
and Intervening*, that Hacking is deeply invested in using a picture of science that tracks the way science actually functions to answer questions that
philosophers ask about the outputs of scientific activity. While the title of
the book might give us the idea that experimentation (essentially the subject
matter of the second half of the book) has something to do with intervention,
Hacking is strikingly (and surprisingly) noncommittal on the point. What he
does say, in passing, seems to indicate that all he is concerned with are the
techniques of experimentation, though he is characteristically more suggestive than definitive.

Discussing Penzias and Wilson's discovery of the microwave background
radiation of the universe, Hacking remarks that "It is sometimes said that in
astronomy we do not experiment; we can only observe." And he adds that
"it is true that we cannot interfere very much in the distant reaches of space,
but the skills employed by Penzias and Wilson were identical to those used
by laboratory experimenters" (1983, 160). But he does not follow up on the
remark, not even to say that *in fact* they were experimenting on the early universe, or that they were not. He is pursuing a different question, one having

to do with which of the two, experiment and theory, has conceptual priority, and which is (or is not) the other's underlaborer. Still, in discussing microscopy Hacking comes a little closer to telling us that experiment must involve intervention. It is when we can *do* things with microscopes that we begin to know how to see with them. "Practice—and I mean in general doing, not looking—creates the ability to distinguish between visible artifacts of the preparation or the instrument, and the real structure that is seen with the microscope" (1983, 191). So what he is telling us, in part, is that even observing requires some intervention into the world, and that rather than being interestingly theory-laden, observation done right (that is, done by manipulating the world with the tools by which we grasp it, and manipulating those tools themselves) overcomes whatever theory may have gone into the making of those tools, or may be lurking in our background belief structure. This has, perhaps, taken us too far in the other direction. For what Hacking seems to be telling us here is that there is, after all, no fundamental observation/experiment distinction—but that both require intervention.

Again, though, it is hard to disentangle Hacking's view about the nature of experiment from his concern with questions of scientific realism, and that debate is not my focus here. Let me take another stab at this: In discussing how scientists create phenomena, Hacking introduces a useful distinction between *phenomena* and *effects*. The distinction is this: "Phenomena remind us, in that semiconscious repository of language, of events that can be recorded by the gifted observer who does not intervene in the world but who watches the stars. Effects remind us of the great experiment[er]s after whom, in general, we name the effects: the men and women, the Compton and Curie, who intervened in the course of nature, to create a regularity which, at least at first, can be seen as regular (or anomalous) only against the further background of theory" (1983, 225). Now, at last, we can see that, in practice at least, experiment is carved off from observation at the active/passive joint. Hacking is an interventionist.

3.5.2. WOODWARD

Woodward has had an enormous impact on how we see the foundations of causation and how we see the role interventionist thinking plays in our coming to know about causal relations. In *Making Things Happen*, Woodward (though ostensibly merely giving an account of one particular kind of explanation, the causal) also presents us with a model for experiment, the kind of thing by which we can gather and control the explanantia that figure in such explanations. Woodward's model of causation, and of causal explanation, is crucially dependent upon the notion of intervention and manipulation. In fact it would not be an overstatement to say that these crucial notions are *constitutive* of his account of causation. Insofar as experiment is supposed to tell us something about causal regularity in the world (and while that's not all we're looking for with experimentation, it is a big part of it), we cannot even begin without finding out what happens to various systems under appropriate interventions. Woodward does tell us in various places that the interventions he has in mind do not need to be performed by any human agent—or by any conscious agent at all, for that matter. What really matters is whether there are "natural process with the right causal characteristics"; if there are, then they "will qualify as interventions" (2003, 128). These processes will exist whenever they are so structured to exemplify the type-level and token-level definitions Woodward gives. The interventions here are used to characterize the circumstance where X is causally related to Y. For the type level, there must be an intervention variable I such that it causes X, and also when I is such as to cause X makes it that no other variables cause X. Also, I does not cause Y indirectly along some causal path not passing through X, nor does it depend statistically on any other cause of Y that doesn't cause Y by means of X. We can then use I to manipulate X to determine whether X is a cause of Y (2003, 98).

This makes the connection between experimental intervention and causation explicit. Now Woodward is well aware, and concedes, that no raw interventionist model of causal explanation can capture our intuitive notion of causal explanation. This is for two reasons, both of which he illustrates (2003, 127–33): First we want to be able to explain causally things we are unable to intervene on ourselves. The moon, for example, functions importantly in causal explanations of the tides, and while I suppose we can these days at least imagine ourselves intervening causally on the moon, we are in no position to intervene on it in a way that could possibly reveal its causal connection to the tides. We would want to be able to ask, "How much would the tides change were the moon's orbit made much larger than it is now?" But we are in no position to intervene on that orbit. Second, certain causal questions cannot be addressed at all by executing any possible, nonmagical intervention. We think we can sensibly ask questions like "were the moon's orbit *right now* much larger than it is, without anything else in the solar system changing, what effect would that have on the tides?" Not only do we think we can sensibly ask such questions, we think we can answer them correctly. But no physically possible intervention would lead to just the right antecedent conditions for answering such a question. While Woodward is aware that his account requires certain idealizations that make it not quite a pure interventionist account, we are still left with some important questions. First, can we be content with an account of causation that takes as primitive a notion (intervention) that is itself explicitly causal? Woodward waffles, in my view, between claiming to tell us about causation and claiming to tell us about causal explanation. The latter appears to be in pretty solid shape at the conclusion of his account; with respect to the former, things are less clear.

Woodward's account of causal explanation is predicated on the view first that "a very central part of the commonsense notion of cause is precisely that causes are potential handles or devices for bringing about effects" (2003, 12), and second that he sees "causal explanation in science as building on and requiring causal knowledge of a more mundane, everyday sort" (2003, 19). And while Woodward will not definitively commit himself to the latter point—one that he takes to be merely suggested by the fact that all cultures understand cause, but few have science (2003, 18)—it is clear that his view is more than suggestive. He claims that "acquisition of causal knowledge must have, at least sometimes, some practical point. There must be some benefit, other than the satisfaction of idle [!] curiosity, that is sometimes provided by these activities" (2003, 18–19). Woodward repeatedly, tacitly appeals to this claim in his argument for the fundamentally interventionist character of causation.

Despite his laudable efforts to remove anthropomorphic considerations from the concept of cause, Woodward continues to conflate the etiology of our concept of cause with the concept itself. He takes the illuminating fact that we can come to causal notions by intervening on and manipulating material systems and concludes that intervention is built into the concept we discover thereby. Our discovery of causation is grounded in our practical interest in and engagement with the world, and some causal relations are both ripe for exploitation by our interventions and understandable in the light of their downstream consequences. But causation is independent of all such notions, and an account of causal explanation should not be grounded in them. The mistake Woodward makes in grounding his discussion is to pass off as an account of a *concept* a story about how creatures roughly like us come to have *acquaintance with* and *mastery over* that concept. Even were we to endorse the proposed etiology of our use of that concept, we still have not learned what the concept itself amounts to on that basis. For that we need to say not only why we use it, but how.[7]

The manipulations we perform *are* a species of causal action; the important task for us as theorists is to detach the part that is connected to our in-

---

7. I'm basically content to follow Putnam on this: a concept is a sign used in a certain way. So to know the concept is to know how it is (should be?) used.

terests in order to find the notion of cause itself in the residue. Such a task may well be at cross purposes to the attempt to even better exploit the concept— but so be it.

3.5.3. THEIR IMPACT

So we have two important examples of theorists about experimentation who are wedded to the notion that experiments are fundamentally, conceptually inseparable from intervention. In each case some care is required to see precisely what intervention amounts to, and there is clearly some wiggle room for either of them to plump for a notion of intervention that doesn't have much to do with the intention/state-change/data model of experiment that began this chapter. However, it should be clear that the influence on philosophical accounts of experiment exerted by these two theorists will tend to embed a more straightforward notion of intervention into the background of discussions of other aspects of experiment.

In philosophy we must resist the danger of thinking that because views are not explicitly articulated, they are not widely held. Brendel, for example—an interesting theorist about thought experiments whose work I will examine later—falls exactly into the camp of those who take it as not worth arguing for, or even mentioning, that experiments require interventions. She makes that assumption in her analysis of thought experimentation, where we will see the interventionist ideal exerting a bad influence outside its proper place.

My aim in this chapter has been to present as well as begin to undermine that ideal. I have offered some arguments here against the idea that intervention is fundamental to the concept of scientific experimentation. I find those arguments compelling. In chapter 4 I will do two things: I will first make clear what role intervention *does* play in experimental practice and what goods it provides the experimenter; and second I will explicitly display (by means of a fable) the blinded RCT as independent of the need for the intervention of any agent. That example will shore up any defects that might remain in the argument of this chapter. The fable I present is possible, thus properly illustrating the conceptual point, but it is extremely unlikely in practice. So I want to be sure the point is not misunderstood. I do not say that intervention is not a practical necessity in our efforts to gain experimental knowledge. Instead, I say that it is *conceptually* superfluous. To really understand the nature of the knowledge experiment gives us, we must display it free of such superfluous elements.

4

Intervention's Goods

4.1. What Intervention Does Give Us

As I have said, I do think there is a role for intervention in the generation of data, but it serves rather to liberate the experimenter than to provide epistemic goods that are otherwise unobtainable. Intervention allows us to choose the information-bearing mapping between proximal and distal system. (Later I will identify that particular kind of connection as an infomorphism between them.) And it does so in two ways: first, it gives the experimenter access to a wide variety of information channels that can serve to connect what amount to proximal local logics of observation state spaces to distal local logics about which we wish to reason; and second, it makes it easier to see that the type claims made about the observed tokens are sound and not confounded with other factors of the situation. Both of these goods—wider access and ready assurance—are pragmatic in nature rather than fundamental or constitutive. Consider: it is rare in nature for creatures like us to have access to novel information channels connecting distal and proximal systems, and it is rare for us to have access to data that are not confounded in various ways. Intervening generates many more such circumstances, and so extends our experimental reach, but it does not play a fundamental experimental role. This relates naturally back to my objection to Woodward's identification of intervention as fundamental to causation and causal explanation *because* it is our original source of understanding of causation. There seems little doubt that intervention is our original means of extracting data from a system that is connected to another system in the right way to bear information about it. And it remains our most reliable access to such data. We must not, however, confuse this fact with the incorrect idea that such connections are reliant on intervention.

There is a third, related function of intervention. It is that in those cases where a subgroup of the distal system comprises the observed proximal-system tokens (e.g., when we take a sample statistic and infer a population parameter from it), intervention helps ensure that observed tokens provide valid inferences about the distal tokens.[1]

Intervention is thought to be important because it gives us control. That control, in turn, allows for counterfactual reasoning, which in its turn supports causation attribution. There is something to this view. Much of the panoply of experimental practice can, in this way, be seen as intended to establish causal regularities. Intervention is the foundation, it seems, of our ability to reliably spot causal regularity. Let's consider once again the blinded RCT (BRCT); but this time our purpose will be to strip away those parts of it that are not essential to its epistemic function. We have a single population. We select subpopulations based on randomly assigned exogenous factors that are, importantly, independent of any endogenous factors. We then intervene on one of these subpopulations (or on several in different ways) and not on another (while at the same time making the nonintervention mimic as closely as possible the features of the intervention that are not under test) which is our control group. These two (or more) populations are then compared with respect to properties of interest.

What are the virtues of the BRCT, in particular pertaining to how it achieves its epistemic aims? The epistemic goods in the BRCT come from the fact that the observed properties of the systems subjected to the test condition and those of the systems not subjected to the test condition differ only because of the test condition. We can achieve this by ensuring that in every other respect, in the ideal case, the two classes of system are identical. But as a matter of principle, the ideal case can never be achieved. The reason is apparently trivial, but it illustrates the real challenge facing designers of BRCTs. No two different systems occupy the same space-time location. Even two neighboring patches of grass are in different spatial locations, to recall one of the examples in Kempthorne's (1992) useful analysis of the connections between statistics and experimentation. We are then forced (as a conceptual matter of a sort) to be content with a lesser standard. The observed properties of the systems differing only in virtue of whether they are exposed to the test condition is achieved by ensuring that the systems are identical in every other *relevant* respect. But how do we achieve

---

1. To anticipate some of the analysis to follow, the tokens are normal in the sense of Barwise and Seligman (1997).

that? How do we, not knowing the full dynamical description of the systems of interest and the proper way to control entirely our impact on the temporal evolution of the systems under those dynamics, determine which respects are relevant to whether their observed properties will differ, independent of the test condition we have in mind?

So the epistemic good of intervention in this case is to guarantee that the effects of the test condition are not confounded with other effects that are not under test. And the trick is to see how to secure that good.

Many solutions have been proposed. A useful place to start is with Fisher's classic *Design of Experiments*. Fisher introduces the notion of significance and of the null hypothesis as ways to *mitigate* against too easily accepting and/or rejecting the hypothesis under test. But he stresses that there is no definitive way to avoid the possibility that the test groups differ in ways beyond those influenced by the test condition itself, nor is there a way to avoid the possibility that our application of the test condition has had effects beyond our consideration.

Fisher uses the by now well-known example of someone who claims to be able to tell whether milk was added to a cup before the tea or after (Fisher [1935] 1974, §5–12). He proposes a test involving eight cups of tea, of which four have the tea added first and four have the milk added first. The probability of guessing the right order on all cups is 1 in 70, so that seems to be a pretty good test. We are not very likely to be mistaken in rejecting our hypothesis that the subject cannot tell the difference when the probability of a successful discrimination is so low. But because the tea cups and quantity of milk and temperature and so forth will clearly differ across the cups, we do not want any correlation between those features and the order in which milk and tea are added to the cups.

Randomization is his only suggested recourse, but as he says, even randomization is not quite a *method*. Rather, "randomization" is a shorthand for an admonition to construct the experiment in such as way as to distribute the possible differences between test groups so that those differences are not correlated in any interesting way with the differences that would arise from the test conditions themselves. Randomization is not enough, of course, since we might well introduce along with the test condition some other identifying marker. For example, even though we add the tea and milk in a randomly determined order, if we also add sugar whenever the tea is added first, that completely undermines the randomizing condition. Now instead of 1 chance in 70 of getting the order right, there is something like 1 chance in 2 if we assume along with Fisher that the subject is likely to perceive the difference in sweetness as constantly associated with either the tea going in first or the milk going in first.

Less obvious than the addition of sugar, but as worrisome for the test, is that, by knowing whether the milk or the tea was added first, the experimenter can somehow tip off the subject. This possibility exists even in cases where the test is not a psychological or psychic one. The well-known placebo effect makes clear that when subjects know that the test condition is present, their behaviors and even physiology can differ. Likewise, there is a connected feature of observers. If the observer of the outcome of some experiment is aware of whether the test condition is in place in a given instance, that awareness influences the observations made in predictable as well as unpredictable ways.

So another epistemic good we want to secure is that we avoid any changes in observed properties in the systems under test that are caused by the presence or absence of the test condition being signaled either to those very systems or to the observer of those systems.

The way this is classically mitigated is to introduce blinding procedures. As far as possible the situation with the test condition present is made identical to the situation without the test condition present, but where that is not possible the correlations between those differences are prevented from being signaled to the test systems by those preparing and administering the test, nor are they signaled to those observing the outcomes.

Can we always be sure that no such signaling has taken place? No. Likewise, we can never be sure that the low probability outcome we use to reject the null hypothesis is not the outcome we in fact find. The procedures we adopt in experimental investigations are, again, designed to *mitigate* the risk of forming incorrect beliefs, but they cannot *eliminate* that risk entirely.

As Fisher ([1935] 1974, 19) puts it:

> it is only necessary to recognise that, whatever degree of care and experimental skill is expended in equalising the conditions, other than the one under test, which are liable to affect the result, this equalisation must always be to a greater or less extent, incomplete, and in many important practical cases will certainly be grossly defective. We are concerned, therefore, that this inequality, whether it be great or small, shall not impugn the exactitude of the frequency distribution, on the basis of which the result of the experiment is to be appraised.

The basic outline here is that we intervene into nature in prescribed ways that help us to mitigate the dangers of making incorrect inferences based on correlations that are not reflective of the true causal situation. Our methods of intervention give us control over the situation that allows us to eliminate any spurious correlations that confound our attempts to find the true causal variables.

But now that we have in front of us the story of the standard procedure, we can ask, "Are there epistemic goods that this procedure secures for us but which cannot be secured in some other way?" And we can also answer: "No." As Fisher's conclusion should make clear, it cannot be a conceptual feature of experimentation that we have made such intervening moves in order to mitigate against confounders. Rather, there is a purely pragmatic question of whether such confounders are sufficiently minor as to allow the distribution of outcomes to bear sufficient information to give us knowledge. There are, to be sure, pragmatic advantages, but as the fable below is supposed to further illustrate, those are the only advantages. The epistemic superiority of intervening is entirely illusory.

4.2. A Fable of Intervention

Once we admit (as I think we should) that there can be experimental knowledge, then I think we should accept the following as a pretty good example of it. A diligent researcher, Marie, who has some experience with radioactivity now wants to find out about some of the effects of, say, X-rays on rats. Marie finds a typical rat and makes a large number of clones of it. She then uses a random number generator to break her new population of rats into several groups. These groups are then randomly assigned identifying tags and a treatment protocol: they are each herded onto a target platform where they are exposed to either no radiation, or one of a number of different radiation amounts, each treatment lasting for the same amount of time; the procedure is perhaps repeated at various intervals. Once the protocol is concluded, the rats are maintained for a specified period while their appetites, general health, etc., are observed. Marie then kills and performs autopsies on the rats. It should go without saying that Marie has been clever enough to hide from herself the actual exposure amounts for any of the rats: their tags trigger the X-ray machine to deliver a certain amount of radiation, the same for each member of a given group, but according to an algorithm involving a random number generated by the computer controlling the X-ray machine. Only after the autopsies are completed and the observations recorded does Marie check up on which rats received which exposures.

Now let us suppose further that Marie has observed that there is a strong negative linear correlation between appetite in rats subsequent to the protocol and exposure to X-rays during the protocol, and that there is a strong positive linear correlation between that exposure and the number of liver tumors observed during autopsies. I think we would have good reason to conclude that Marie has discovered a pair of causal effects of X-rays on rats, discoveries that clearly count as giving her experimental knowledge of these effects.

This all works, we might think, because of the protocols established by Marie for segregating the rats, for intervening in their natural life course by the introduction of radiation, and for doing so according to a randomization and blinding routine. In short, Marie's intentional intervention, tightly connected to her deliberate plan, is the crucial element in generating this experimental knowledge. But what, precisely, about her intervention is so crucial for generating this knowledge? Let's come at this a little sideways.

First, can Marie give *us* experimental knowledge by relaying to us her procedures and results? I take it that a great deal of the edifice of our scientific knowledge is built on the assumption that she can. Of course, we may wish to replicate her results in order to be sure that that she hasn't made any mistakes, or that she hasn't been extremely unlucky and observed an extraordinarily low probability situation.[2] But suppose that has been done. Perhaps another diligent researcher, Pierre, has intervened in his own lab on his own samples of rats, and he has also found these strong linear correlations. When both have communicated their results and methods to us, do we then have experimental knowledge of the correlations, and indeed of the causal link between reduced appetite and X-rays and between tumors and X-rays? We could demand more . . . perhaps that Eve and Irene, two other diligent researchers, communicate *their* results. But at some point we have either to say about this case "yes, we now have experimental knowledge" or to stop claiming that we have any experimental knowledge at all that we didn't get by performing an experiment ourselves.[3] I claim that, if the correlations do hold of rats generally, then Marie's initial experimental report suffices to give us knowledge of them. Could we be in a better position in some sense by getting Pierre's reports? Probably. Pierre's reports should probably increase our confidence in Marie's reports, as should Eve's and Irene's. But Marie's reports should

2. We'll turn to replication shortly, but it is not relevant for appraising this episode.

3. I'm pretty sympathetic to skeptical worries generally, and I do not believe that it is simply irrational to deny that experimental knowledge can be communicated, nor to deny that it is impossible generally. I do claim, however, that anyone who thinks that the Newtonian–Boylean scientific revolution in Europe inaugurated a practice that has been pretty successful at giving us experimentally based scientific knowledge should agree that Marie (or Marie together with Pierre [or at least Marie together with Pierre and Eve and Irene]) has been able to generate and communicate experimental knowledge about these rats.

suffice if her samples are large enough and her spread of treatments extensive enough. As far as the evidence presented is concerned, Marie's report and her data are surely sufficient to establish good grounds for believing the results, and for assuming the results are correct. That sounds like good experimental knowledge.

Ahh, but is Marie trustworthy? Perhaps one would want to establish this independently. I have been assuming all along that she is, but perhaps to secure our trust Marie can just record the proceedings (including a data dump from the computer that is generating the radiation exposures) and share those recordings with us at the end. Again, it is perfectly possible that Marie can fake the tapes, and so it is possible that later we will come to find that she didn't really perform the experiments as advertised—in which case we had good grounds to believe in these correlations and still did not really have knowledge of them, because we didn't have the information we thought we had. This worry, however, accomplishes nothing more than to remind us that nearly all knowledge *claims* are defeasible. It is possible to conceive of scenarios that undermine nearly any knowledge claim one can imagine. While this is an interesting point for the foundations of epistemology, it is not really germane here. For we are attempting to isolate the role intervention plays in generating experimental knowledge, not attempting to find out whether experimental knowledge is possible at all. Let us stipulate that Marie is not deceiving us, and that we still have (accurate) recordings of her entire procedure. Again, I claim that we now have experimental knowledge of the correlations Marie observed.

Now, let's probe this a little bit. Given that we have the recordings, do we need Marie's written account of her observations and of the autopsies she performed on the rats? Can we not in some sense perform those observations ourselves? I don't say that we don't require training in the evaluation of rat appetites and in the identification and enumeration of tumors. But given such training, we have in front of us the means to make those observations ourselves. So even were Marie to die before reporting on the experiment, and were Marie's notebooks to be lost, we would still have experimental knowledge.

But does Marie even need to be present for the autopsies? Perhaps she has robots that are sufficiently programmed to produce sections of the rats' livers. Indeed, is Marie herself necessary for these proceedings at all? Suppose we simply stumble upon recordings of a robot facility that clones rats, randomly exposes them to X-rays in accordance with the above protocol, feeds them for a while, then sections their livers—and produces recordings of the proceedings. What stops us from taking this as giving us experimental knowledge that X-rays cause lower appetites in rats and that they cause liver tumors in rats? The probability of this is, of course, vanishingly small. But there is no conceptual difficulty here.

What does this fable show? It shows at least that the intentions of agents are not required in order for an experiment to be conducted. Does it show that intervention is not required? That depends crucially on what one means by an intervention. If one means only that to get causal knowledge some signals are required bearing information about what happens when the purported cause is present as well as when it is absent, then generally yes, experimental knowledge requires intervention. But one means that someone sets out to intervene in nature and then does so, then no, experimental knowledge may be had without intervention. Most importantly, if we are supposed to understand intervention as more than the fact that some cause produced some effect on systems we're interested in but also, for example, somehow was used to give us control over the causal evolution of various systems, and in turn that that control is somehow constitutive of our epistemic grasp of the situation, then intervention is not at all required.

What then are we to make of various accounts of experiment and of causal explanation (for example Woodward's) that are based on the notion that all causal knowledge and explanation derives from interventions, either actual or possible (or in Woodward's case merely notional)? I think these accounts require fundamental revision.

Indeed, the order of reasoning above, where control gives counterfactual knowledge which in turn gives causal knowledge, is epistemically backward. For our capacity to skillfully exercise control *relies* fundamentally on our possession of causal knowledge in the first place, be it explicit or implicit. While it is true that the production of data can be greatly facilitated by, and can even in some cases rely crucially on, our interventions, the fact remains that this knowledge comes not because we ourselves produce those data, but rather (as Hume said and as the Renaissance medical scientists knew) because we are able to sift those data whatever their source.

4.3. Doing versus Knowing

The above fable should help us to see clearly that the strength of BRCTs (and of intervention in general) does not lie exclusively in what they do. What they do is important insofar as it allows us to find situations through which we can know things: randomly generating some variables gives us access to situations without certain kinds of confounding; blinding gives access to situations where, for example, the observer's knowledge does not impede the flow of information about the hypothesis under test; etc.

The conflation of experimenting and intentional intervention is complemented by another conflation: experimental knowledge is in many cases the product of manipulation, unmediated by theoretical activity. Hacking's *Representing and Intervening* (1983) makes just such a conflation. It is a very important book for bringing issues in experiment to philosophers of science, and Hacking correctly points out (echoing Franklin [1986] in his equally important book, *The Neglect of Experiment*) that experiment and experimental practice had largely been ignored in many philosophical accounts of science in favor of theoretical issues; further, that this neglect was particularly puzzling given that empiricism is the official position of the vast majority of contemporary philosophers of science, and that experiment is a key feature of empiricism. Given my contention that only experiment gives us knowledge beyond token observations of proximal systems, the failure of philosophers to pay sufficient attention to how experiment functions was a serious oversight. Happily in the light of these books and other work by Hacking and Franklin and others, philosophers in the last several decades have devoted concerted attention again to experimentation. So I do not want to give the impression that I find Hacking to be misguided in his approach. His book was written to free us from the idea that science is principally about the language of science, and whether or not one believes that we were ever really prisoners of that idea, the book trenchantly demolishes it. Science really is about the world, and not simply our talk about the world.

But the subsequent attention to experiment generated by the book has served to reinforce the perception of a sharp divide between experiment and theory on the one hand and between experiment and other observational tasks on the other. Hacking's focus is on establishing that experiments can have what he calls lives of their own, that these lives are oriented around the production of effects, and that effects and productions are the central features of experimental practice. In the context of microscopy he says that "practical ability breeds conviction" (1983, 191). I will not object to his characterization here. Rather, I have shown that the various items of interest to Hacking—the effects, the techniques, the abilities—are best seen as aids to generate, in various instances, the data supporting the key inferences that constitute our experimental knowledge. As a practical matter these aids are crucial, in the same way that as a practical matter diagrams in Euclidean geometric proofs are crucial, and yet those diagrams are not the proofs[4] and the source of

---

4. To be sure, it is true that there are systems of proof that are themselves diagrammatic, and for which the proofs are sequences of diagrams. That's a separate point. In Euclid's *Elements* the proofs are generally the arguments, and the diagrams as such perform no logical function.

experimental data is not the experiment. What intervening into nature allows us to see is that some proximal system is of some given type. But there things end, unless we adopt the further inference that some distal system is itself of some type. It is that further inference that makes our practice experimental, that constitutes the trial.

Hacking says that he "shall insist on the truism that experimenting is not stating or reporting but doing—and not doing things with words" (1983, 173). And while we can certainly join him in rejecting the assimilation of experimenting to doing things with words, we should simultaneously resist taking as a truism that experimenting is doing in quite the way he understands it. In attempting to resurrect the theory/observation distinction, Hacking implicitly appeals to another distinction on which that "truism" relies: the distinction between natural and human experiment. That distinction is without a difference.

I have argued that a focus on intervention as a key element of experimentation is the wrong way to analyze what is distinctive about this part of the sciences. Instead, what we should focus on is the role experimentation plays in the generation of scientific knowledge. While experiments are surely about the material systems they involve and not about words describing those systems, they are not thereby removed from the knowledge generation aspect of scientific inquiry. Instead experiments play a crucial role in the generation of scientific knowledge because they are fundamentally directed to such generation, and any adequate account of experimental practice will involve accounting for that role.

The principle purpose, or justification, for appealing to intervention as basic to experiment is that it allows us to control in some measure the course of nature—to twist the lion's tail. We confront nature, according to Kant, not as a petitioner but as a judge who compels nature to speak. As arresting as such analogies are, and as important as they were for the early experimental philosophers in establishing experiment as a legitimate source of knowledge, they do more to obscure the conceptual foundations of empirical knowledge than they do to clarify them. Of course experiment is doing, and more than doing with words alone. The problem with assimilating experimenting to stating or reporting or using words generally is not on the side of misunderstanding experiment, but on the side of misunderstanding knowledge. Knowledge is not principally a feature of language; rather, it is a feature of the informational state of the knower. Knowing is not doing things with words any more than experimenting is. But understanding *that* does not advance our understanding of experiment.

I think we should see the requirement that experiment be an intervention into nature in the same way that we now view the requirement that meaning be defined in terms of verification conditions. Certainly we cannot fully understand what a statement means if we have no idea what difference it would make in the world were it to be true. Similarly, we cannot fully understand how information flows in an experiment if we don't know anything about the material conditions governing that flow. But just as it is a mistake to call the meaning of a statement its method of test, so too is it a mistake to call what we mean by an experimental result an intervention that would produce that result.

One crucial danger of assimilating experiments to interventions is that it obscures and directs our attention away from other possible avenues of experimental knowledge. For example, given that we cannot, probably even in principle, intervene in large-scale cosmological systems, we might think that we cannot have experimental knowledge of those systems. As we will see in the chapter on analogue experiments, the fact that we cannot intervene in cosmological systems is not a hindrance in principle to our securing experimental knowledge of those systems. Similarly, thought experimentation involves no interventions into material systems—indeed, it involves no material systems—but as we will see, it can produce experimental knowledge. Before we can appreciate the close connection between all branches of experimental inquiry, we need to move beyond the interventionist paradigm and focus on the key feature of experimentation: the extraction of information from one system to learn about others.

Simply put, the point to intervention is to secure important, auxiliary epistemic goods of two sorts: First, that we actually have the information indicating that some token system of one type is also of some other type. And second, that this information is not misleading, so *generally* tokens of that one type are also of the other type, and not that we have chosen a particular token of a special subtype of the one type that is also a token of the other type. (We find, say, that all things that are of the type *sugar* are also of the type *soluble*, not merely that this particular sugar cube is.)

4.4. The Production of Phenomena and the Generation of Knowledge

Boyle knew an important fact about intervening: that it does not matter how phenomena were produced, but it does matter what those phenomena are and what we can say about them with confidence. Of course, that confidence will most often be higher when we have done the producing ourselves, but in few other areas of life would we confuse that kind of convenience with a conceptual feature of the thing itself. Find a ten-dollar bill on the ground, or get it from a friend, from the US Mint, from a known counterfeiter: none of this is relevant to what it is to be a ten-dollar bill. It is relevant only to our confidence *that* it is real. Experimental knowledge works the same way.

To look forward a bit at this point, and to emphasize the importance of knowledge to the experimental enterprise, I want to preview the differences between two accounts of Hertz's cathode ray experiments, mine and Jed Buchwald's, which will be discussed at length in chapter 7. Buchwald focuses, as does Hacking, on the doing: what were Hertz's tools? what can these tools be used for? what, generally, could Hertz *do*? But the first thing to find out is what knowledge Hertz aims to produce with these tools. It is not that Hertz's intentions are relevant to the constitution of his experiments. Rather, Hertz's aims are a good guide for our own understanding of how his experiments function to generate for him that experimental knowledge. We trust that Hertz's understanding of how the machinery functions is sufficiently accurate that his aims lead him to properly exploit the capacities of the devices he uses. His aims tells us where to look ourselves, and once we find his principal aim (establishing qualitatively new knowledge about the electrostatic and magnetostatic properties of cathode rays) we can begin to trace the way that data produced by his device can be understood as bearing information that can generate knowledge claims concerning cathode rays generally.

The crucial advantage that intervening gives the experimentalist is freedom to choose the system about which one wishes to become informed. While intervening is not conceptually different from the perspective of epistemology, it does afford us control over the source of our experimental information. As I will clarify in part II, the key to experimental knowledge is information flow through a channel whose main conduit is the proximal experimental system. Creating that system then gives us flexibility over the distal system that is the source of the information we desire. As are so many constant background features of the world, the proper function of intervening can be obscured by the role the *output* of that intervening plays in our investigations. The phenomena produced as a result of our intervention are conceptually indispensable for producing knowledge about the source of those phenomena. Thus it can seem that the intervening is itself conceptually bound up with producing experimental knowledge. As we have seen, though, it is not. And this has implications from two sides. "What kinds of experiment can there be when the phenomena are not the result of a material intervention?" contains two salient questions: "What experiments are performable without intervention?" and "What systems other than the material are apt for intervention, and can such interventions produce information-bearing phenomena?" The former is addressed below in the chapter on natural experimentation; the latter in the chapters having to do with simulation and thought experimentation.

We will see explicitly in the case of thought experiments the confusion that results from our tendency to unreflectively assimilate experimentation to intervention.

4.5. Law: Method versus Substance

In *The Diffident Naturalist*, Sargent (1995) makes clear an important misunderstanding on the part of those who take the notion of scientific law as arising from thinking of the universe as fundamentally governed by hidden agency. She says that we should take the analogy with law to be methodological (nature is put to trial to find out truth) rather than substantive (nature is not to be understood as law-like in operations or itself to obey some law). Common law provided Boyle with epistemic resources (Sargent 1995, 42 ff.). When we understand the case in this way, we understand that the nature of experiment is not to be found in any system of intervention or type of observation. Rather, we understand that the nature of experiment is in the method of controlling the generation of knowledge given our material and cognitive limitations.

Sargent makes perfectly clear that Boyle was opposed to attributing any essential significance to the artificiality of man's intervention.[5] The point to intervention is expediency and efficiency. Of course Boyle was insistent that we not be too quick to make inferences about what powers nature has on the basis of too few and too easily found observations, but he clearly saw that the epistemic advantages of intervening over gleaning were not to be overstated.

Despite his adoption of a justificationist standard, Boyle did understand something about the generation of experimental knowledge that we seem to have forgotten. He understood that it is a merely contingent feature of our place in the world that we need to intervene into it to gain that knowledge. By taking our intervention into nature to be itself a perfectly natural class of event, rather than something sui generis, he answered the charge that when we intervene we see something other than nature at work, that we see instead merely the results of our own handiwork. But this line of thinking is easily inverted. For when we remove human agency from our analysis of experimental knowledge, we have not thereby changed anything about the nature of that knowledge. An observer who could see all happenings in the world would find more than enough opportunity to infer whatever causal laws are really operative. Given our own limited station, we must instead construct sufficient happenings to make those inferences ourselves. But we should not

---

5. And we should also invert this reasoning to see that we do not lose our control variable when we remove human agency.

forget that the reason we do so has nothing to do with the nature of experimental knowledge and everything to do with the rarity of the conditions that are appropriate for the generation of that knowledge.

We have seen in this chapter that intervention, while an important part of our experimental practice, can in principle be done without. It gives us a great deal of freedom and control, but provides no sui generis epistemological advantages. Despite the standard accounts we have all been exposed to, there is simply no necessary role for intervention.

We turn now to consider another such standard component of our accounts of experiment: replication. As in the case of intervention, we will see that it plays an important but not a necessary role in the generation of experimental knowledge. I will make this clear by first analyzing what goods replication can really be expected to provide, then connecting that to the origins of our commitment to replication, and then considering the so-called replication crisis in the social sciences. The framing of the current situation as a crisis in *replication* highlights both our confusion over what is at stake in the crisis and what function replication plays in our experimental practice.

5

Replication

Replication seems to have played an important and central role in the development of contemporary science and the knowledge generated by its means. Yet the nature of that role is not very clear. Replication also has an important ceremonial role in contemporary scientific practice, as the one true arbiter and validator of experimental claims; but it is also unclear what substantive role (if any) replication plays in contemporary science itself. An important reason for this lack of clarity is that the nature of replication is itself not very clear—neither what counts as a replication of a scientific experiment, nor which experiments should be replicated and why.

A central focus of this book is the meaning of experimental knowledge claims, and relatedly what is necessary in order to have experimental knowledge. The foundation of that epistemology is that experiments give knowledge by facilitating the flow of information. If replication is to have any part to play in the epistemology of experiment, then, it will have to contribute to facilitating that flow. In these next few chapters I will be developing an account of replication that suits it to its role in the production of experimental knowledge.

It might not seem that understanding experimental replication and its function in the generation of experimental knowledge would entail any difficulty. So let me present the following as motivating, indirectly, the worry about how replication functions.

I remember in college doing an exercise with some fellow students as part of our laboratory studies: We would look through the objective of a microscope, squeeze an atomizer bulb, and, when we saw moving dots and circles, we would repeatedly flip a switch on a device next to the microscope. This action appeared to cause the circles and dots to reverse their direction of motion. We would pick out one of the dots and note the times it spent moving in each direction between flips of the switch. After several reversals, the dot would escape our view and we would start the process over. We were told we were measuring the amount of charge that spread over oil drops as they escaped the atomizer, and thus indirectly the charge on the electron.[1] Readers familiar with the history of experiment may recognize this description of Millikan's oil-drop experiment. The history of this experiment is fascinating indeed, and Franklin in particular in many of his works has given excellent accounts of its significance and of Millikan's methodology and treatment of his data. My interest at the moment, however, is not the experiment itself so much as what is included in claims like "in college we replicated Millikan's experiment," and in claims that some experiment or other was or was not replicated. Evaluating such claims and then developing a clear understanding of replication will tell us a great deal about experiment in general, and in particular will shed light on the interesting interplay between the questions some experiment is being used to answer and the proper identity conditions on experiments. It will all come down to an account of the knowledge-generating capacity of the cluster of activities that are singled out as the experiment itself—that is, how those activities together do or do not allow information to flow between researchers and the systems they are investigating.

In his essay *The Conflict of Studies*, Isaac Todhunter (1873, 16–17)[2] offers his own charming (if dated) appraisal of experimental replication.

> Experimental Science viewed in connection with education, rejoices in a name which is unfairly expressive. A real experiment is a very valuable product of the mind, requiring great knowledge to invent it and great ingenuity to carry it out.
>
> When Perrier ascended the Puy de Dôme with a barometer in order to test the influence of change of level on the height of the column of mercury, he performed an experiment, the suggestion of which was worthy of the genius of Pascal and Descartes. But when a modern traveller ascends Mont Blanc, and directs one of his guides to carry a barometer, he cannot be said to perform an experiment in any very exact or very meritorious sense of the word. It is a repetition of an observation made thousands of times before, and we

---

1. If charge is quantized and each time the charge magnitude is different on the different drops, then one will find after many trials that only one value for the charge divides all those magnitudes without remainder.

2. Thanks to Andy Blitzer for bringing this to my attention.

can never recover any of the interest which belonged to the first trial, unless indeed, without having ever heard of it, we succeeded in reconstructing the process for ourselves. In fact, almost always he who first plucks an experimental flower thus appropriates and destroys its fragrance and its beauty.

To take another example. We assert that if the resistance of the air be withdrawn a sovereign and a feather will fall through equal spaces in equal times. Very great credit is due to the person who first imagined the well-known experiment to illustrate this; but it is not obvious what is the special benefit now gained by seeing a lecturer repeat the process. It may be said that a boy takes more interest in the matter by seeing for himself, or by performing for himself, that is by working the handle of the air-pump: this we admit, while we continue to doubt the educational value of the transaction. The boy would also probably take much more interest in foot-ball than in Latin grammar; but the measure of his interest is not identical with that of the importance of the subjects. It may be said that the fact makes a stronger impression on the boy through the medium of his sight, that he believes it the more confidently. I say that this ought not to be the case. If he does not believe the statements of his tutor—probably a clergyman of mature knowledge, recognised ability, and blameless character—his suspicion is irrational, and manifests a want of the power of appreciating evidence, a want fatal to his success in that branch of science which he is supposed to be cultivating.

Todhunter here hits on something important: experimental knowledge is not created anew when the actions of the experimenter are repeated. He may well be overstating the modal claim about how scientific beliefs *ought* to be formed and how hearing about a result is better than seeing a performance of the act that first generated the result. But I will say this much: the only way to get knowledge experimentally is to see that some system has some property and to come to believe on that basis that some other system has it too.[3] Whether knowing experimentally is better or worse than knowing experimental results by testimony, I don't know. I don't have much commitment to that sort of claim. But I do want to know what it is to know experimentally. I think Todhunter is right to this extent: if I recreate a situation as it was said to be, only in order to see that things happened as they were said to, and come thereby to believe something about some distal system based on my newfound confidence in these earlier events, I do not thereby generate experimental knowledge so much as evaluate favorably the earlier knowledge claims. The information borne by those happenings cannot be transmitted to me by them

---

3. As I will put this when I come to my detailed theory of experimental knowledge: the one system bears information about the other.

when that same information is already present in the prior signals that, by assumption, I already received.[4]

All of this has at least these consequences. First, simply repeating the various procedures of a prior experiment does not count as replicating the experiment for any useful understanding of the term. Second, there is thus at least a potential puzzle about what should count as replication. To solve that puzzle will require knowing what replication is for, and what role it plays in the generation of experimental knowledge. It is not the other way around, where knowing what replication is would tell us what role it plays.

Whether or not replication does, or even should, move from the ceremonial position it now occupies into a position of true power, students of science should understand how it works, and what does and does not count as the replication of a scientific experiment.

5.1. The Contemporary Context

A great deal of progress has been made over the last couple of decades on the mechanics of replicating experiments from the point of view of evaluating the standards for the sanctioning of data produced in one experiment by means of those produced in another. But the foundational story that replication has something crucial to do with the production of experimental knowledge has received little critical evaluation. Fidler and Wilcox (2018), for example, in their very helpful *Stanford Encyclopedia* entry, continue (quite explicitly) to use "replication," "reproduction," and "repetition" interchangeably. This marks out that their interest, and broadly the interest of the philosophy of science community, is somewhat orthogonal to my own. Their main focus is to understand the culture of replication (or lack thereof), the crisis in replication, and what reforms in our statistical methods this crisis heralds. Underlying their account is the persistent and erroneous belief that replication, because it is justificatory, is an essential ingredient in the production of experimental knowledge.

Steinle (2016) as well, though more concerned to distinguish between replication and repetition, still folds them together under reproduction, and continues to insist that the main role of reproduction has been the securing of conviction in results already produced. Be that as it may, this has little to do with the production of experimental knowledge as such, chasing as it does

---

4. What we might say is that this re-creation has removed an impediment to the original information causing my belief.

the red herring of justification. Tetens (2016) claims in the same volume even more forcefully that reproducibility is at the heart of science, and he retains a heavy focus on justification and a very strong sense of experiment as fundamentally interventional.

I do not suggest that my overall view of replication is sharply at odds with any of these three, or with other contemporary work on the issue. Indeed, like them I view replication as deeply embedded in the practice of science, and believe that its use should be expanded. But my focus here is not on the meta-scientific question about how the details of statistical analysis and so forth can be brought to bear to solve the so-called replication crisis, the focus of so much contemporary work on replication. Though I reject the justificatory context of experiment, there is a role for replication studies with that focus. For replication does have a role to play both as proxy monitor for information flow (a measure of whether we know that we know), and as an aid to understanding the persuasive discourse that is at the heart of much of what goes on in the sciences. These are both important for understanding how science functions as a discipline. But these two purposes are less relevant to the conceptual analysis of experimental knowledge that is the focus of this book. Therefore, little of that exciting technical work will appear here.

5.2. Coming Up Next

After setting up the problem of replication a little more, I will in the next few chapters turn to a real, but minor, controversy over experimental replication. The issue is whether Thomson or Perrin or both replicated Hertz's experiments to establish the charge state of cathode rays. This seems to me an easy case, but it highlights the importance of asking the right questions when assessing replication claims. I next consider a less controversial case familiar to students of contemporary physics: Weber's gravitational wave experiments. The status of the replication of Weber's experiments by other physicists is not controversial at all; what is controversial is how the episode closed. This case prompts an introduction to Collins's idea of replication as essentially contested, and as pertaining solely to controversial results and attempts by the community of experimentalists to bring to a close episodes involving such results. It also prompts a reexamination of the debate about that idea. I will agree that Collins hits on some important insights, but I will argue that ultimately the view is shortsighted. My conclusion from the Weber case will be that the use of replication in science has been fundamentally misunderstood.

Subsequently, I examine the experiments that prompted the banning and/or tight regulation of saccharin as a sugar alternative. What we learn from this, and the prior, examination is that the key role of replication in experimental practice is as a tool of calibration. I further explore the calibrating role of replication by means of the results of Seok et al. (2013a, 2013b) on murine models of inflammation. It is simply not the case that replication is a means of establishing experimental results. Instead, replication validates experimental *practices*: it tells us that information could have flowed between the experimental situation and the larger world, not that it did so.

But that in turn makes it mysterious how there could be a replication crisis in science. Isn't the whole point to the crisis the realization that without replication of results we never have the experimental knowledge we think we do? While that is the lesson many want to draw from the crisis, I show that what the crisis *really* teaches us is both that social science experiments are generally woefully underpowered and that the role of replication remains poorly understood. The latter problem I am attempting to remedy in these chapters; the former I leave to the social sciences themselves. Generally, though, this crisis is not confined to the social sciences, and I try to extend the analysis to show that the lesson of crises involving replication is that calibration is more central to and more difficult in the sciences than is commonly thought. What failures of replication illustrate is that we very often just don't get information about our subjects of interest, even when we're lucky and stumble upon a scientific truth. Why? Because often our experiments are simply not properly calibrated to be apt sources of information.

Curiously, seeing things in that way makes our interest in such episodes as the replication of Hertz's cathode ray experiments more understandable, because they become about the experiments themselves and their aptness for knowledge generation rather than about the idle topic of whether some great scientist was or was not wrong on some occasion. But they also make understandable and believable a point that is complementary to the one I began with: that redoing the activities of an experiment is not sufficient to replicate it. The complementary point is that redoing those activities is not *necessary*, and indeed experiments can be replicated without any new events being generated. I illustrate this with the series of replications of Kociba et al.'s studies of dioxins and liver cancer (1978). There nothing was done other than changing the protocol that turns records into observations—but that change of protocol amounts to recalibrating the information channel sufficiently to replicate the original experiment.

Generally, I will be arguing that replication is itself a kind of regulator of information flow, one that functions by giving new researchers access to a channel of information from some distal system via some proximal system that is of the same type accessed by prior researchers. While replication can generate experimental knowledge when a prior experiment did not, replicability itself has little directly to do with the capacity of a given experiment to produce knowledge. What I will make clear is that replication is not directly useful for generating experimental knowledge nor, surprisingly, for validating it—not exactly. Its target is not the *results* of a scientific experiment; it cannot show either that they are correct or that they are not. Rather, replication is for the evaluation of prior experimental activities' suitability to secure the information that would be necessary for them to have generated experimental knowledge. Making that clear will, in turn, make clear that the principal function of replication really is calibration—the tuning of apparatuses (and the checking whether they are tuned) to detect information-bearing signals.

6

The Received View of Replication

There is something unclear about the nature and purpose of experimental replication. The only thing that *does* seem clear is that many are convinced that replication is the ultimate sine qua non of experimentation, even more so than intervention. Why is that? What is it about replication that gives it its air of importance?

The story about the importance of experimental replication runs like this. What really got the scientific revolution off the ground was the rise of societies of experimentalists, and their members' insistence on the repeatability of experimentation. The weekly meetings of the Royal Society, for example, comprised a demonstration of some experimental fact, followed by discussion of that fact among the witnesses to it. Because these witnesses were multiple and because they were witnesses as well to the presence of others who were themselves helping to witness the fact, these demonstrations really did amount to a generation of novel experimental knowledge. The idea was that it wasn't enough that a lone researcher would be able to produce some effect or other (even were it the redoubtable Hooke back in his rooms preparing for the meetings); it was crucial that the effect could be reliably and repeatedly produced on demand, and in front of witnesses. Otherwise it was simply no effect at all. Presenting the narrative as I have just done indicates that replication, one among many means of securing certain epistemic goods, has been elevated above its station, to the hallmark of experimental practice—so that no practice can be said to produce experimental knowledge without the stamp of replicability.

6.1. The Beginnings of Empiricism?

Even taking that narrative seriously as an account of the origin of our attention to replication, there are obvious questions that arise. First: What, on that
view, is the purpose of replication? What role, if any, does it play in the epistemology of experimental science? There is the historical version of the question, of course: "Was replication used by Boyle and company as the arbiter for
when experimental knowledge had been generated?" Perhaps counterintui-
tively, the answer seems to be "no." For while it was important for generating
widespread confidence in a result, the public witnessing of the production of
experimental results had little to do with whether that result could be known.
But doesn't that fly in the face of the self-understanding of those early prac-
titioners of the scientific method? Didn't Sprat, for example, best known as
the author of *The History of Royal Society of London*, think of repetition as
the generation of the scientific fact itself? It seems so: he tells us that some
one or two members are sent out to perfect and perform an experiment, but
then they are to report to the entire assembly and "bring all the History of its
process back again to the test. Then comes in the second great Work of the
Assembly; which is to judge, and resolve upon the matter of Fact. In this part
of their imployment, they us'd to take an exact view of the repetition of the
whole course of the Experiment" (1667, 99–100).

What Sprat seems to be talking about, though, is not how one person
may or may not gain experimental knowledge from some trial. Rather, he is
propounding the idea that we are to see the assembly itself as a kind of experimenter and sifter of the data generated by the experiment. He explicitly com-
pares the assembly to a single inquirer who, beginning with good intentions
and initial diligence, may well come to be deceived on the occasion of hitting
upon a causal story that strikes upon the imagination. By contrast, the slow,
fractious deliberation of the society cannot be undermined in this way (1667,
105–7). Sprat consistently speaks about the society itself as knowing things,
discovering things, adding to its store of knowledge. Thus those experiments
that are repeated before the company have little to do with replication as we
tend to understand it today, and more to do with the one body, the Royal
Society, itself practicing an experiment, and then performing it.

What about Boyle, then? Doesn't he require replication as a condition on
experimental knowledge? In the preface to the first part of "A Continuation
of New Experiments Physico-mechanical" he describes repetition as having
the point of providing either greater success or confirmation of previous experiments. However, this is always in the context of helping *someone else* to
come to have knowledge of his own (Boyle's) results. It is, in a manner, am-

plifying the signals that have already been generated. Boyle does not speak about these repetitions as adding epistemological weight, or as providing an independent route to the knowledge generated by them. Rather, the discussion is always on the way such repetitions can make it easier for others to see and come to know his results.

In the context of the development of contemporary science, and especially the dueling standards represented by Boyle and Hobbes, Shapin and Shaffer (1985) some time ago worried about this issue of what counts as replication, what standards one could bring to bear to decide whether something was replicated, and what purpose was served by the replication of scientific experiments. Their focus was on the production of experimental facts. Principally, they were concerned with what I would characterize as discrete observations that result from some intervention into nature. However, their discussion also involves the debate between Boyle and Hobbes over the reach of experimental claims. The issue might be phrased like this: is it possible to learn laws of nature when experimenting, or is it only possible to learn discrete facts of an observational nature that must then be interpreted in the light of true knowledge of natural law? As Hobbes puts the case, one can only advance in scientific knowledge on the basis of a knowledge of the true laws of motion— laws that cannot be discovered from experiments but must be already known in order properly to interpret those experiments.

While the analysis they offered has been widely influential, Shapin and Shaffer seem not to have entirely understood the significance of replication studies and the role they played in the development of contemporary experimental science. In brief, their view is that there is a social category that is fundamental to the development of our science. That category is the matter of fact, and the main development of Boyle, and of the Royal Society more broadly, that led to contemporary science was the technique for turning isolated experiences into these matters of fact.

Boyle's description of his thinking about replication is in sharp contrast with Shapin and Shaffer's understanding of replication as "the set of technologies which transforms what counts as belief into what counts as knowledge," an understanding situated in and arising from their background understanding in which the (scientific) "fact is a constitutively social category: it is an item of public knowledge" (1985, 225). In their chapter on "Replication and Its Troubles," Shapin and Shaffer uncritically accept Harry Collins's (1985) account of the experimenter's regress, and then attempt to understand how Boyle's complicated practice of negotiation is supposed to provide an exit from the regress. I'll address and reject Collins's arguments below, but even accepting his general notion of regress, there is little in this historical episode to show that the facts—in particular about the spring of air—offered by Boyle were ever in any serious need of auxiliary support. As Boyle repeatedly makes clear, the only real question is how to *explain* that spring, not whether there is such spring. The majority of Shapin and Shaffer's evidence in this chapter bears on disputes about the *quality* of the vacuum obtained by various rival pneumatic philosophers. Those disputes are then somehow equated with worries about Hobbes's dismissal of the entire edifice of vacuum technology.

The entire episode seems to me to be entirely overwrought. Hobbes took the notion of vacuum to require the complete absence of any matter in the vessel, and none of his adversaries cared about that at all. That Hobbes was worried about replication tells us little about replication's significance for experimental knowledge. Far from playing the role of arbiter of belief, replication seems rather to play the role of publicizer and disseminator of scientific results.

It also, though, is clearly a means for checking on the proper functioning of experimental apparatus *before* setting off to seek new experimental knowledge. Wootton (2016) provides a useful counterweight to Shapin and Shaffer on this. He rejects, quite rightly in my view, the claim "that replication is always problematic, and that in the end what counts as replication is always decided by the intervention of authority" (2016, 349). I don't know that there are many who would make this claim, which he attributes to Pinch and to Collins. That seems not quite right, and we do need to be careful here about exactly what is being asserted by the various proponents to this debate.

Wootton also points to Shapin and Shaffer, who as we have seen are channeling Collins's account of replication. But there is no claim that replication is *always* problematic. Rather, the theory developed by Collins is that replication is only of interest when the facts are in dispute, and in each such case the resolution of the dispute centers around coming to agreement about what constitutes replication in that case. As we have seen, replication, understood in Collins's (and by extension Shapin and Schaffer's) sense, is not merely repetition, but is itself a kind of social fact. These scholars are not denying that there are natural kinds in the neighborhood (e.g., answers to questions like "Was the air pressure in the interior of Boyle's air pump within 1 percent of that in Torricelli's tube?"), but they are reserving the word "replication" for when the question "do our scientific standards count these experiments as the same?" is answered, "Yes."

While there is clearly some confusion here over the nature of social facts, and what it would mean for replication to be such a fact, there is also a certain lack of clarity about the roles replication was playing in the early modern period. The principal two are advertising[1] and calibration. Regarding the latter, for instance, Wootton observes that before they would start to produce novel results, members of the experimental community would reperform Torricelli's experiment (2016, 349, e.g.). Why? He is not explicit about this. But the discussion makes clear that it was both to secure their mastery over the technique and to check that their device was reproducing known effects. There was no serious doubt about the effect itself, and a failure would have been seen as indicating a broken device or an incompetent investigator, not a refutation of Torricelli. Even without consensus about how to *explain* the result, the result itself was considered a stable check on the apparatus that should be done before an experimenter set out to perform novel experiments (cf. Wootton 2016, 336). The fact that those engaged in experimental practice endorsed rather than resisted such procedures makes clear that our understanding of the significance of replication is simply backward: replication doesn't validate experimental knowledge; rather, experimental knowledge can provide a background against which replication tells us about our own skill as experimenters or the aptness of our tools for performing experiments.

6.2. The Foundation of Empiricism?

That replication as we understand it had, for the early experimental philosophers, anything like the importance that is generally claimed for it is uncertain at best. But let us leave aside for a moment the historical origins of our commitment to replication, and see whether there is something about it that is essential to our own practice. After all, though Boyle and company got things started, their views are not determinative of the actual epistemology of science. Perhaps then it remains plausible that replication provides some important epistemic benefit after all. Perhaps when we replicate another's experiment we somehow generate novel warrant for our belief in the result of that experiment, and thus we generate novel support for the experimental knowledge claim. This accords with the platitude one often hears, and that we will see below in the discussion of the "crisis in replication," that the scientific enterprise is founded on the possibility of reproducing the results of others' experiments in order to provide an independent test of their results. Generically, one might say, a result does not count as a scientific result unless, and until, it has been subjected to the independent testing of others.

---

1. This is both to generate credit for the original experiment and (perhaps more importantly) to recruit new members to experimental practice.

This role for replication is simply to check the veracity of claims that some experiment has been performed, and has produced results as reported. That is important, both as a check on fraudulent claims and as a check on whether these claims result from some error. So to say that an experiment is replicable may be to say that the narrative of the experiment and of its result is correct. Notice, though, that in this understanding replication has no role in generating scientific knowledge directly. At most it tells us that it is acceptable to form beliefs about the world based on that experiment, or perhaps to have our beliefs about the world sustained by it. In this view, a replicated experiment would be a verified experiment, and replication would be simply one among perhaps many methods of verification. And this sense of replication is pretty clearly a kind of Type II calibration, and not itself a generator of experimental knowledge.

Before going any further, I need to clarify who counts as the knowing subject in the case of experimental knowledge. One often hears (and I suppose I often say myself) that we know something or other on the basis of experiment. But in what sense is it true that *we* know such things experimentally? Not many of us have gone through the trouble of generating novel knowledge of nature experimentally. I believe, for example, that TCDD (a dioxin) is a promoter of liver cancer in female Sprague–Dawley rats, that the presence of gravitational fields bends light beams, and that human agents suffer from the endowment effect (giving extra value to a thing just because we own it). Indeed, I think I probably know these things. I also think *we* know them, and that they are all experimentally generated items of knowledge. But I am not sure that I myself know them experimentally. There are a few people who know them experimentally; the rest of us know them (if we know them at all) by testimony. Testimony is a fine way of getting to know things, and that's good given how much of what we know comes from testimony, but it is not experimentation. When someone knows something experimentally it is necessary that the appropriate observations be used in causing belief. My, and our, knowledge may well derive from experiment, in the sense that it is derived from the experimental knowledge of researchers or groups of researchers, but it is not true that it is known experimentally by us, even if it is knowledge deriving from others' experimentation. I just want to flag here that there is potential confusion, in much the way that we may be confused by claims about, say, mathematical knowledge. I know a lot of mathematical things that have been proven, but I know many fewer by proof. I am content to suppose that knowledge can be transmitted by report or testimony, but I am not sure that knowledge by proof, or by experiment, can be.

So, how does this help us to understand what is going on in replication? Let me put it like this. Suppose the place where I get all of my scientific beliefs is in the pages of Nature, and suppose further that if I read something in Nature I come to believe it. In that case, as a knowing subject, I have an interest in making sure that the things that find their way into scientific journals like Nature are true. That is, I want to be sure that Nature is properly calibrated as a channel of information about the claims that appear there. So we might think that replication is itself a method for calibrating venues that report on experimental claims. That would be a *kind* of epistemic benefit—not to the maker of the experimental claim in the first place, but to the rest of us who do not have access to the experimental signals themselves. To be sure, should I (or anyone) perform an experiment and come to believe something thereby, the fact that someone else reperforms or replicates that experiment can shore up my belief if my results are confirmed and undermine that belief if they are not. It does not, however, provide any necessary epistemic support. If I know the experimental claim on the basis of my experiment, that fact need not change either on a confirming or on a disconfirming replication of my experiment. It might well be rational to consider changing my belief in the latter case, but acknowledging that possibility is a far cry from acknowledging that replication is somehow the foundation of experimental knowledge.

Replicability (or the more demanding standard of actual replication) is neither necessary nor sufficient for experimentation to produce knowledge. Like intervention, it provides value, but not a value that it alone can provide. Rather than belaboring this point, I will turn now to some case studies that clarify the calibrational nature of replication.

7

Hertz and Cathode Rays

There continues to be a minor controversy concerning whether or not Perrin and Thomson separately replicated Hertz's attempts to measure the electro-static properties of cathode rays. Hertz, of course, concluded incorrectly that they were charge neutral. It will be instructive to examine this case again in order to further clarify what is at stake in questions of replication. The case will not take us very deeply into the nature of replication at first, but it will show us something about how to characterize experiments so that we can pose properly questions about replication.

I think this case can be used to give a compelling view of how what we want from replication is connected to our understanding of what counts as replication. A plausible view is that Hertz's experiments investigating the charge on the cathode ray were replicated by Thomson and by Perrin, even though the design, construction, and operation of their machines differed somewhat from Hertz's (and from each other's). My analysis of the case will highlight the kinds of good that replication gives us and at the same time show how Thomson and Perrin secured those goods with their experiments. In particular, I will show how the knowledge made available by Hertz's device (had it been set up properly, or had he interpreted it properly) is the same made available with Thomson's and Perrin's devices. This is so because all three men were asking the same question, and because the same flow of in-formation from the properties of cathode rays through particular instances of the rays in these devices, in conjunction with the manipulations performed on the total systems, answers that question. These experiments ask the same question in the same way. That is why, if Thomson's and Perrin's experiments show what we think they do (that cathode rays are charged), then a proper running of Hertz's experiment (including removing the interference from the

Faraday cage without undoing its job of mitigating stray electric fields in the lab) would find their charge-bearing feature.

7.1. Did Thomson and Perrin Replicate Hertz?

It seems obvious at this point that both Thomson and Perrin did replicate Hertz's cathode ray charge-catching experiment. I discussed this some time ago (Mattingly 2001). The issue in my opinion involves saying clearly what Hertz was asking about in his experiment, and then showing that Perrin and Thomson were each asking that same question and doing so in appropriately analogous ways. Hertz wanted to find out whether cathode rays carried an electric charge. The problem is that they're very hard to handle. He had already, he thought, shown that the cathode beam itself could not be deflected by electric fields, although it could be deflected by magnetic fields. So there was, he thought, a puzzle. His solution to the problem of measuring the charge on the rays directly was, essentially, to catch them in a bucket. He inserted his cathode ray tube into a metal cup that was attached to a very sensitive electrometer. He put the cathode and anode together at the head of the tube, and made a hole in the anode so that rays could pass through it into the region of the tube surrounded by the charge catcher. His idea was to allow the cathode rays to flow, and that if they carried a charge, that charge would induce a charge on the surface of the metal cup that would be detected by the electrometer. Hertz was unable to measure any charge on the rays.

Perrin developed a similar device, with a slight but important difference. His charge catcher was inside the tube itself. It was also inside the anode. Perrin's anode was a box at the far end of the tube, and the charge catcher was inside it but electrically isolated from it. Perrin allowed the rays to enter the charge catcher and detected charge.

Thomson, by contrast, used a very different device. It was similar to Hertz's in the sense that the anode had a circular hole in it to allow rays to pass into the main tube. But Thomson's detector was off axis, so he had to use a magnetic field to direct the rays into the bucket. When he did so, he also detected charge.

Thus we have three experiments set up to measure the same thing. One fails. The two that succeed seem somehow the same, and yet somehow different from the one that failed and from each other. Is there replication here? There is certainly no repetition. How do we answer the question in the first place?

Jed Buchwald (1995) provides an interesting method for sorting out the right way to think about experimental replication. He considers Hertz's cathode ray experiments and subsequent experiments by Perrin and Thomson and concludes that Hertz's experiments were not replicated by either Perrin or

Thomson. His basis for this conclusion is a detailed appraisal of Hertz's experimental resources—that is, an appraisal of what Hertz could do. How does he find that out? He goes through the various pieces of apparatus to find out what purpose each was serving in the original experiment. And he finds something interesting. Past the anode, Hertz inserted a piece of metallic gauze. What could it be for? What experimental resource does it provide? Buchwald concludes that the gauze is to make pure the rays from the cathode, to eliminate spurious tube current. This, he thinks, is an important resource, pointing to an important thing that Hertz could do, and by extension needed to do: purify rays.

Buchwald is not wrong that Hertz included this gauze, but he is wrong about its purpose. Hertz's idea for measuring the charge is straightforward, but there is a complication, and one that Hertz repeatedly complains about in his published work: his lab was an electrostatically noisy place. In particular, his source of electricity was often generating stray fields. In order to isolate the experiment from this influence, he surrounded the machine with a metallic shield. In order for Hertz to find out the charge status of the rays, he needed a space that insulated the rays from the fluctuations of electromagnetic fields in his lab, especially those at the cathode ray origin, and the Faraday cage (for that is precisely what the metal box around the experiment is) electromagnetically provides this. The problem is that if the charge bucket were completely shielded, Hertz would not have been able to get the cathode rays into it. So, to allow rays in, but to mitigate the effect of the laboratory fields, Hertz used a piece of metallic gauze at the point of entry.

Perrin's apparatus also includes a Faraday cage. It has a much larger opening that Hertz's, and it doesn't have the gauze over the open end. But in Perrin's case the gauze is not necessary. In part this is because Perrin's detector is sufficiently far removed from the cathode and the source of any electrical interference. Moreover, because the Faraday cage is within the tube, Perrin has control over the rays' entry into the cage and then into the charge catcher. He uses a magnetic field to move the rays out and back in, and sees charge accrual only when the rays go in. This control over entry into the catcher allows Perrin to further mitigate the possible influence of fluctuations in the electromagnetic field. Thus Perrin can ask and answer Hertz's question, in Hertz's way. Thomson as well has both this distance from the cathode and control over entry of the rays into the charge catcher. He also asks and answers Hertz's question, in Hertz's way. The only problem here is Hertz's setup. His gauze has an unfortunate side effect, one not anticipated by Hertz but one he could have understood easily: it damps the charge by absorbing many of the electrons of which the rays are composed. There were still plenty of glowing rays *visually*, but the charge density was seriously reduced by the gauze.

Buchwald is simply mistaken in his claim that the gauze was for ray purification, and his mistake follows from his focus on doing. We can see this easily when we see Hertz repeatedly talking about having pure rays in circumstances where there is no gauze at all. Purity is simply not an issue for Hertz, and it is trivially achieved simply by elaborating the rays a distance from their source. If instead of starting by trying to find his uses for equipment one focuses on what Hertz was attempting to know, we can see that Perrin and Thomson were after the same knowledge. What did Hertz want to know? Well, he wanted to know whether the rays were charged. Beginning there, and attending to the things Hertz complained about as interfering with his ability to know, we easily find the purpose of the gauze and the other material resources he brings to bear. This is the key to understanding experimental replication: a focus on knowing. Of course we should pay attention to what experimentalists are doing—their doing is the foundation of their knowing—but we do not want to mistake the method of securing assent to an experimental claim for the experimental claim itself.

The information theoretic account in part II of this book will provide resources to help experimenters avoid that mistake and to clearly situate various attempts to replicate an experiment as first making knowledge claims, and then attempting to secure assent to those claims rather than proceeding the other way around or failing to consider explicitly the distinction between asserting and justifying.

If our goal for an experiment is to establish that we can know something on the basis of that experiment, and our goal in replication is to evaluate the knowledge claims that arose from the original experiment, then we must focus on exactly what can be known from the experiment, rather than on what the experimenter can do. And we must focus on the way what we seek to know is to be known. It might now seem that we are opening the door to the question of what the experimenter can do. However, I do not think that is the case. On the model we are using here—one that characterizes experimental situations in terms of how information flows within them—the particular manipulations one researcher can apply do not, at least in the first instance, bear directly on the question of whether and in what way knowledge claims are established or on whether those claims are established in the same way from situation to situation. Instead, once we have an understanding of the way information would flow in the experimental situation, we can see what might obstruct it: for Hertz, lab noise has been explicitly theorized as an obscuring information block, to be removed by the gauze; impurity of rays is for him no such problem.

Here, in outline, is how I will characterize experimental situations: There is some distributed system. The various parts of the system are connected by information channels, and each of the parts has both its own internal logic and a set of maps that transform the tokens of that logic into tokens of the logic of the distributed system as a whole. Each part also has a set of maps that transform the tokens of the distributed system into tokens of the local logic. These maps can be used to develop maps from the local logics to each other. I will have a great deal more to say about this later, but the point for now is that these systems of maps encapsulate the kinds of questions we can use the experimental system to ask and answer.

When we characterize Hertz's experiments in terms of the questions they were built to ask, we are able to judge that there was replication. While there is something odd about saying that Perrin and Thomson replicated an experiment that Hertz got wrong, we can see that Hertz's device was capable of asking the very question we have identified—are the rays charged?—but that it muted the answer in a way that Hertz was prepared to understand, and that he could have reversed. Thomson's and Perrin's experiments asked the same question in the same way, but did not so mute the answer. What could one know on the basis of these experiments? The charge-to-mass ratio, for one thing. And indeed, surprisingly, one can find a good value for that quantity in Hertz's data. Even as muted as the charge was, the device functioned to bear the information that the rays were charged; it is just that the charge-to-mass ratio calculated based on the data he collected seemed beyond the realm of the possible to him, especially without a much clearer registration of the fact of caught charges.

Hertz's published work indicates that he would recognize the charge-damping character of his device after very minor consideration. Faced with the Thomson and Perrin results, he would immediately see that their experiments replicated his and showed that the rays are charged, and he would recognize further that, by either removing his filter or increasing the electrostatic sensitivity of his detector to observe charge at the level he himself calculated would be in the device even after attenuation, his device would confirm their results. We cannot be sure, of course, because Hertz did not live to see these experiments performed. However, my analysis of the situation makes clear that Hertz's understanding of electrodynamics would make such a fact obvious were he confronted with it.

7.2. Knowing versus Doing

The focus in this chapter so far has been on the tools of experimentation and their role in adjudicating claims of replication. Some (Buchwald, for example) think a principal goal of judges of experimental replication claims is to account for what experimentalists are able to do with such tools. I have argued that it is better to ask what they can know with those tools. My approach gives a satisfactory answer to questions about the replication of Hertz's cathode ray experiments. It also anticipates part II of this book, where I ground replication in the information theoretic account of experimentation.

Before moving on, however, I will consider one more view of the Hertz episode, provided by Chen (2007), who suggests that there is a kind of hybrid way of looking at questions of replication. It's a neat idea. Chen suggests that experimentation should be understood as a kind of means–end reasoning. In the case of Hertz's experiments his idea is that I am focused on the end, and Buchwald is focused on the means. Thus Buchwald is partly right to orient his analysis in terms of the tools Hertz has in front of him (his means), while I am partly right to focus on the ends Hertz identifies (what he wants to know). Chen's view is basically that to understand replication is to coordinate the means available to the experimentalist with the ends they have in mind. That's consonant with what I wrote in my original Hertz paper, and with the information flow account of experiment.

Of course experimentation is governed by means–end reasoning. In that sense Chen's point is clearly correct. What Chen does not, I think, appreciate is that my focus on Hertz's knowledge-gathering aims *led* to an understanding of his means for accomplishing those ends, and to the correct understanding of which *apparent* resources for accomplishing those ends were really resources of a very different kind. Chen's continued focus on the gauze as a tool for elimination of tube current is a result of continuing to begin with what *appear* to be means (as identified by Buchwald) and attempting to integrate those smoothly with the end of measuring the charge on the cathode rays. The end of the gauze is made clear by Hertz—it is to eliminate interference. Working backward from the end that Hertz clearly identifies for us, however, as I did in my original piece, makes clear that the gauze is helping to secure that end not by filtering current but by mitigating against stray fields generated by the device itself, and by the rest of Hertz's laboratory. Moreover, by not seeing how the gauze was supposed to function to attain Hertz's ends, Chen continues to misread Perrin's experiment. Perrin of course had no method of eliminating tube current, but tube current is not an impediment to gaining information about the charge state of the rays. Perrin *does* protect his charge catcher from stray fields generated by the device by moving it away from the source of the rays and from any possible stray fields. So Perrin can now manipulate the rays to keep them out of the charge catcher, and their entry covaries exactly with the accumulation of charge. Again the point is that Chen, like Buchwald, begins with the gauze, finds a possible use for it, and then attempts to align that use with Hertz's knowledge acquisition aims. As a result Chen and Buchwald miss both the fact that Hertz does not have that use in mind and the fact that its real use is clearly identified elsewhere in the experimental report. Instead, we should begin with Hertz's discussion of the experiment and follow along where we are led by his analysis of how information is supposed to flow.

Do we think of observations of the cathode rays as generating experimental knowledge? Of course we do, in conjunction with other features of the situation. But how, on my view, does the knowledge they generate move from merely observational knowledge to experimental knowledge? Is not the proximal system on which we are intervening in this case also the distal system about which we are drawing conclusions? In these experiments observations are made of cathode rays, and cathode rays are what we find out about. So isn't there something funny about the definition of experiment I have been using? That's a reasonable worry, but not, I think, serious. For the tokens we observe of cathode ray phenomena allow inferences to full-fledged type knowledge, knowledge about all cathode rays. Sometimes our background knowledge is such that there is essentially no gap between token observations and type generalizations—and sometimes our machinery is such as to facilitate that seamless transition. But without the inference from these particular rays to rays generally, we have done no more than observe; on the other hand, with that inference we have used our observations to experiment.

8

Replication as Contested

The Hertz example doesn't really tell us much about what constitutes replication, or what its real purpose is. For that it will be more useful to examine a case where the nature of and point to (attempted) replication is up for debate. I will argue that much of what replication is and should be is calibration, not the routine checking of others' results, as many who see a crisis of replication seem to think. I will take "calibration" in both its senses: that of preparing systems to function as apt channels for information flow, and that of checking that systems are functioning as apt channels for information flow. The first pertains to the generation of experimental knowledge itself, the second to the evaluation of that knowledge. It is this second sense of calibration that is most relevant to understanding replication's role in experimental practice.

I will begin with the use of replication to quell debates in experimental practice over controversial claims, both those concerning effects and those concerning the significance of those effects for our experimental knowledge. I will then consider another important attempted replication that showed clearly that no knowledge of gravitational waves had been been produced by Joe Weber's observations in the late 1960s and early 1970s. This episode made it clear that Weber's device simply could not produce data about those waves. The replication attempts were via other devices, constructed in good accord with his protocols (i.e., constructed as versions of his devices), to check whether such versions could themselves produce gravitational wave data. I will evaluate case studies by Collins and by Franklin and Collins. I will endorse Collins's understanding of what brought the episode to a close, but little else in his argument.

In connection with this episode, Feest (2016) provides a welcome exception to the tendency to fold together the production of scientific knowledge with the justification of knowledge claims. Her view of the debate between Franklin and Collins is that analysts have failed to appreciate the distinction between these two modes, and on this point I think she is right. She claims that attempts at experimental replication very often point toward an attempt to clarify how rationality standards can be brought to bear on what she calls the "*investigative process* itself, in particular as it pertains to determining the applicability and scope of novel concepts" (2016, 41). Feest's work then fits nicely with my view that much of replication is about both Type I and Type II calibration. Part of the function of calibration must be to sort out which propositions, so to speak, are of the kind that experimental procedures bear on—and key to *that* will be clarifying the conceptual terrain. I recommend that work, then, for another view of what goes on in the interaction between Collins and Franklin over the so-called experimenter's regress. Here, however, I am going to focus directly on how implicit epistemological commitments steer much of the discussion into a dead end.

8.1. Collins's Internalist Epistemology

Harry Collins has emphasized as strongly as anyone the importance of the disconnect between the significance scientists *say* replication has for the foundations of science and the actual role it plays in their disciplines. "Thus, though scientists will cite replicability as their reason for adhering to belief in discoveries, they are infrequently uncertain enough to need, or to want, to press this idea to its experimental conclusions. For the vast majority of science replicability is an axiom rather than a matter of practice" (Collins 1985, 19). He has noted as well that replication is most needed in times of deep confusion in particular fields. These occasions bear some resemblance to what Kuhn deems science in crisis, when normal rules do not apply and scientists are in the midst of establishing and validating new methods of generating knowledge—in this context, experimental knowledge.

However, Collins still takes replication to be central to the experimental method. It is a "vital idea. Replicability, in a manner of speaking, is the Supreme Court of the scientific system" and "corresponds to what the sociologist Robert Merton (1945) called the 'norm of universality.'" Indeed, Collins calls it the "scientifically institutionalized counterpart of the stability of perception," saying that its "acceptance . . . can and should act as a demarcation criterion for objective knowledge" (1985, 19).

By invoking the stability of perception, Collins identifies the many issues that are connected to replication in science with various aspects of the general problem of induction, both in Hume's version concerning how we are able to project predicates, and in Goodman's version concerning how we know which predicates are projectable. For Collins the key question is how we can know when we have successfully replicated some experimental result. I believe his answer is that we can, but that it is a kind of socially constructed knowledge, by agreement and by authority—items he would like added to the standard repertoire of epistemic tools. He makes this even more explicit in his paper on "The Meaning of Replication and the Science of Economics," where he explicitly invokes "external" explanations for the "internal" progress in science (1991, 131). His view seems to be that we can never from a purely epistemic standpoint fully validate our knowledge claims, and that we are in a closed epistemic circle, or a regress of reasons, that we can only escape by appeal to something beyond considerations that are internal to scientific method.

As illuminating of the social character of the scientific enterprise as his investigations are, much of his epistemological discussion suffers from the attempt to identify knowledge with the capacity to persuade, and his discussion of regress rests on the need to provide decisive rejoinders to the skeptic, or to those scientists who believe that we have not fully or appropriately replicated their work. But we do not need to silence the skeptic in general, or those who are skeptical about a specific instance of scientific replication (or validation). We must *reply* to the skeptic, and that we can do, but we can never silence her, nor should we. Knowledge does not rely on the absence of any possible objections, nor on ruling out of court skeptical doubts. Instead it relies on uncovering, building, and employing channels that really do connect us to the information that we want, and coming to believe on that basis. Challenges from skeptics about the workings of these channels, when met, provide even more knowledge, specifically about *how* the channels provide knowledge. When their challenges have not yet been met our knowledge is not thereby undermined: either we didn't have it to start with because the channels don't work, or we did because they do work, so we still have it but don't yet know exactly how. Unlike the traditional skeptical portrayal, our knowledge is not fragile and in constant jeopardy all the way up until our response to the infinitely removed, final "but how do you know?" stage; instead our knowledge is stable and increases at each stage of the dialectic.

Collins's account of how replication functions recalls again the problematic addressed by the founders of the Royal Society. Their idea was to build absolute assurance regarding matters of fact into the very process of experimental science. Only in that way, they thought, could critics as diverse as Hobbes and Shadwell be silenced, and their enterprise made safe. But such methods are not necessary. Science works, and the way it works is by continual calibration of its tools, not by providing decisive rejoinders to its critics.

One of my controversial assertions in this part of the book is that on a fundamental level replication has very little to do with scientific knowledge claims. We will see, however, that my assertion is entirely consistent with replication being a crucial part of the scientific enterprise. Replication as such, when it is of most use, tends to provide second-order epistemic goods: that we are right to believe some claim (or not); that some method of belief generation is also knowledge-generating (or not). These types of good, especially the latter, are deeply connected with progress in science because they both control what beliefs we take to be generated by given information-bearing channels, and control what information-bearing channels we tend to open (i.e., what experimental apparatuses we construct) in order to generate our scientific beliefs. They are not, however, at all connected to whether a given belief is itself an item of scientific knowledge.

Collins insists that "replication is the establishment of a new and contested result by agreement over what counts as a correctly performed series of experiments" (1991, 132). But agreement has little to do with the real meat of the case. What replication does is evaluate the methods, procedures, and ultimately the results claimed to be associated with a previous experiment. It does this by determining whether those methods and procedures were, in the first place, adhered to, and then whether they are appropriately calibrated to produce those results by allowing information to flow from the distal system targeted by the researchers through the proximal system with which they had to deal.

So replication is twofold, and needs to do two classes of things that are not clearly possible to accomplish using a single experiment: checking that implementing the required methods and performing the required procedures would yield the previous results pushes the replicator in the direction of closely mimicking the previous series of actions; and evaluating the capacity of those procedures to yield information about the target system itself pushes the replicator toward a novel method that both yields that information and illuminates the way in which the earlier experiment either could or could not do so. Thomson's and Perrin's replications of Hertz's cathode ray experiments are models in how the tension between these two aims can be resolved. As I showed, the appropriate analysis will center on understanding how information was supposed to flow, how it did or did not flow, and how subsequent experiments bear on those issues. From this perspective, replication is not so far away from calibration.

8.2. Replication and Calibration

As a general matter, calibration either produces devices that are proper conduits of information, or shows devices *to be* proper conduits. In the latter case, if a Type II calibrating replication shows that the original device was properly calibrated, we already have its information, and any beliefs caused by it were already knowledge; if it shows that the original device was not (or could not be) properly calibrated, we get neither the information nor the knowledge claimed for it. In that sense replication does not give experimental knowledge of the original system. The very act of replication may, to be sure, coincide with a new experimental act that itself gives such knowledge; but it is not *replication* giving the knowledge. The interesting case of Type I calibrating replications arises when we have signals in the channel that *could* carry information, but there is something wrong with the channel; fixing that can both calibrate the channel and produce new experimental knowledge of the distal system by means of the information already produced but somehow trapped within the experimental apparatus (in its extended sense). What we see is that a principle purpose of replication studies is best characterized as (re)calibrating of the original study.

I believe that replication is not itself a generator of experimental knowledge about the distal system targeted by the original experiment. Rather, it is a method of calibration in the second sense—a kind of technique to generate conviction that we know, or knowledge that we don't know.

Collins (1991, 131–32) says:

> Repetition should not be confused with replication. Replication is the establishment of a new and contested result by agreement over what counts as a correctly performed series of experiments.

Replication then, according to Collins, is only appropriate in the context of establishing contested results, in which context questions of validation and verification of going procedures are not relevant. He thus rules out of the replication court most standard scientific practice and undertakes to adjudicate only claims to have established fundamentally novel techniques, procedures, standards, or what have you. A crucial difficulty with Collins's view is that it makes replication essentially contested by fiat. It also gets backward what is at stake in some paradigmatic cases of replication controversy. In the case of Hertz and cathode rays, for example, even those who disagree that Hertz's experiments were replicated by Thomson and Perrin do not assert that Thomson's and Perrin's experiments were incorrectly performed. Rather, the assertion is that those experiments do not mimic correctly *something* about Hertz's experiment. It is that something that we as theorists need to identify in order to have a clear notion of what replication amounts to.

Collins's view, while going farther than I would want in making replication essentially contested, does conform nicely with my own view of what can constitute replication in such situations. So in the attempts to replicate Hertz's cathode ray results, multiple different techniques, with their own status in the arena of scientific acceptability, are brought to bear to settle the question of how Hertz's procedures should themselves be taken to bear on his claimed results. While the techniques do not reproduce his methods, and moreover invalidate his result, they do properly settle the question: they rule that his techniques do not suffice to establish his results. What they do is reproduce what were information-bearing channels that would be acknowledged by Hertz as such, at least according to his understanding of the technology inso-far as we can assess that understanding by reference to his texts.

Further, such an account of replication also conforms to my novel struc-tural claim that token identical experimental interventions may well be in-stantiating type-distinct experiments as well as token-distinct (but type-identical) experimental replications. This is so because along the way, these studies also produced first-order experimental knowledge of the charge state of cathode rays. As I argued above, that is distinct from their calibration/replication task.

8.3. Collins's Version of Replication

It is challenging to sort through the various murine metaphors in Collins's book to find out what he takes to count as replication precisely. The exact connection between our scientific endeavors and his mouse scientists is ob-scure in the extreme. However, we can say this much: There are two versions. The first, called the analytical, is rejected after a brief examination, but Col-lins never gives us a theory to consider. Instead he considers in outline the idea that replicating experiments should be a balance between being closer to identical to the original experiment (in order that their data bear more strongly on those originally produced) and further from identical (in order that they produce novel information). Collins notes that there is an episte-mological tug-of-war in store for the theorist using this method and there-fore abandons it (1991, 38). It remains obscure what type of *analytical* notion he has in mind there. In any case, he immediately replaces it (1991, 39) with what is called a "sorting schema"—that is, a series of operations observers of scientific activity could use to see what it is that counts, according to the methods in play by scientists, as a true replication of some experiment. This so-called empirical procedure considers all the activities on earth as a whole and then classifies them in a quasi-Linnean sorting procedure that first elimi-nates "all activities not to do with the subject" and then continues until finally it eliminates every activity that doesn't count as replication. The key step is, apparently, step five, where we eliminate activities that do count as experiments, but do not count as competent copies of the original. This step, Collins claims, defies attempts by those operating within science itself[1] to produce a rational theory of replication (1991, 42). That is, the step in deciding of two experimental procedures whether one is a properly done version of the other is supposed, somehow, to be beyond appraisal from within science. Little in the way of direct argument for this claim is made, however. So I will turn instead to an episode at the heart of Collins's understanding of experimental regress: the replication of Weber's gravitational wave experiments.

8.4. Weber and Gravitational Waves

In the case of Joe Weber's gravitational wave experiments there is no controversy over whether they were replicated by various teams at the time. That is generally acknowledged. There is a controversy, however, over what makes those replications, and what brought the episode to a close. That controversy is also over the nature and role of replication itself.

General relativity implies the existence of gravitational waves in our universe. Generally, any acceleration of a massive body will cause gravitational waves. These are not to be confused with gravity waves, which are simply waves that are caused by gravitational fields. The waves in the ocean count as gravity waves, and those are easy to observe. Much more difficult and complicated to observe are gravitational waves. The most obvious difficulty comes from the fact that gravity is many orders of magnitude weaker than electromagnetic forces, which are responsible for the vast majority of the things that we see. So any experiment measuring gravitational waves needs to be very sensitive, and very well insulated from any other influence (sound waves, radio waves, seismic waves, etc.). Even such detectors will not be sensitive to the kinds of gravitational waves typically present in our region of space-time. Instead we will only be able to observe gravitational waves that result from very energetic events like the collision of two black holes, or neutron stars.

There now seems to be good reason to think that gravitational waves have been observed in the laboratory. The Laser Interferometer Gravitational-Wave Observatory (LIGO) collaboration announced on February 11, 2016 that on September 14, 2015 gravitational waves had been detected for the first time in history. The extremely sensitive detector is really two devices separated by just over 3,000 kilometers. At each location a laser beam is split at the

---

1. He talks about terrestrial efforts as opposed to the science-fictional mice that are supposed to be performing the classification.

intersection of two 4km-long perpendicular tubes, sent down those tubes, reflected, and recombined back at the intersection. The changing interference patterns can be used to detect the minutest changes in the relative length of the tubes. Many things other than gravitational waves can cause the tubes to oscillate in length—electromagnetic fields, seismic waves, fluctuations in temperature, etc.—but when those changes are consistent between the two widely separated locations, we can infer that a gravitational wave has passed by.[2] As I would put it: those synchronized changes bear the information that a gravitational wave has interacted with the tubes and caused the oscillations (given, as we think is the case, that nothing else could have been responsible).

The LIGO group, however, was not the first to *claim* that gravitational waves had been detected. On February 8, 1967, Joe Weber (1967) communicated the result of a two-year study of gravitational waves to *Physical Review Letters*, stating that the discovery of gravitational waves by this study could not be ruled out. A year later (1968) he was making a stronger claim: his detectors, he said, were clearly responding to a common external source that might be gravitational waves. The following April (1969), he claimed to have ruled out any plausible external sources other than gravitational waves. In September of that year he submitted a paper claiming more than consistency. He says: "These results are evidence supporting an earlier claim that gravitational radiation is being observed" (Weber 1970, 276). While Weber, who died in 2000, never gave up his claims of gravitational wave detection, by the early 1970s his claims were roundly rejected by other physicists. What happened? Why did other scientists reject his results? Why did he continue to maintain them? And what does any of it have to do with whether or not gravitational waves were, or even could have been, detected by Weber's device?

I'll begin with the fascinating reminiscence of Allan Franklin regarding his rapprochement with Harry Collins (together with editorial comments by Collins) over the role of sociological factors in the development of scientific knowledge, specifically with respect to the Weber case (Franklin and Collins 2016). More than any other student of the experimental sciences, Collins has attempted to understand exactly how this episode developed, progressed, and eventually closed. His principal method of research for his analysis of the Weber case in "Changing Order" (1985) is interviews with scientists, but he says about his subsequent treatment of the case that "my large book, *Gravity's Shadow*, which was published in 2004, uses every resource I could get my hands on including all the published materials I could read plus private

---

2. More details can be found at the LIGO website; this information was drawn from their fact sheet: https://www.ligo.caltech.edu/system/media_files/binaries/300/original/ligo-fact-sheet.pdf.

correspondence painfully extracted from Joe Weber" (Franklin and Collins 2016, 98). Collins calls his basic orientation "methodological relativism," which is the position that "to do good sociology of scientific knowledge it is vital not to short circuit the analysis by explaining the emergence of what people count as the truth by the fact that it is the truth" (2016, 98). So far so good, but then he adds the astounding non sequitur that "therefore, in the course of the analysis of what comes to count as the truth of the matter you have to assume there is no truth of the matter" (2016, 98). Of course such a claim has nothing to do with the methodological position he describes.

Allan Franklin has also studied this case, using the more traditional methods of history and epistemology of science. He soberly "advocates an essential role for experimental evidence in the production of scientific knowledge" and insists that "one can decide what is a correct experimental result independent of that result by applying what he calls the 'epistemology of experiment'" (2016, 99).

The background of this interesting document is that Collins and Franklin have had rather sharp disagreements over the years both on how to conduct analyses of the production of scientific knowledge in general, and on Weber's claims to have observed gravitational radiation in particular. However, in recent years they have come closer intellectually, and the document cited is meant to reflect the things in the case they agree about, and those they still disagree about.

One thing that is universally accepted, though, is that almost immediately after Weber published his results, other physicists attempted to observe gravitational waves using devices built on his model. They were not successful.

Weber did have what appeared at first glance to be quite compelling data. Those data did require explanation. That explanation, however, had nothing to do with gravity waves. As emphasized by Franklin, Garwin's letter in *Physics Today* made it perfectly clear that selection bias (the physicist's version of p-hacking) could easily produce what appear to be extremely unlikely coincidences out of data that were known not to involve gravitational influence. Garwin outlines results by James Levine showing that a judicious sampling routine could result in a six-standard-deviation "effect" (Garwin 1974). Weber never seemed to understand the power of such selection bias, insisting that he did not rely on the computer and that his team's results were arrived at using "real-time counting and pen-and-ink records" (Weber 1974, 13). He was not able to appreciate that Levine's pseudoeffect is not a computer artifact, but instead the result of a decision about how to evaluate signals, or that it simply reflects the fact that if we decide that a message has been sent with a certain content (that there's a gravitational wave source, for example), then we can generally look at a string of symbols and "find" that content within the message. Basically, by choosing how to cluster various signals (corresponding to spikes in the background noise), Levine was able to find apparent correlations in data that were known not to be correlated in that way. That analysis completely blocks Weber's analysis—even were there gravitational waves in the device, he could not have known it using his methods.

Something else to notice here is that Collins's view of what it is to accept a scientific claim is completely inadequate (Franklin and Collins 2016, 104). He says that he is

> always asking, would someone determined to believe in the reality of Joe We-
> ber's claims be forced to reject them by "this" or "that"? Among the "this's"
> and "that's" are the counter-experiments. I argue that if you were determined
> in that way, the experiments would not prove to be decisive though, of course,
> they would still be important evidence. Joe Weber would have been much,
> much happier if others' experiments had supported his own but the other ex-
> perimental results did not, and could not, force him or his allies—of which
> there were a few (see Collins (2004))—to give up.

This kind of thing is precisely the problem with Collins's "relativism." There is a persistent ambiguity between the truth of the knowledge claim itself, the question of who believes the claim (individuals, communities, etc.), and the cause of that belief—whether an entrenched prior conception, signals bearing new and relevant information, or signals that don't bear that kind of information. The way these things are discussed within a justificationist framework makes it extremely difficult to get any kind of handle on what's going on.

It is difficult to evaluate any of these questions when they are not clearly separated. That is the most frustrating thing about Collins's analysis. Did Joe Weber observe gravitational waves? No. The follow-up experiments done by other groups establish that his device could not be calibrated to transmit information about the kinds of event he inferred from the signals he received. And Levine's analysis, by contrast, showed that the device was apt to generate spurious signals that appeared to carry such information. But Collins mixes up that kind of fact question (what information about Weber's device was generated by these replication/calibration attempts?) with a very different kind of question (is it possible to force someone to retract an empirical claim on pain of irrationality?). We know from Duhem, from Quine, from Kuhn, from Lakatos, etc. ad exhaustion, that the answer to the latter question is "no." It is never possible to force someone to give up an empirical claim on pain of irrationality as long as they give up sufficiently many other empirical claims.

That's just a red herring. We are not really interested in the latter question in any case, at least not when our interest is in the nature of experimental knowledge claims themselves. What we do want to know is what information could have been transmitted by Weber's device, and what could not. As soon as we see that Weber is engaged in confirmation-chasing we stop (as we should) considering him as engaged in a scientific inquiry. The scientist Q, quoted in Collins (1985), sets out to destroy Weber's credibility by performing some trivial experiments that would give them a standpoint to decisively show other scientists that Weber's results were nonsense. Collins focuses on the remark by Q that refuting Weber was *itself* no longer doing physics (1985, 94). He thinks this somehow supports the idea of an experimenter's regress. Again, the idea is that there is a regress if I cannot force you to drop your experimental claim. Collins's view seems to be that because Q was no longer doing science, then science is not enough to resolve the matter. But that's completely backward. The science was already clear, but those not in the gravitational community might be misled by the fact that Weber was still going around giving very confident talks, based on what Collins rightly points out are his rationally defensible beliefs. They're rationally defensible, but not driven by the signals he's receiving from his experiments. Of course Q was no longer doing science, *because* Weber had already dropped out of the pursuit of scientific results. Showing that had nothing to do with the replication of his experiment—the science on that was already clear, as Collins's own interviews with the scientists involved make clear.

The reason Collins and Franklin continue to disagree about what brought the episode to a close is that they disagree about what the episode *is*. Collins takes the episode to be resolved when Weber's credibility is so fully undermined that no scientific venue will take his claims seriously. Franklin, by contrast, sees it as resolved when the evidence suffices to settle the question. They are looking at two entirely different episodes. The first is the episode of the scientific community moving on and no longer entertaining the possible, but nonactual, experimental knowledge of gravitational waves presented by Weber. The second is the episode of there being experimental knowledge that Weber's devices did not carry information about gravitational waves. These episodes are causally related in interesting ways, but they are different, and so are their timelines.

Of course we must evaluate the evidence from our own perspective, and we could well be wrong about whether Weber did or did not detect gravitational waves. But that is not relevant to understanding the nature of the enterprise. In any situation involving persuasion, both the beliefs of various agents and the facts about the case will be present, but that is no reason to identify them with each other. What Collins with his talk of an "experimenter's regress" remains unable to appreciate is that epistemic forcing is just not a real feature of our cognitive situation. It is always rationally possible to hold on to any consistent sets of beliefs if we are willing to give up enough other beliefs. That is simply uninteresting from the point of view of understanding what constitutes scientific knowledge.

Returning now to the Weber case itself, we can see that the replication experiments are all attempts to determine whether or not Weber's device is capable of generating the kind of information he claimed for it—that is, calibration in the second sense. Are these other experiments directly relevant to the question of whether Weber did or did not have experimental knowledge of gravitational waves? Of course not. They are useful to us in our own attempts to assess his knowledge claims, and they can be useful as reports in regulating our own beliefs about the gravitational wave events. However, as with all calibrations in the second sense, they do not generate first-order knowledge of the event in question; rather, they generate knowledge *about* our state of knowledge of that event.

8.5. Collins's Claim

Collins's argument here is meant to ground his attempt to establish that the breaking of an experimenter's regress is seamless from the point of view of epistemology and that therefore epistemology comprises everything that goes into whether or not scientific results are accepted. The heart of this argument, it seems to me, is the claim that "the point of contentious experiments is to discover things that should not happen in theory," and the auxiliary claim (suppressed) that the theory of the experiment is somehow coextensive with the theory targeted by the experiment (1991, 138). Both of these claims seem false. First, the point of any experiment is to generate knowledge, independent of what theory does or does not say. Some experiments do seem to generate knowledge that conflicts with theoretical claims, or other experimental claims, but to relegate controversial experiments to the status of theory negators is to minimize and trivialize them. It also allows for a straw account of the so-called experimenter's regress.

Much of Collins's confusion is present in his discussion of the experimenter's regress in *Changing Order* (1985). The experimenter's regress arises from his generalization *without argument* of certain conclusions that he reached about how one evaluates the building of lasers. The proposition in question is this: "Proper working of the apparatus, parts of the apparatus *and the experimenter* are defined by the ability to take part in producing the proper experimental outcome. Other indicators cannot be found" (1985, 74). This proposition is subsequently applied, without any attempt at a generalizing argument, to the case of gravity wave detectors. We are supposed to believe that because it is difficult to know whether something is a laser until it lazes, we cannot know whether a gravity wave detector, a device evaluable using basically Newtonian laws, is properly constructed without being dragged into some infinite regress that can only be escaped by some appeal to the authority of the scientific community.

In addition to the hasty induction, there is a clear conflation in the first place between what it is to be the device and what methods one will use in order to evaluate whether some apparatus counts as such a device. This is part and parcel of the old internalist epistemologist's confusion between knowing and being licensed or justified somehow to make knowledge claims.

Collins is writing about scientific *persuasion*, and how it operates, and that is very important and interesting for us to understand. What he claims, however, goes far beyond all of that and into the authoritarian nature of scientific epistemology itself. Using his examples this way is fallacious. Yes, science is done by people, and some of our methods of belief generation do not track the flow of information. But even in such cases we can have those beliefs *sustained* by appropriate information, and that gives us knowledge even when we are initially persuaded by bad reasons.

As Collins's own interviews suggest, when we consider the scientific status, the *experimental* status, of Weber's claims about gravitational radiation, the people in the relevant community really did have knowledge of what was going on *from the data* rather than from public announcements. They were properly situated with respect to the information flowing from the experiments and did not require some nonepistemic escape from a nonexistent epistemic regress. Right in the middle of his arguments about a critical mass of experimenters contradicting Weber's claims, Collins reports this remark from "another scientist": "It was clearly a case where Weber had tripped himself up because of his data analysis and I felt that it spoke for itself and that those few people who knew about it were enough. But 'Q' did not feel that way and he went after Weber ... and I just stood on the sidelines covering my eyes because I'm not really interested in that kind of thing, because that's not science" (1985, 93). This tells me there is a public flow of opinion, especially in communities on the periphery of the gravitational wave community, and in that flow the critical mass of scientists whose results contradicted Weber's gradually coalesced into a public consensus, a consensus that could seem to have been reached on the basis of some version of an appeal to authority. There is another flow of opinion that is generated and sustained by the information that the various experimental studies make possible; only very ordinary and finite epistemic regresses arise there, and they are stopped by ordinary epistemic means. We see that Weber's results are an artifact of his particular binning procedure, and not derived from any information carried by the experiment's signals. The end.

In addition to these other worries about Collins's accounts, we also know that theory is rarely brought under experimental scrutiny in quite the way Collins envisions. Almost never is the theory governing the experiment under test when the experiment is being performed. Collins's examples *themselves* do not function in that way. Instead these experiments are being used to evaluate some claims or to generate some knowledge, and their adequacy to that task is called into question. Then the resources of the scientific community are brought to bear on that question, at the same time that these follow-up experiments are being performed. Despite the fact that the two processes are running in parallel, we should not see them as wrapped up together in some seamless impenetrable epistemological bubble. Rather, we should see them as they are: experiments evaluated, as all experiments are, against a background of established theory and practice, a background which can change during the evaluation, but which never loses contact with our epistemological standards.

9

Misleading Replication

I now consider the changing status of saccharin as a suspected carcinogen. My analysis provides support for my conclusion that replication is best thought of not as repetition or even principally as confirmation, but as calibration. This episode may clarify how replication studies provide calibration rather than first-order experimental knowledge. We will see in part III of this book that murine studies are generally not adequate to produce knowledge about human biology. Indeed, we will see that animal studies as a whole cannot be said to provide any robust experimental knowledge about human biology. I will argue that this is largely a problem caused by the lack of appropriate information-bearing maps between animal systems and human systems. Here I will preview that conclusion by illustrating it as a failure of replicability, understood as an evaluation of how well these animal models are calibrated for detecting features of humans, and sometimes of murine mammals generally.

The saccharin case is a conceptually important and financially costly example of experimental failure. I am looking at it not because I think our experimental methods are generally flawed (I don't), but because when things are going well it can be difficult to see how they function. Because experimentation is control over the flow of information, those times when it does not flow (but seems to) are going to illuminate better how that control can and should work by illustrating how information bottlenecks in various instances.

Saccharin, an artificial sweetener, was removed from many beverages and prominently labeled as a carcinogen on others because there appeared to be experimental knowledge that it would cause bladder cancer in those who consumed it. Its causal power was considered small but not negligible, and consequently the use of saccharin as a food additive in the US, Canada, and other countries was either banned outright or reserved for those whose health was deemed to be at much greater risk from consuming sugar than from saccharin.[1]

Let's look at why saccharin was banned in one way or another in many countries in the 1970s even though, as far as anyone now knows, it is not harmful to humans. The answer is not hard to find. Studies of rats showed that in high concentrations in the diet saccharin caused bladder cancer in those rats. Typically, those who *do* doubt saccharin's danger do so for terrible reasons. One standard response by the public to such studies is a kind of blanket skepticism: the concentrations in the diet are so high that *of course* it would cause cancer—anything would at so high a dosage. This is, of course, completely wrong. Many many substances are subjected to similar tests and, even at concentrations that no human could reasonably ingest, the test animals are just fine. Or at least they don't get cancer from the test substance. Blanket skepticism of that sort, though common, is silly.

Some things cause cancer in high doses, and some do not. And there have been many studies of rats fed diets containing varying concentrations of saccharin showing that it does cause cancer. Perhaps the most extensive is Schoenig et al. (1985). In that study some Sprague–Dawley rats were put on a saccharin diet before mating, and then some of those were taken off the diet after giving birth and some were left on until their offspring had been weaned onto one or the other of the saccharin diets being studied. These diets involved saccharin in varying percentages of total intake. With very great sensitivity, these diets caused bladder cancer. Table 9.1 shows Schoenig et al.'s results for cancer incidence.

As the table makes clear, there is a strong linear dependence of these cancers on saccharin intake. Yet saccharin does not, after all, cause bladder cancer in humans as far as we know. Despite all the experimental results showing its dangers, saccharin turns out to be as benign in humans as any other foodstuff. So what went wrong? It is a curious feature of this case that very few people seem aware, even today, that this conclusion about saccharin was so far off base, and most seem to regard saccharin as harmful to human health. Indeed, I only learned that it was not harmful because I was planning to study

1. For this episode see especially Schoenig et al. (1985), the 2003 California EPA Reproductive and Cancer Hazard Assessment Section Office of Environmental Health Hazard Assessment "Evidence on the Carcinogenicity of Sodium Saccharin" final report, and "Information Document on the Proposal to Reinstate Saccharin for Use as a Sweetener in Foods in Canada" by Health Canada's Bureau of Chemical Safety Food Directorate Health Products and Food Branch. These give a good overview of the considerations that initially led to the strict curbing of saccharin use, as well as the problems with the experimental evidence behind it.

TABLE 9.1 Incidence of primary neoplasia in the urinary bladder of female rats

Group (test compound; % in diet)	No. of rats	Percentage (and no.) of rats[a] with		
		Transitional-cell papilloma	Transitional-cell carcinoma	Transitional-cell papilloma or carcinoma
NaS				
0.0 (control)	324	0.0 (0)	0.0 (0)	0.0 (0)[b]
1.0	658	0.6 (4)[c]	0.2 (1)[c]	0.8 (5)
3.0	472	0.8 (4)[c,d]	0.8 (4)	1.7 (8)
4.0	189	2.1 (4)	4.2 (8)[c]	6.3 (12)
5.0	120	3.3 (4)	9.2 (11)	12.5 (15)
6.25	120	10.0 (12)	6.7 (8)	16.7 (20)
7.5	118	15.3 (18)	16.1 (19)	31.4 (37)
5.0 (through gestation)	122	0.0 (0)	0.0 (0)	0.0 (0)
5.0 (following gestation)	120	3.3 (4)	6.7 (8)	10.0 (12)
NaH				
3.0	118	0.0 (0)	0.0 (0)	0.0 (0)

Note: NaS = sodium saccharin; NaH = sodium hippurate.
a. Each animal was included only in the incidence for the most severe hyperplastic or neoplastic lesion observed: i.e., carcinoma > papilloma > papillary hyperplasia > nodular hyperplasia > simple hyperplasia. Rats with severe autolysis of the bladder epithelium and rats that died before the first bladder tumor was observed (study month 15) were not included in the denominator used to calculate tumor incidences.
b. The background incidence for total bladder neoplasia at IRDC is 0.8 percent.
c. Including one animal with grossly visible bladder calculi.
Source: Schoenig et al. 1985, 483 (Table 5)

the case of saccharin as an example of *successful* experimental knowledge generation.

The problem has to do first with the kind of experiments used to establish the result in the first place: murine models of human disease, in particular rat models. These models were completely inadequate for the generation of knowledge about human bladder cancer. Here is our current understanding of the impact of saccharin on human bladders: there is no reason to think that any amount of saccharin is carcinogenic in humans. There were in this case several very high-quality interventionist experiments with very good power. Rats were placed on high saccharin diets over their entire lifetimes. Their blood and organ function was assessed periodically throughout their lives, and various kinds of tissue biopsies were performed on them once they were dead. The causal impact of saccharin was quite clear. These rats were getting bladder cancer *from* the saccharin. Yet no humans have. Moreover, there is no particular reason, chemically, to think that they would.

We do in fact have studies showing that saccharin is clearly carcinogenic in rat bladder tissue. But these experiments are themselves contradicted by numerous other rat studies. So were the first experiments poorly performed, and were their data therefore unreliable? No. That's not the problem. In fact, there are two main things going wrong, but these things cannot be illustrated by the failure of replication studies understood in any way related to repetition. The studies are solid. They are properly performed. Performing them again a large number of times would not change anything. The first problem is that the earlier studies mistakenly rely on the rats in the study being good models for human disease, even though we have no really strong reason to think that generally rat metabolism bears any information about human metabolism that is of any biomedical use. The second problem here is that, for some reason, the rats in these studies appear to be metabolically untypical of rats generally. How so? What was finally discovered is that the tumors in these rats develop only doing organ formation, so only in very young rats. Rats, apparently, secrete a chemical compound to sequester and eliminate the saccharin (something like pearl formation in oysters), and the sequestered saccharin nodules can induce cancerous tumors in the bladders of some strains of young rat during bladder development. There is simply no mechanism in humans that is analogous to this. So the real situation is much more complicated than can be helpfully illuminated by a model of the epistemic role of replication built around the verification of the procedures in the original study. These rats are themselves poor models for rats generally. Indeed, rats generally are no more susceptible to bladder cancer from saccharin than are humans. The lesson here, in part, is that without that metabolic pathway being generalizable, the experiments were unable to ask questions about human bladder cancer. Such failures of questioning ability are typically relegated to discussions of external validity, but they should never be far removed from discussions of replication generally. In fact, the way to understand what we normally mean by replication is this: Sometimes studies are performed that seem to yield experimental knowledge. There is a trivial need to be sure that the experimenters are competent, honest, and not unlucky. But the main task of replication is to assess the capacity of the experimental procedures to generate the experimental knowledge that is being claimed. The former task evaluates the observations themselves: do these procedures result in these data? The latter evaluates the experimental protocol itself: could those observations bear information about the distal system of interest? That latter task, replication properly understood, is an instance of calibration.

We see that subsequent studies of rats, in conjunction with our understanding of human development, show that the original experiments were simply not capable of giving the knowledge claimed for them. Do we thus, on the basis of these replications, have knowledge that saccharin does *not* cause human bladder cancer? That is a trickier question. Let me leave it aside here and simply note that this case raises a number of questions about the nature of replication that need to be answered before it is fully understood.

Once again, the real problem is that sometimes rats (and murine mammals more generally, and other mammals even more generally) are impossible to calibrate as apparatuses that support the transmission of information about human biology. Indeed, we will see later that "sometimes" is misleading. Rather, it turns out that animals are simply terrible models of human disease in any probative context. In other words, when we really don't know about some disease course, and we are trying to find out about it using animal models, those models are as good as useless. We'll come to that. But here I just wanted to consider the case of saccharin from the perspective of replication understood as calibration. What it showed is that while all of the early studies on saccharin were correct in the data they produced, those data did not carry information about human metabolism—and that it was further studies on the precise timeline and etiology of the disease that made that clear, studies that revealed the exact way the data carried the information it did. Here attempts to replicate were able to illuminate the calibration failure of the original experimental systems.

This is not at all unusual in cases of nonhuman animal models of human disease. I will conclude this chapter with a preview of an example of calibration failure that I will treat more fully in part III. This is one of the most interesting calibrations in recent years: Seok et al.'s (2013a) study of the aptness of murine models of human inflammatory responses. Their examination of the connection between inflammatory response pathways in murine and human systems established (in conjunction with other work in this area) that murine models are not, after all, apt conduits for information flow about human inflammation in particular, and probably not for human biological processes in general. They showed that the correlation between various processes that are activated in mice in response to inflammation were totally uncorrelated with the human response—and across a very wide range of inflammatory triggers. What we will see from their calibration efforts is that many experiments that generate knowledge about murine systems, while successfully generating experimental knowledge about species-wide features of mice, or rats, or what have you, fail to generate knowledge about the systems they are being used to try to probe. The problem here is not the experiments at all. It is simply that these experiments are appropriate conduits for information flow from distal intraspecies systems, but not from distal interspecies systems. Many failures of external validity conditions are of this sort, and this corresponds exactly to what is often discussed under the heading of "conceptual replication" in the social sciences. What is most interesting to me is that examples of this sort are in line with my understanding of replication as calibration, and that failures to replicate knowledge claims in the sciences are providing a direct negative appraisal not of the claims themselves, but of the capacity of the original experiment to have produced and/or detected the relevant information-bearing signals.

We have now seen in outline two examples of replications that are best understood as calibrations of prior experiments, as well as some episodes that are clearly calibrations (albeit failures) that function in much the same way. I think it is not unreasonable to conclude that the main epistemic good produced by replication is, in fact, calibration.

Replications can also calibrate in the first sense—that is, they render the original intervention, or device, or protocol capable of bearing the appropriate information. We will see a clear case of that when we turn to replications involving no new data. First, though, I will make an effort to say clearly what I understand replication to be.

What Replication Is, Finally

In this chapter I am going to consider more concretely how we should understand the expression "experimental replication."

10.1. Replication as a Relational Notion

Nothing is intrinsically an experimental replication of some other experiment. In this respect replications are similar to explanations, as are experiments. All three of these are complexly relational, and they require a specification of the context in which they are being produced. Experimental practices do or do not give knowledge *of* something *to* someone. Since experimental replication is itself an experimental practice, the same is true of it. That is where I will begin. What I will show is that when these practices target a system of interest in the world at large, the knowledge generated about that system is *experimental*; when the target is itself some prior experimental activity, it is appropriate to call it experimental replication showing that the prior activity either could or could not have generated experimental knowledge of the sort claimed for it.

To be sure, there will be cases when some practice both generates experimental knowledge of a system targeted by a prior practice and generates knowledge of the aptness of that prior practice for generating knowledge. What that shows is that the same practice (set of activities) can be fit into both a replication context and an experimental knowledge generation practice. That's just how relational notions work.

So we might find ourselves in the situation of 1, where experimenter 1 observes happenings at proximal system 1 in order to generate experimental knowledge of the distal system. Experimenter 2 observes happenings at

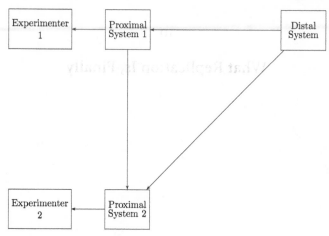

FIGURE IO.1. Schematic of experiment and replication

proximal system 2. Were the system there properly calibrated, she could well be doing two things at once: finding out about the calibration state of the first experimental situation (that is, whether or not it was capable of generating experimental knowledge of the distal system); and finding out about the distal system. These are contingent doings, based on her state of knowledge and the way the information channels connect her with the various systems. The point is just that we should see her role, qua replicationist, as monitoring what is going on at the experimental system centered on proximal system 1. There the arrows represent a flow (or possible flow) of information from one system to another.

In the rest of this chapter I will be offering some general considerations about the kind of thing that replication is, and what it does. Together with the case studies leading up to it in the preceding chapters, these considerations will constitute an argument that replication, while an important part of what we do in science, is far from a sine qua non of the method. It does not give us experimental knowledge; instead it can in some cases give us the knowledge that we have experimental knowledge and in other cases that we do not.

10.2. Experimental Identity Conditions

In a strict sense, of course, an experiment cannot be performed more than once, because it is a spatiotemporal event with its own identity; we simply cannot demand replicability in the strictest sense. It seems a hopeless task to attempt to detail the appropriate identity conditions necessary before we could, with any justification, call two experiments the same. In any case we don't want to allow the likes of, say, Pons and Fleischmann to keep pushing back against attempted replications with demands for stricter and stricter approximations to identity. What we recall from the cold fusion experiments of the late twentieth century is that no matter how many times, to no matter what specification, someone attempted to reproduce the results of Pons and Fleischmann, who claimed to have generated nuclear fusion at low temperatures in their lab, no results were found. But rather than conceding that there was a flaw in their method, that their experiments were not capable of transmitting the information they thought, they simply insisted that *their* experiment did generate these results and that others had simply not replicated them properly. Of course that is nonsense. Their experiment *has* been replicated and found wanting.[1] Similar remarks apply to Weber, as we saw above. Demanding strict identity of experiment is clearly too strong.

But if strict identity is too strong a standard, what standard then is appropriate? It is not enough to say that some experiment or other was performed and its results either agreed with earlier results (so it is a replication that confirmed the earlier results) or did not agree (so it is a replication that disconfirmed the earlier results). That is far too weak. We need to show enough similarity to give a compelling reason to accept some experiment as replicating another.

Homeomorphisms are often suited to do the heavy lifting in such circumstances. We would want some kind of function from one device to another such that the function on the first device's components results in the appropriate version of those components in the target device, and such that the function from the original device after carrying out relevant operations results in the same state of the target device, as if we first applied the function to the original device and then carried out the new device's version of the operation. For example, I might build a model of Millikan's device for measuring the charge on the electron. I would have an atomizer with some oil in it, a variable strength electrical capacitor, a microscope, etc. Then whenever Millikan sprayed electrostatically charged oil droplets into his apparatus, I would do the same. Whenever he adjusted the power in his capacitor, I would do the same. And so forth. Then when Millikan finds a candidate drop and suspends it in his view for a series of oscillations of the electric field, I would do the same. Finally, Millikan and I would have performed "the same" experiment. Now this condition is a lot better than strict identity, but it's still not really

1. For more on this episode see Gieryn (1992).

possible in general. Most of the time there is no such connection between the manipulations of one experimental procedure and those of another. Which oil drops Millikan caught is a highly contingent matter, and the odds are long against anyone trying to replicate his experiment getting any reasonable approximation of the sequence of observations Millikan had.

But wait—surely we can simply say at this point that we need not find drops of the same size with the same speeds and same charges as Millikan's. That level of repetition can't be necessary. Can it? Of course not. But we know that because we *already know* what is and is not relevant in this case. Where there is dispute, when there is a question about the necessary conditions, when replication is most relevant to our scientific aims, we won't generally know these things. All in all, saying with any precision what counts as two instances of an experiment is hopelessly complicated at best. So much the worse then for anyone who wants to understand from the ground up what counts as an experimental replication.

Kempthorne (1992) makes this obvious (but in some sense often neglected) point with respect to agronomy experiments. As he points out, no matter how carefully one tries to match two plots of land, they are different, and no matter how nearly overlapping in time two interventions are on the same plot of land, those times are distinct and the situation has changed. One can of course say that the significance of these differences is minor and not worth considering with respect to the questions being asked by the experimenter. But to assert that convincingly requires that we have already in place a notion of what counts toward replication, and also that we know enough about the system of interest to say what is and what is not a significant difference with respect to that notion. All of that is to say that we don't get for free the concept of replication by asserting that one experiment is simply a new instance of another; it is our account of the notion of replication that *underwrites* that assertion. What Kempthorne concludes is that we must have a method for overcoming the failure of replicability in order to mitigate the essential differences between individual interventions. His method is a kind of statistical sampling routine. He claims this kind of routine is necessary because the methods on offer (principally, in his view, variations on and expansions of Fisher's methods of partitioning) will, generally, lead to clumping of the features of various systems in ways that produce spurious correlations and intervention effects. We needn't endorse Kempthorne's view to agree with him that the problem of deciding what counts as a replication is real.

Surely, though, we do understand what it is to replicate an experiment, even if we have trouble articulating it precisely. So let me try again. Note that in addition to the issues surrounding identity conditions on experiments, we must also understand what goals we have for replication in the first place and how we are to evaluate whether they have been achieved in any given case. Generally, though, what we are looking for when we replicate another's experiment is an evaluation of the way the experimental observations bear on the experimental knowledge claim made as a result. What is the evidence that is generated in experiments? What are the data? Is there a natural, or objective, or robust method for characterizing and classifying different experiments as instances of the same experiment (or experimental protocol, or intervention, etc.)?

In the last analysis experimenters should not be looking to replicate experiments—these are localized 4D objects in any case and cannot be exactly reproduced; instead they should be looking to appraise whether observation reports (that some tokens t_i are of type τ_i) generated by an earlier experimental process support their associated experimental knowledge claims (token p being of type π bears the information that token d is of type δ). Replicating the conditions will give a distinct data set and so provide evidence about the right function of phenomena and interference to attribute to the data. Woodward (2010) is especially illuminating on this point.

My conclusion from all of this is that one experiment replicates another when it answers the same experimental question in the same way. That formulation is a little vague, but shortly I will ground it in my information theoretic account of experimentation.

Why might one want to replicate the experiment rather than the results of that experiment? One reason, of course, is to settle priority claims. If some result is established experimentally, then the performer of the experiment is credited with the discovery. But should it turn out that the result was not established by that experiment, that in fact the experiment is not capable of establishing the result, then naturally priority is lost. Another reason (and one I take to be more significant) is that we want to understand how the world works and which claims bear on which other claims, and we therefore want to know of a given experimental procedure whether it is capable of supporting the claims that are made in its light. Of course there is also the worry that different methods may give different results, that there is only one method available, that other results need checking with this new independent test method, etc. And all of these rely crucially on our ability to determine of a given experiment whether or not its results are spurious.

It is a complicated business to sort out which local logic of the initial experimental situation forms the target of attempted replication. There is of course the variety of beliefs that the initial experimenter(s) had in mind. What did that person or group take to be the point of the experiment? What knowledge was supposed to be produced therefrom? What would count as success in achieving those aims? Then there are the various apparatuses used to establish observational facts about the proximal systems being investigated. How do these function? What would be different were they slightly different, or markedly different? How secure is our knowledge of the behavior of these apparatuses? How robust are our conclusions about the observations made employing these apparatuses should that knowledge fail in various ways?

And then, crucially, there is the question of how we connect up the proximal observations with the distal system that is the object of our experimental knowledge claims. What material circumstances are special about the experimental situation and threaten or secure the basic experimental knowledge claim that p being π bears the information that d is δ? Sorting out all of these issues will allow us finally to assess the nature of the connection that supports the flow of information, the infomorphism (if any exists) between the proximal and distal systems. And once that is done—once, that is, we have a sense for the way in which information of a certain sort can flow from the distal system through the proximal experimental system—we are ready to entertain the question of whether some other experimental system can be used to replicate the results of this experimental system. In this book nothing as formal as a complete state-space description of the logical structure of experiments will be attempted. Nor is that necessary. What is necessary is to see how this story can be told, and to see it in such a way that the knowledge we have of various experimental systems can be displayed as appropriately revealing the ways the logics of different experimental systems can be mapped onto each other. The point is to give conceptual guidance for picking out the significant issues and ignoring others that are merely distracting. The analysis is meant to function much the same way formal logical analysis functions in the background of rigorous but natural language–based philosophical argumentation—we ground, and guide our discourse in the light of the results from formal study, without carrying out, in each instance, a complete formal analysis.

10.3. Independence

There are two things one might mean by the independent testing of a scientific result. The first has to do with the observations detailed in the report. The second has to do with our confidence that those observations bear properly on other systems—that is, whether there is an appropriate infomorphism between the systems. Replication of results is no reason to think we had experimental knowledge in the first instance; it may be just a reason to think that we redid the original experiment. Either we calibrate the original device or we know nothing new about it.

Support for observation claims: When the experimenter tells us that some system has certain properties (that a given percentage of rats developed tumors; that students in the classroom made certain market exchanges; that so many detector trigger events took place), we want to know whether these things really happened. But more than that, we want to know what these things tell us about the state of the systems under observation: that the tumors arose after exposure to a suspected toxin; that the trades the students made were made subsequent to students' obtaining sufficient information about the relative worths of the things traded; that the pictured stars really were behind the sun when photographed. Many attempts to replicate appear to focus on establishing the fact of some effect, or on further clarifying the parameters associated with the effect. For example, in psychology we want to establish that these particular men who wear red are more attractive to women than some other men are *because of* the red clothing. In medical science we wish to establish that ingesting antibiotics is causally relevant to recovery from certain diseases. Whether or not some system has some property (or some group of systems is such that each member has some property), then, is generally in question in attempts to replicate an experimental matter. This claim ties in nicely with many of the original discussions of this by Fisher ([1935] 1974), for example. We want to be sure that what we are looking at is not a spurious statistical feature, or a feature caused by unusual actual agency.

Support for claims of relevance to other systems that were not part of the experiment: We want to know whether other rats will get tumors in such circumstances; we want to know whether red-wearing men are generally more attractive; we want to know that the interactions in the interior of the device will produce similar reactions in particles elsewhere. And more, we want to know that they arose by the right causal pathways—that they were not merely *subsequent* to the initial happenings, but were *caused* by them. The first of these concerns the attribution of type characteristics to the proximal system, and the second concerns the relation of those type designations to distal systems. The latter is crucial, and has everything to do with the derivation of experimental knowledge from the experiment. Indeed, the major question for the theorist about experimental knowledge is how such observations in experimental circumstances can bear information about the features of distal systems. So while this question identifies crucial features of the information theoretic account of experimental knowledge, it has relatively little to do with replication as such. Replication is focused more on the issue of whether or not the observations made in the experimental setting accurately reflect the properties of the things observed in that setting.

So replication is best understood as directed toward the proximal systems involved in the original experiment. Even so, there are two different kinds of thing one might mean by the replication of another's experiment: simply checking whether the alleged empirical claim does or does not hold; or checking whether the apparatus used really did, or was capable of, generating those observational data.

10.4. What Is Experimental Replication?

I have come to see "replication" as an unfortunate word to use to try to express what is going on in the evaluation of prior experiments. Etymologically, it has the nice sense of folding back, of turning an experiment back on itself to scrutinize its *bona fides*. Its present use, though, carries the sense of redoing, or of making a copy of something. But what is the point of that in the context of generating experimental knowledge? Is it that we think experimentalists might be lying? Of course they might. But then, so might be all those from whom we accept knowledge claims. I will not deny that sometimes there is fraudulent research and that methods for eradicating it might involve checking that the experiment as described in the research report has the outcome claimed for it. And only in those cases where the report was honest would that amount to redoing the original experiment; when the report was dishonest it would be performing a new experiment altogether. None of this, however, is what is typically meant when we hear that repeatability is the foundation of our empirical sciences.

To understand what is going on in experimental replication, we first must be clear on the goal. There are several possible things that we might be after: (1) we may, as above, wish to determine whether some experiment really was performed as described; (2) we way wish to find out whether the observations made on some system were correct; (3) we may wish to find out whether those observations, even if correct, were typical of the system being observed; (4) we may wish to find out whether observations of that sort suffice to establish experimental knowledge claims of the sort made; or (5) we may wish to know whether those experimental knowledge claims made are correct. There may well be others. Notice that on this list the first two have little to do with experimentation as such. They are just run-of-the-mill considerations having to do with observation. Notice as well that the last two have nothing to do with repetition. Number 4 is a second-order question about the adequacy of the original experiment for establishing some claim, and number 5 is simply a question about nature itself, a question that presumably can be answered in many ways that are unrelated to the original experiment.

That leaves number 3 as the only candidate case in which repetition is replication. Even there, however, it seems spurious. For if all one is really doing is taking more data, so that the total data from proximal systems more accurately reflect features of the distal system of interest, then the right thing to say is that the experiment continues. If there were not enough observations made in the original situation to bear information about the system of interest, then why think the experiment was entirely done in the first place? If we recreate for all intents and purposes the original setup of the experiment and take data as before, in what sense is there a new experiment? What conceptually separates that from, say, various runs of the same experiment? Nothing, I would say.

In my view, to replicate an experiment successfully is to exhibit the appropriate identification between the information flow in the replicating and replicated experimental situations. This way of understanding things focuses on the heart of the experiment itself—the turning of data about the proximal system into knowledge about the distal system. For those purposes, the key to replication is to make sure that the experiment was able to do what it was advertised as doing, and to do so by mimicking in some ways the nature of its purported information channel. This is what I mean when I say that replication is folding the experiment back on itself.

To replicate an experiment, then, is something along the lines of modeling the proximal system in the original experimental situation sufficiently to evaluate whether this fact, that its initially being of type $\tau_i'(t_i')$ entails that it is subsequently of type $\tau_f'(t_f')$, bears the information that the distal system initially being of type $\tau_i(t_i)$ entails that it is subsequently of type $\tau_f(t_f)$ when $t_i' > t_i$ and $t_f' > t_f$. In English, the replication is a check on whether the first experiment was capable of generating the knowledge claimed for it. It is an entirely separate question whether the replicating experiment is *also* an experiment on the distal system itself; it might well be if, for example, the original experimental situation could not provide the information in question, but the new one does. Then it replicates in the sense of giving experimental knowledge that the original was not properly calibrated, as well as itself giving experimental knowledge of the distal system. This is the only way in which replication bears conceptually on experimental knowledge, and it does so in its calibrating role.

As I said, precise re-creation of the original experimental situation is relevant only when that situation is too underpowered to generate experimental knowledge on its own and we simply need more data. And, as is pretty obvious anyway, in any other case precise re-creation will not give new knowledge, for if the original was not properly calibrated a precise re-creation of it will not be either. Instead what we must have, in order to shed light on the original knowledge claim, is an experimental situation calibrated to give knowledge of that original situation.

The more exactly we redo, or copy, a prior experiment, the less we are able to check the aptness of that experiment for producing knowledge. I would say, though I don't think it does a great deal of work, that experiments that merely amount to the repetition of prior practices are more like continuations of experiments than repetitions of them. This is a point on which I agree with Harry Collins. One can certainly learn that a prior practice was underpowered if the same activities generate different results. There is also a persistent question here of whether, through replication, we seek to check experiments or their results. The two senses are at odds with each other—or at least, when we redo an experiment and get the same result, that typically does not tell us that the result is correct unless we were concerned only about the atypicality of the original sample (as in an underpowered experiment) and allay that worry by the newly taken sample(s) (as would be the situation should the two experiments together have sufficient power).

In the former case there is no need, strictly speaking, to repeat the experiment itself, or anything like it. One could just as well perform another experiment that decisively answers the question of whether the original claim was true or false. Suppose you come to me with the observational claim that there are seven jujubes in the room next door. You may have performed a complicated chemical assay of selected samples of gas in the room and concluded on its basis that there are seven jujubes in there. However, I needn't worry about whether I can perform the same experiment as you in order to replicate your results. I can simply, piece by piece, put everything in the room through a sieve the size of a jujube and count how many jujubes appear during the procedure; or if the room is empty enough I can just glance around and count how many jujubes I see; or I can, one by one, remove every object from the room, noting as I do whether or not it is a jujube, and increasing my jujube counter each time something is a jujube; or I can perform any of a myriad of other operations that have as their end result the determination of how many jujubes are in the room. Once I am done I have either confirmed—hence replicated—your result or I have disconfirmed it—hence shown it to be nonreplicable. Let's call this replication of results rather than replication of the experiment. Replication of that sort is very useful for the way it bears on experimental claims we are thinking of accepting. If for example some third party is wondering what number of jujubes it is right to believe are in the room, it's probably a good idea to wait to consider a number of efforts to establish the number (especially for complicated or difficult-to-establish results). But again, such replication has nothing at all to do with whether we did or did not know the correct number after the first investigation. Increasing our confidence when we are correct is often a good thing; changing our views when we are incorrect is generally also a good thing. Both of these can be facilitated by the replication of results or its failure, but there is simply no conceptual connection between experimental knowledge and such replication.

10.5. Hertz Again

What is the interest of Hertz's case, beyond the fact that some great experimentalist was dramatically wrong about his claims, and that others asking the same questions but using structurally appropriate information channels (with respect to the question being posed) got the right answers? Does it matter whether there was any replicating going on? The insights in the last few chapters should help us to see what is of interest here.

The Hertz episode provides us with a case of calibrating replication as well as a fascinating look into the mind of one of the most powerful experimental thinkers in the history of Western science. The rich detail and frank description of Hertz's frustrating attempts to generate knowledge of the cathode rays from fractious and delicate equipment is well worth an in-depth examination. This is all the more true for students of the history of science who wish to evaluate the status of Hertz's knowledge claims. When it comes to the charge on cathode rays our questions about whether Hertz's experiments were replicated by Thomson or Perrin are about more than simply assessing the experimental knowledge claims that Hertz was making from the point of view of what was established by Thomson or Perrin. It is merely a boring historical fact that Perrin and Thomson, in establishing their results, showed that Hertz's were wrong. The exciting question is whether the structure of the information-bearing maps the various researchers were exploiting connected in the same way their proximal observations to the distal systems they were studying, and whether that shared structure of information flow allowed one or another of them to provide calibration for the others. As one can see by returning to my exchanges with Buchwald, it requires a careful and close reading of Hertz's texts to determine the basis for those claims and to properly evaluate the information-transmitting capacities of his devices. It is of note that Thomson and Perrin can be seen as having both Type I and Type II calibration for Hertz's device. That is, they assessed his original calibration with respect to the claims that he was making, and found it wanting; but by offering a clear demonstration that the rays are charged, they also provided the calibration needed to allow information about the mass–charge ratio—information frozen if you like within the data—to flow in the light of their own new information.

Seeing replication as directed toward calibration, then, justifies what would otherwise be a puzzling interest in Hertz's cathode ray experiments. Either they're worth studying or not. If they are, it can't be because of any *real* worry about their epistemic status. Hertz's results were just plain wrong, however whiggish it might be to say so. If they're to tell us anything worth knowing about scientific experimentation generally, it will be in what they say about the connection between calibration and replication. Their interest lies in what we learn by seeing where (and if) Hertz's calibration failed.

Once we see that Thomson's and Perrin's devices were tuned to the same information-bearing types of signals that Hertz's device was, we then learn several important things. First, knowing that his device was responsive to these signals, we have a clear reason to look in Hertz's reports for the residue of those information-bearing signals—and we find it. That leads, second, to learning that in fact the signal was attenuated *by Hertz's own apparatus*, and it gives the clue that locates that attenuation in the gauze. Hertz, then, by his appropriate attempt to make his device responsive *only* to signals bearing information about the charge state of the rays, filtered out the vast majority of those signals—the very opposite of what happened with Weber, who in essence amplified the noise in his device in just the right way to make its signals seem information-bearing when they were not.

To see Hertz's experiment as worth replicating, to see it *as an experiment* on the electrical properties of cathode rays, requires that we see the Faraday cage as misapplied. Otherwise we are speaking only of a historical curiosity that Hertz, a famous physicist, happened to believe, on the basis of a device that was incapable of showing it, that cathode rays are not charged. To be charitable to Hertz, we must see that his analysis of the experiment was flawed in order to see that it was correctable and that the device could (and largely did) answer the question asked of it.

11

Replication without New Evidence

I am now going to make a totally outlandish claim that I hope will seem obviously true by the end of this chapter: *experiments can be replicated without ever generating new evidence.* This claim is illustrated with an analysis of experimental investigations of human cancer risk by way of data on rat cancer risk.

Much of what we think we know about the human metabolism is at least informed by the results of experiments on rats and other nonhuman mammals. But those experiments can fail to yield experimental knowledge of humans, and only attempts to replicate their results—attempts that fail—can show us this. We have seen already a case in point: research on the relationship between saccharin and bladder cancer. Attempts to show that ingesting saccharin causes human bladder cancer fail. While it is true that some rats really do develop cancer on ingesting saccharin, it turns out that the phenomenon is not general, even in rats, and not generalizable at all to humans. This in some measure highlights the importance of replication studies, but it is more germane to, and will gain more traction in, the more extended discussion to follow, in part III, of analogical experiments.

An example of rat studies that is more interesting from the perspective of replication proper is a *series* of experiments on rats investigating whether or not dioxins cause liver tumors in rats (and of course one performs these experiments to find out about cancer in humans as well). I claim that two well-known "reanalyses" of the data from the celebrated Kociba et al. (1978) study count, on any acceptable view of replication, as replications of the original experiments using the selfsame biopsy slides as the original. This case is nicely analyzed by Heather Douglas (2000) as part of her arguments attempting to establish the ineliminability of nonepistemic values in science. A critical analysis of those arguments is not appropriate here, but I do want to expand on some of her treatment of those experiments.

The original Kociba study took place over the course of two years, and generated data on 236 rats. These rats were broken into four treatment groups with different levels of exposure to 2,3,7,8-Tetrachlorodibenzo-p-Dioxin (TCDD), administered orally. During the course of the study the rats were periodically examined physically (for the presence of tumors) and had their blood checked for various substances. A key clinical result for my purposes, however, is that at the conclusion of the two-year study, sections of various organs were dyed and preserved on slides. The sections of liver preserved on these slides were used as the basis of Kociba et al.'s conclusion that TCDD is a cause of liver cancer.

One curious thing about this case is that since the original experiment was performed, several different groups of researchers have gone back to the original biopsy slides and, applying to some extent explicitly and to some extent implicitly new standards of appraisal, they have made conclusions differing from Kociba et al.'s regarding the prevalence of cancer indicated by the slides. Here is my contention in a nutshell: these reappraisals are experimental replications of Kociba et al.'s original experiment. What I think happened is that Kociba et al. established a channel that allowed signals, possibly bearing information (at least about the susceptibility of Sprague–Dawley rats to liver cancer in the presence of TCDD), to flow into those slides. Those slides now *themselves* are fossilized versions of those signals, and observations of them, if calibrated correctly, may allow that information to flow into the observer, who is then performing a novel experiment. I myself am not a biologist, and I have no idea at all what the right observational protocol would be to properly calibrate the (extended) device, but the right protocol should give us knowledge about the connection between exposure to dioxins and liver cancer in rats.

One might object that these new studies of the slides are simply the same experiment reanalyzed. At this point, though, let me recall the earlier discussion of experimental knowledge, an account that will be further clarified in part II. What distinguishes these studies from each other as different experimental performances is the state of knowledge of the researchers, and the community at large, on the occasion of the investigation. In the same way that structurally identical interventions and observations can sometimes count as different kinds of experimental activity (calibration, or replication, or novel effect generation), so too can different sequences of activity that are *numerically* identical up to the observation stage.

Consider this: when you and I carry out and attend to the same operations in the lab, we may well be performing different experiments. You may be merely calibrating a device whose operations you know well, and which you have already used successfully to generate new knowledge. I, on the other hand, may well be unaware of those earlier investigations and yet be pretty confident of the operational state of the device, so I may be finding out novel things about nature at that moment. The purpose of my experiment is different from the purpose of yours. If all that is true (and I hope that what has gone before has convinced you that it is), then the dosing of rats, the killing of those rats, the performing of their autopsies, and the preparation of slides all taken together do not yet suffice for the performance of experimental investigation of the role of TCDD in rat cancers. Those slides must be observed. Information must flow all the way into a knowing subject before an experiment has been performed.

This complements the point made in the discussion of intervention. There I argued that the sifting of the data is at the heart of the experiment—it is what transforms those activities into experimental knowledge. I argued there that intervention itself serves the purpose of giving us freedom to choose the experimental knowledge we want. Here I am going to argue that the way the data are used to find out about what was going on in other experiments determines whether or not there has been a replication.

Here the question that confronts us is whether some of these experiments, these new occasions for getting information out of those slides, are replications of the others. It is relevant in answering this question that the researchers in the later experiments were aware of the results of the earlier experiments. This awareness bears on their state of knowledge. What about the scientific community itself? What is its state of knowledge of the role of dioxins in rat cancers at the time each of these experiments is performed? To answer this question definitively requires that we know the role dioxins play in rat cancers, for that bears on the nature of the information in these signals. That's one issue. The second issue is to decide whether the beliefs about rat cancers that were formed in the first experiment were properly formed. That will require, in addition to knowing whether the material conditions of the experiment were sufficient for generating knowledge about the role of dioxins in rat cancers, knowing what the correct standards of evaluation are. The question that the researchers are addressing using those standards is whether various rats have tumors, and if so, how many. The answer to that question then informs the further question about the causal link between those tumors and dioxin exposure. To know how it informs that question requires that we know about the statistics of the experimental situation—the numbers of rats, their relative exposure levels, etc. Those issues can be set aside for the moment. What is at issue in the three experiments we have in front of us is the question of whether these rats have tumors and how many, and that issue arises from a dispute about what the right standards are for evaluation of the various rat organs.

I argue that, in fact, the subsequent observations of the original rat organ slides by new researchers did indeed amount to replication of the experiments. Which parts of the original experiment were in doubt? Only (as far as I can tell) the transition from the material conditions of the experiment to the inference about the causal role of dioxins in generating tumors in the rats. That is to say, the right way to look at the reevaluation of the slides is as the new experimenters rerunning the experiments with (in their view) better equipment that was calibrated more properly to the question of whether the rats did indeed have tumors. (This equipment, we might say, comprised their eyes and their professional judgment about what does and does not count as a tumor.[1])

But can I really be saying that two different researchers, interacting with the same experimental apparatus, and considering the same data—even perhaps at the same time—can be performing two different experiments? Yes. More than that, I am saying that they *must* be performing different experiments. Only by thinking of experiments as demarcated by their knowledge-generating powers can we get a good conceptual grasp of how they can generate knowledge at all. And once we realize that, then the two activities of two different researchers each evaluating the data must count as two different experimental events.

Richard Fry generated a worry about this view: Suppose I am testing some hypothesis, and suppose I have a hundred slides of rat livers. After examining fifty of them, I'm convinced of my hypothesis. Looking at the other fifty should, one would think, still have to count as experimental-knowledge gathering. But if my justification goes up to 100 percent after the first fifty, it would seem, on the view being articulated here, that the experiment ends right then. But that seems like a weird result. Indeed, it's more than just weird, because while it ends right then as far as I'm concerned, if someone else, being more skeptical, isn't convinced until the seventy-fifth slide, then doesn't it follow from the view that it keeps being experimental for her until then? So is the whole thing even experimental at all for people with some justification for the contrary conclusion? This might seem like an unstable place to stand, and I am sensitive to the worry.

1. See in this connection the story that Bogen and Woodward (1988) and Woodward (2010) tell about data and phenomena.

Fry's own suggested answer is that there's no such thing as being 100 percent justified. In other words, simply acknowledge that justification is open-ended at the top, at least as regards experimental knowledge. That does seem to be a possible response to the worry. We should, of course, adopt the view that insofar as we are good empiricists, who update our probability assignments in accordance with Bayes's theorem (even if we're not Bayesians), it is good practice never to saturate our prior probabilities. That is, prior to testing our various hypotheses it is bad practice for us as empiricists to take the probability of some empirical hypothesis to be 0 or 1; instead it should be assigned a value somewhere between those extremes. Then we will never have occasion to assign a probability of 0 or 1 to any empirical hypothesis, since no finite amount of experimental evidence will ever move one of my probability assignments all the way to 0 or 1—and finite evidence is all we ever have. But it is true that we sometimes do assign, *effectively*, a probability of 1 to empirical hypotheses, and it's probably sometimes even OK that we do so. Here one could recall Popper's analysis: certain propositions, while never really being fully confirmed by the evidence, simply drop out of the game of scientific testing. We don't have inductive proof that they are true, because there is no such thing, but we simply make such things presuppositional and move on. While I am, in some sense, continuing to take data every time I eat bread, I am not really testing the hypothesis that bread nourishes rather than poisons. The fact that it does so (*pace* gluten-free crusaders) is simply no longer at issue for me—I have removed that question from the game of science. But that is not relevant to a test that is still live.

That's one option. A second possible reaction is more specific to the case of experimental science. Say that I am subject to a variety of biases, as are we all. In particular I am apt to read data in ways that conform to my prior expectations about things. I am both more likely to see things that confirm my expectations and more likely to give them heavier weight than I am things that don't. I do not do this on purpose, even in the middle of a heated debate. However, my intentions are not the only feature of my cognitive profile that is at play when I observe. Knowing this, and wishing to mitigate the effects of my biases and my bias-promoting cognitive architecture, I will adopt various protocols to help with this. These are the same protocols that are part of the standard construction of experiments: random assignment of treatment; blinding; double-blinding when the subjects are themselves potentially aware to what's going on; and so forth. These protocols will generally not be broken until all the data are analyzed and reported on. And so I am unlikely to find myself in the position of abandoning my experiment because I'm fully convinced. This is probably what in fact keeps many experiments going to the end.

My own response, though, is to say that none of this is directly relevant on my epistemology. Because my epistemology has nothing to do with justification, any apparently weird results that arise from justificatory concerns are not mine. As I will elaborate in part II of this book, I adopt Fred Dretske's information theoretic account of knowledge. There the only question is whether my beliefs are appropriately information-responsive. Thus my own certainty is not probative here. What is probative is whether the signal itself raises the probability to 1. I can't know that it has, generally, and so I have recourse to various rules of thumb that mimic internalist epistemological views. I use statistical analyses to generate satisfaction conditions on which I am prepared to make various bets. But these are not where the knowledge conditions lie. On what I take to be the right epistemology for empirical science, knowledge is belief caused by information, so it doesn't much matter what my own appraisal of the probabilities is—if I believe (or accept, perhaps) the result because information flowed, then I know.

This discussion does, however, serve again as a way of driving home the point that different experiments can go on, one right on top of the other, so to speak. If you and I are working on a medication for some serious illness, and I am worried about its long-term efficacy and you are worried about whether it has any short-term benefits, you may well terminate your experiment before I terminate mine, even though we share all material circumstances identically up until you break protocol at some given point because the experimental system has produced all that the data that are required to bear to you the information that is relevant to the knowledge claim type you are seeking, and yet for my interests more data are required.

In the case of the rat trials above, the right thing to say, it seems to me, is that the subsequent researchers were attempting to replicate the earlier experiments. Replication, though, is not (as we saw) merely doing the same experiment. It is in part answering the question of whether the earlier experiments could have produced the results they did, and perhaps also the question of whether they in fact did so. From that point of view, it presents no difficulty that all the previous data generation is shared among the three groups. What is, and remains, a serious issue for the philosophy of science is determining which of these three observational episodes produces appropriate knowledge of rat tumorigenicity in the face of dioxin exposure. While I am not prepared to answer that question here, seeing the episode in the way that I have displayed it may well make it easier to perform the needed analysis of the state of knowledge of the various researchers—and of us as well.

12

The Replication "Crisis":
A Calibration Problem

There is no *replication* crisis in the experimental social sciences (or in any other science, as far as I can tell). There is a crisis, but it is one *revealed by* (attempts at) replications, not caused by a *lack of* (successful) replication. Generally, replication is not itself a generator of experimental knowledge about the distal system targeted by the original experiment.[1] It is better thought of as a kind of technique by which we can know either that we know, or that we don't know. That is, it is better thought of as a kind of calibration of prior experiments. That is the funny thing about replication. It is not really about the original *results*; it bears most directly on the connection between prior experimental activities and claims about results made on their basis.

In what remains of part I of this book I will address the so-called replication crisis in the experimental social sciences, and recast it as revealing that in those sciences experiments are typically (or at least more often than they should be) calibrated incorrectly and thus cannot produce the knowledge claimed for them. This discussion will further support my contention that replication is principally an exercise in calibration; neither confirmation nor refutation, it is an evaluation of the aptness of a device (or lab or protocol) for giving knowledge. I will show that the standard narrative underlying the "replication crisis" talk derives from a flawed understanding of the role replication plays in experimental investigation.

Periodically, some discipline that is traditionally seen as a softer science than physics or chemistry will awaken to the thought that its progress as a scientific discipline is being held up because of how poorly it fares in the

---

1. Though as we shall see when we consider replication without new data, it can certainly function in that role alongside its calibrating role.

replication game. Library science, for example, is no stranger to this phenomenon. Christinger Tomer (1992, 468) lays out the problem for library science
explicitly in his review of *Replication Research in the Social Sciences*:

> Of these flaws, none is more important than the failure of research in library
> science to transcend the fragmentation and discontinuities characteristic of im-
> mature disciplines. The worth of research investigations, the value of their
> contribution to the body of objective knowledge, is determined by the con-
> firmation or refutation of their findings and by the extent to which the in-
> vestigations and their findings can be replicated. As P. A. Lamal writes in the
> essay entitled "On the Importance of Replication," replication of research "is
> necessary because our knowledge is corrigible" (p. 31). In library science, com-
> plaints about the inability to build a substantial, coherent body of professional
> knowledge may be attributed more or less directly to the *infrequency* with
> which research is systematically replicated.

Interestingly, the substance of Tomer's review is that the entire volume on
replication is a disaster: there is, he claims, nothing of substance other than
the occasional article that makes clear that editors in the social sciences are
strongly biased against the publication of replication research. But Tomer's
own failure in the review is twofold. He adopts uncritically the idea that lack
of replication is an important cause of a dearth of secure knowledge in library
science (and other social sciences), and he also embraces the idea that those
sciences that *do* have a substantive body of such knowledge are, somehow,
consistently better at producing replications of the work in the field. Neither
is true. Instead, and as Harry Collins came close to pointing out, the strength
of a discipline can in part be measured by how *little* it replicates. If replication
really is, as Collins says, like the Supreme Court for empirical science, then
constant appeal to it should be seen as an indictment of the strength of the
laws and methods of a discipline rather than a signal virtue.

But isn't it true that there really is a crisis in scientific replication? Lamal
and the other contributors to that volume are not alone. In the last several
years there have been many calls to strengthen the culture of replication.
These calls are especially forceful in the social sciences.[2] However, worries
about replication are not absent even in physics. What are we to make of the
fact that physics experiments are now years-long collaborations that probably couldn't be done again even were there political will to pay for it? The
delicacy and complexity of these monstrous accelerators and their detectors
make any realistic attempt to redo what was done impossible. And the results
of these investigations are foundational and are meant to tell us whether we

2. See the Saltelli and Funtowicz (2017) article in *Futures*.

need an entirely new conception of physical theory or can simply continue on as before. Presumably, if we need replication anywhere we need it here.

The massive scale of the devices and the collaborative teams, and the extended times required for some experiments in physics might make replication failure there seem to be a different kind of issue than in the experimental social sciences. In those latter disciplines there appears to be a capacity to replicate, but there is no incentive for researchers to do so. However, there is no reason to single out the social sciences in this. Even in those subdisciplines of physics where replication does remain feasible, there is no incentive to do it. The *culture* of experimental replication appears to be broken even when it would be materially feasible to replicate. Recalling that, as Ioannidis (2005) and others have shown, the number of positive results published in the scientific literature must, as a plain statistical matter, be much greater than the number of correct positive results, we should be worried about and should be interested in weeding out such spurious results. But there appears to be no mechanism in place for eradicating them from the scientific canon. Saltelli and Funtowicz (2017) argue that the crisis is real, and they offer some possible remedies. All of these, however, are reliant on a change in culture about how studies are reported in the first place.

Despite our standard teaching methodologies in science and in philosophy of science, where we emphasize very strongly the point that no results should be accepted as correct without independent test, the plain fact remains that all of us are more or less prone to fallaciously accepting the first word we hear on a subject. Too often that first word is a positive result that must later be rescinded. Chabris and Simons (2010) voice such worries in "The Invisible Gorilla," where they bemoan the addition of spurious results into the popular conception of scientific results. Such a situation would be bad enough did such results not also infect the scientific literature itself. But of course they do. While this is principally a problem for popular accounts, it does appear to cause its share of mischief in the sciences, where initial, incorrect studies continue to be cited, sometimes for many years. An idea one sometimes sees floated is that experimental studies should be accepted for publication by journals in advance of observing the effects being probed, provided only that certain conditions on professional standards of research are met. Let me here voice my support for another idea:[3] before publication of results, some number (at least two? more? I'm not sure) of independent replications should be performed. Research reports detailing the original experiments and the

---

3. This idea is in some respects similar to Kahneman's (as outlined in his 2012 open letter to *Nature* concerning priming effects), but it is more general.

replications would be published together. In the cases where replication fails to confirm the original study, the first public word on the subject would be that we know either that the effect does not exist or that it is not known to exist. But in the cases where the replications succeed, those of us who have not done the experiment would have much better ground for accepting the results into our structure of belief.

Is such a thing possible? I'm sure that it is. But will it really happen? I doubt it. There has already been significant pushback against the idea of increasing the profile and academic respectability of replication research. It is true that there are now a few journal options for replication research, but the venues are limited and the number of studies in these venues is very small: the *Preclinical Reproducibility and Robustness* gateway of F1000 Research, started in February 2016, had twenty-six articles as of April 2018; the *International Journal for Re-views in Empirical Economics*, started in April 2017, had four published replications (all by Joachim Wagner, all published in volume 1); *AIS Transactions on Replication Research*, founded in 2014, had thirty publications as of April 2018 (most but not all of which were replication studies); etc. No amount of argument that something should be prestigious will make it so, and replication is simply a low-prestige activity. The problem, of course, is that the current state of academic and professional science is oriented toward the positive result as the unit of success, despite the lip service we pay to the contrary. Until that changes, replication will never take on the significance it merits as the arbiter of the validity of scientific results. Additionally, those whose research is called into question by failure to replicate consider replication efforts to be attacks rather than well-intentioned efforts to get at the truth. Perhaps that would not be so were replication studies a standard part of the generation of scientific publication; but bootstrapping our way to a culture of scientific activity in which it would be is a daunting prospect.

But so what? What we have here is in fact not a replication problem, but a power problem. If we did have a culture of replication, the main result (at least in the social sciences) would be to tell us what we already know: that our methods in those disciplines are not capable of generating the knowledge we are claiming. That, again, is a calibration problem—it shows us that the methods we take for granted in these fields are too weak to generate useful information flow.

To understand what replication really amounts to and what it takes to do it is clearly of importance generally in the empirical sciences. But in the case of social science research (including psychology and economics, for example) the situation is more fraught with difficulty, and the need for clarity is therefore even greater. The systems studied by physicists, and chemists, and even biologists are much simpler and more stable in important respects than are the systems studied in the social sciences. In the latter cases the importance of human agents is so great that their unique and difficult-to-predict features cannot be ignored. This difficulty of prediction will have implications for the inferences that can be drawn from any experiment or series of experiments.

Generally, a replication study for a given experiment should be directed either at (1) assessing the correctness of the original researchers' claim that some device, or protocol, or what have you really did (or could have) generated the observations claimed for it, or (2) assessing whether the device, or protocol, or what have you suffices to generate the knowledge claim made on the basis of these observations. Failure to appreciate that there are these two purposes and that the latter is of critical epistemic import results in the kind of confusion exhibited below.

In his column, "Psychology's Woes and a Partial Cure: The Value of Replication," Henry Roediger, past president of the Association for Psychological Science, discusses varieties of "replication attempts: direct replication, systematic replication, and conceptual replication" (2012). He says, "When someone uses the phrase 'failure to replicate,' they almost always have in mind (or should have in mind) direct replication."

In *Nature*, speaking about the limitations of conceptual versus direct replication, Ed Yong quotes Brian Nosek, who studies social psychology at the University of Virginia in Charlottesville: "It is the scientific embodiment of confirmation bias," says Nosek. "Psychology would suffer if it wasn't practised but it doesn't replace direct replication. To show that 'A' is true, you don't do 'B.' You do 'A' again" (Yong 2012, 300).

But how obvious is this? I don't even see how properly to scan Nosek's claim. He must mean something like "to see whether doing 'A' has some outcome, you don't do 'B.' You do 'A' again." That of course is one way. Another way might be to look at a film of someone doing "A" and note the result. Or one could use a strongly confirmed theory of "A"-like stuff and just figure out on its basis what would happen were we to do "A." But that is not at all the point to replication studies in social science. We are generally pretty convinced that doing exactly "A" again would result in essentially the same outcome. The questions we typically have are either "how exact do we have to be in replicating 'A' to guarantee that same outcome?"—when, for example, we are concerned that the original study was underpowered and we want to use more and different test subjects in the same protocol—and "what does the fact that doing 'A'-like stuff produces 'C'-like outcomes tell us more generally?"—when, for example, we are worried that some peculiarities of "A" situations don't ramify to a larger context (from Harvard economics students to average

US consumers, perhaps). There are also cases where the power of the study is so small that the possibility of an aberrant result in the first study is too high to take the result seriously. But in that case doing the exact same study *again* simply leaves us with two poorly powered studies. No particular result of such a study is illuminating. What we should be telling ourselves in such a case is that the original study is not capable of giving us knowledge, and we should just do a different study. Could that study be a much higher powered version of the original? Sure, but no matter what we find out from that study, even if it is the same as the original outcome, we got no experimental knowledge from the first one. If the only point to replication is to get more subjects into the study, because generally these studies are underpowered, then what we should be learning is that the real crisis comes from calling the original studies "experiments." They are better thought of as idea factories, or advertisements to get someone to do an experiment. Activities that, as a class, are not capable of generating experimental knowledge ought not, in my opinion, to be thought of as experiments. If on the other hand we have some different question in mind, then blanket statements such as Nosek's are of little use.

That is why it is better to begin by asking what the target of the replication attempt is. Do the scientists whose work is being replicated make broader claims based on their experiments, rather than merely pointing out that certain actions were performed and that certain observations were made? If so, it matters what the scope of such claims is when we consider the experiments as targets of replication. Not much has been done to clarify what direct replication would even be, exactly. The closest one sees is something like "the same methods" to test "the same question," and it is left undefined what it would mean for methods to be the same. Indeed, the debate in psychology over whether direct replication is necessary or desirable is muddled, because there are arguments made against the worth of direct replication that amount instead to arguments that direct replication is not possible. The worry articulated by such folks is that by moving to any new setting, the original moderating conditions that allowed the original researchers to pick up on the effect will be absent. But if not all relevant features of the informational situation are present, the new experiment is clearly not directly replicating the original. So, this argument continues, the only way to confirm experimental claims is for the original researchers to continue taking data.

There will be other issues as well. For human agents, a crucial difficulty is deciding what feature was actually being measured in any given observation. If all we have to deal with in the case of economics experiments, say, is whether or not some trades were performed or some purchases were made, we don't have a very clear idea (absent extensive auxiliary data) what the true economic behavior of the agent actually was. For in economics experiments generally we can observe only that agents report a willingness to behave in certain ways in a real market; we cannot observe real market behavior, nor can we directly observe agents' psychological states to be sure their behaviors are being caused by what we think. Such issues are made very clear, and are addressed clearly, in Kahneman, Knetsch, and Thaler (1991) in their tests of the endowment effect. They are at significant pains in the article to show two things: that the agents involved understood the tasks they were to perform, in the sense of understanding the structure of the economic system of which they were a part; and that the agents involved were motivated to perform according to basic economic presuppositions, in the sense of maximizing the value of their trades. So they first showed that the agents they were observing were not motivated by other factors that would make their reported willingness to sell higher than their reported willingness to acquire—that they were cagey traders, for example, who always overstate the former and understate the latter. They needed to secure these two features or the experiment could not have established that the agents involved, by and large, succumb to an endowment effect. Only against the background of a settled observation does it make sense to consider whether or not some other observations will bear on replications of the experiments involving those observations.

The real problem is that for psychology experiments, as for all experiments on complex agents, finding the right information-bearing maps between the proximal and distal systems is not an easy task. Simons (2014) is quite right about this, and also quite right that the difficulty of generalizing our results from one study or from a suite of different studies is in separating the signal from the noise—or as I prefer to put it, separating the information in the signal from the noise in it. The question, of course, is: how do we make that separation? Notice though that much of the difficulty of generalization will be traceable to external validity conditions. As such it will not bear very strongly on questions of replication, in the sense of confirming results. It will to some degree, of course: the inferences that are licensed on the basis of any data are naturally constrained by what is known already by those making the inferences. That, however, is a general feature of experimental inference, not a special problem for the determination of whether some attempted replication does or does not succeed.

So we haven't seen much evidence that there is some kind of crisis of replication as such in science, but we have seen that there is a crisis having to do with the soundness of the methods used to draw conclusions in the experimental social sciences. While it does seem correct that not enough replication is happening in disciplines such as psychology, economics, library

science, and other social science disciplines, and that consequently many
spurious results continue to avoid disconfirmation, that seems not to be the
urgent problem. Instead the prevalent failure to replicate results indicates a
crisis of method. The standards for what counts as experimental knowledge
generation in these fields are simply inadequate. When we do see a strong
push toward replication of research in such fields, what we find is that much
of the published work does not stand up to scrutiny. Why is this? As I mentioned, Ioannidis (2005) has provided significant data and analysis of exactly
this question. His conclusion is that in many fields the strength of the experimental studies is very low with respect to the sources of possible error, and
journals are too ready to publish the results of such seriously underpowered
studies. Stated so baldly, the conclusion is nearly analytic when one sees how
frequently replications are refutations. But the considerable virtue of Ioannni-
dis's analysis comes from the detailed measures he provides for the sources
of error and bias, which measures allow him to quantify what proportion of
studies we should expect to fail of replicability.

The failure of some high-profile results to replicate does not indicate that
we need more replication. Instead it tells us that we have good reason to think
that the disciplines where those studies were generated are methodologically
unsound. Lack of replication studies generally is not a problem for knowledge
within a discipline. Indeed, if it *were* a significant problem for a particular
field, then that would tell us that generally the cause of belief (for researchers) or acceptance (for the discipline understood collectively) of experimental knowledge claims in that field is not the information that those claims
are about. That sort of problem would amount to a serious methodological
failing for the field. What replication failures are telling us about these fields is
that their tools are not properly calibrated to give knowledge. Sometimes, of
course, even in a properly functioning discipline there will be claims accepted
where that acceptance is not caused by the appropriate information-bearing
signals. But that should be rare if we have a discipline where we are good at
getting ourselves in the way of such signals and tend not to accept things
otherwise.

The theory of replication outlined in this chapter is compatible with the
thought that replication should be rare but that in cases where we really do
have experimental knowledge replication will be generally available. As I said
above, I will not endorse Shapin and Shaffer's (1985) view that replication is
what turns belief into knowledge; indeed, it is my basic presupposition here
that belief already is knowledge when it has been caused in the right way. It
may well be that sometimes experimental practice is insufficient to generate
the information-bearing signals that would be necessary to give knowledge,

and in those cases multiple sources of observational data might together bear that information. However, in those cases it will sometimes be appropriate to reserve "experiment" for the generation of the entire collection of data and its subsequent causing of belief. Even so, there will sometimes be signals that bear information, and these signals will suffice to cause beliefs in one or another researcher, and perhaps different beliefs in different researchers. In those cases it probably makes sense to think of there being experimental replication.

Conclusion to Part I

13.1. When Do Experiments End?

I have in some measure given my answer to this question throughout part I of this book. Experiments end when some agent (even presumably a complex social agent like a research group) comes to believe (or accept) an experimental claim on the basis of signals (comprising all manner of different possible objects of observation) or decides that there is not enough information in the signal to cause such a belief. They end in experimental knowledge when, in the former case, those signals bore the information that the distal system is the way the experimental claim says. The work Peter Galison (1987) was doing in "How Experiments End" is, I think (and have argued implicitly, at least), best understood as trying to explain what is going on when we assess the goodness of others' experimental claims, or what is going on when groups of researchers come together to try to decide what the facts are on the basis of these many operations. The answer here is about our knowledge about our knowledge, rather than about first-order knowledge claims. The latter case, however, bears directly on making it so that these signals cause belief or making it so they don't. In such situations people are still experimenting, and the experiment ends when beliefs stabilize around the relevant claims. Of course, as I have argued throughout this first part, there may be many overlapping experiments going on at once. They can start with any events in the world independent of the intervention by an agent, but they end when agents come to believe.

13.2. Transition to Information

I have so far considered a number of issues traditionally associated with the analysis of experimental knowledge. I have advocated for nonstandard views of these matters, but they are principled views that follow from my presupposition that the key to understanding experimental knowledge is to see experimentation as controlling the flow of information from distal systems (the systems experimenters are trying to find out about) to proximal systems (the systems experimenters are actually observing through their encounters in the laboratory or in the field). A main takeaway from the discussion is that the calibration of experiments as conduits of information is a crucial component of that understanding. Part II of the book will sharpen the discussion by presenting an explicit theory of experimental knowledge grounded in information flow. This theory will back up my claim that the results of part I follow from seeing experimentation in the way I will describe.

Information and Experimentation

14

Introduction to Part II

In part I, I considered the significance for the conceptual foundations of experimental knowledge of calibration, intervention, and replication. Of the three, calibration was found to be of fundamental conceptual significance because its presence or absence determines whether information can flow through some experimental apparatus, and thus whether such apparatus is apt for the generation of experimental knowledge.

I was there working with intuitive notions of information, information flow, and knowledge. In this part I intend to make much more precise the nature of the claims I was making in part I, by rehearsing some standard issues surrounding the concept of information and then revisiting those issues in greater depth. I will be adopting the communication model of information outlined by Claude Shannon. While there are a number of competing models for how to understand information, especially as it relates to the foundations of knowledge, I am most content with Shannon's. In my opinion, it does everything one could want from such a model, and introduces no complicating novelties. Many people seem to believe that information, especially as conceived in communication theory, cannot provide an appropriate foundation for our knowledge, but I am not one of them. It seems to me evident that we come to know by being informed, and informed in just the way that communication theory says. I won't offer much in the way of novel, direct arguments to support this view, but those developed by Shannon (in Shannon and Weaver [1949] 1963) and later by Barwise and Seligman (1997) seem sufficient, and I defend those here. What *is* novel here is the theory I develop of experimental knowledge based on information, and that theory will show by its *own* adequacy the adequacy of the conception of information I have chosen.

Over the next several chapters I will develop that theory by first explaining what is special and unique about experimental knowledge, and why information flow is a good way to capture what happens when that knowledge is generated. I then turn to a very rapid overview of Shannon's communication theoretic account of information, followed by a similarly rapid overview of Dretske's information theoretic account of knowledge. I respond on Dretske's behalf, but in my own way, to certain objections that have plagued his view, in particular resolving a central problem for learning by deductive inference on his account (the so-called closure problem). That done, I broaden Dretske's account by resorting to Barwise and Seligman, who show how to apply his theory to distributed systems. Having set the stage in that way, I then make use of these three familiar (at least to some) conceptual pieces to construct my own account of experimental knowledge. That account will be turned to good use in part III, where I resolve some vexing issues in the philosophy of experiment.

The Basic Features of Experiment

15.1. Specialness of Experiment

What separates experiment from other forms of knowledge-generating activities? What does experimentation have that distinguishes it from all other forms and makes it the hallmark of the scientific method, at least officially? The short answer is "nothing." Naturally, though, I cannot stop with the short answer, given how controversial it will be and how contrary to good sense it must sound. Nevertheless, the short answer is correct. In this section I will offer a longer answer that makes it sound less absurd; what remains of the book is an even longer answer that will make it sound obvious. The longer answer begins by asking another seemingly simple question: what do we do when we find out something we didn't already know? That is, what happens conceptually when we learn something? Clearly we go from a state where we did not know to a state where we do. Since this isn't really a book in contemporary epistemology, I won't be worried about the fine structure of the concept of knowledge, and I certainly will not attempt to respond to every puzzle that field has generated over the last several decades. I will make things easy on myself here and take it that one knows when one believes correctly, where to believe correctly requires both that what one believes is true and that one believes it in some appropriately good way.[1] What that appropriately good way must be, nobody seems to be very clear on, or at least nobody seems to have convinced a plurality of the other epistemologists. Most contemporary epistemologists do, however, seem to think that this conception—believing the truth in the good way—is generically what we mean by knowledge. All

1. My own preferred explication, as I will discuss below, is (not surprisingly) that to know is to believe because of information. But nothing said here should hinge much on what particular conception of knowledge one has.

we really need to agree on at this point is that in going from a state of not knowing to a state of knowing, one has come to believe something that one did not believe before, and to do so in the good way (or one continues to believe something that one did believe before, but does so now in the good way, rather than the less good way one believed it before). So far this is all very vague, but it is enough to move us forward.

We can ask, for any conception consistent with the above, "OK, but how can I go about getting more empirical knowledge?" Surprisingly, there seem to be only a very few items on the list of possibilities. I can, by using the tools of logical inference (both inductive and deductive, and maybe mathematics as well for those who think it's separate from logic), note that some stuff I believe conflicts with some other stuff I believe, and make a choice to modify what I believe by either removing, adding, or modifying some things that I believe or my grounds for believing them. When this goes well, and the beliefs that are already knowledge dominate the beliefs that are not, then that conceptual reordering can produce new knowledge. Or I can, by some perceptual experience or other (including, for example, hearing someone's testimony[2]), come to believe something new, and when I do that in the good way, I can come to have more knowledge. That's not much, but it does seem to cover it. To be sure, there are other ways to change what I believe: I can be hit in the head with a rock, and suffer sufficient brain trauma to change what I believe about some things. I could, I suppose, simply decide on a whim to believe something because I want to. (I don't *know* that I could, but many people have reported to me that they can, so I won't dismiss it out of hand.) I think it's physically possible that somehow an external agent of some sort (the world itself, space aliens, a person with telepathic skills) could change my brain state in such a way that I come to believe something new, and even perhaps something true. (In my view this is just a generalization of the rock case and adds nothing new to that case, but some may disagree, so I don't mind adding it.) This seems to exhaust the possibilities; but in none of these cases is there any new knowing going on, because in none of these cases do I come to believe in the right way to know the things that I now believe. The things that cause me to believe in the right way must be more than just causes;

---

2. Testimony is an interesting case. Like other sorts of knowledge, it may be said to begin with some observational event (the hearing of what someone is saying); however, we come to believe what was said not on the basis of the speaker having the property *said x*, but rather on the basis of being connected already to the speaker in some kind of epistemic community. I use my observations to come to know what she is asserting, but my belief in the thing said comes from accepting her testimony. Testimonial knowledge is a crucial source of our empirical knowledge, but it would not count as experimental.

their belief-causing powers must be properly connected to the truth of what I come to believe—and none of those other causes are.

So how does this bear on the question of experimentation? Well, if experimentation is a way to acquire knowledge of nature, then it must operate either by restructuring the logical space of my beliefs about nature, or by changing what my beliefs about nature are through the generation of perceptual experience, or a combination of the two. It seems then that there is nothing in the nature of experimental knowledge itself that is *conceptually* different from any other kind of knowledge. So any difference must lie in the particular methods by which either the logical or the perceptual restructuring of what we believe takes place, or in the nature of the beliefs themselves. That is, the difference must lie either in the way experiments generate knowledge or in the structure of the beliefs themselves.

It is my view—and the view much of this book is dedicated to spelling out and defending—that the hallmark of experimental knowledge is the kind of believing involved rather than the means by which this believing comes about. That is, experimental knowledge is distinguished by the way certain beliefs are connected to each other and not by the way those beliefs, and the observations that underlie them, were generated. I will argue that experimental knowledge claims, insofar as they differ from any other kind of empirical knowledge claim, are generically of the form "token observations that proximal systems p are of type π support the inference that distal systems d are of type δ." Experiments and all their panoply of apparatus will be in service of producing such token observations of proximal systems and ensuring that they do indeed support such inferences about other distal systems. The basic contention of this book will be that the best way to understand how this works is by thinking of experimentation as controlling the flow of information between distal systems and our minds through the proximal systems with which we deal.

15.2. Experimental Knowledge

I will now return briefly to a consideration of the role of induction. Understanding how experiments give knowledge comes from understanding what inductions are being made and against what background presuppositions they are being made.

Even asserting that some token is of some type on the basis of an observation (e.g., "this sample of material masses 3 grams") is complicated conceptually. Either the assertion is sharply restricted—to the precise time of observation, for example "at 6:26 I noted that the sample was in the scale's pan, and the reading was '3gm'"—or much more is going on than mere sensory registration. This point is not new. It is in fact the basis of all skeptical worries about the senses and their capacity for giving knowledge of the world. As Hume has it, the mere having of a sensory impression has, by itself, no informative connection with any other such impression, nor really with any further facts about the world. The point Hume makes is, in some sense, unassailable—just as was Popper's rephrasing of the point. While I am not advocating a skeptical position, I focus on this case to make the important point that even basic observations are deeply structured and function only by exploiting the basic inference of experimental knowledge.

The main reason why it is worth stopping to consider the nature of induction, and our reliance on it in generating scientific knowledge, is that what goes wrong in scientific experiments, what prevents them from giving scientific knowledge, is generally that some inductive step fails. It is not induction itself that fails; rather, the way the various steps of the experiment are connected with our background dependence on the reliability of induction is improper in some way. I will, in what follows, develop a general story about how experiments function to allow appropriate appeals to induction, and thus to generate knowledge (against the background presupposition that empirical knowledge is possible).

My view of induction is that at various occasions when we generate knowledge claims, there are relevant inductions that can and should be made, and others that should not be made. The relevant inductions are those supported by the construction of the device being used and the background knowledge of those using the device. The class of allowable inductions for any circumstance is nested in the sense that using any of those inductions will not take us out of the background practice that supports them, while performing any different induction would.

My task will be, on the one hand, to display scientific experiments as methods for exploiting the inductions that form our background knowledge generation practice and for generating new relevant inductions and, on the other hand, to show that the same methods of exploitation and induction generation are available for a wider class of practices than is normally counted as lying within the class of scientific experiments. By showing first how the latter methods function, and the conditions under which they do function, I will then be able to show that the former methods meet these conditions as well.

15.3. Performing and Informing

I want in this section to say something about how I understand the activity of experimentation generally, and how I think of it as a knowledge-producing activity. The production of effects and the entire panoply of exploratory experimental interventions comprises what may be called the performative mode of experimentation. There is another equally important mode of experimentation, which may be called the informative mode. This mode is the use of observations to assign properties (law-like relations, type designations, etc.) to systems that have not been observed: for example, concluding that effects observed in one system would be observed in other systems if the latter systems were to be observed in the same manner. The intersection of these modes, of course, is at the point of observing that some system does have some property or other in some circumstance or other—i.e., that we can say of proximal system p that it is of type π. The performative mode, and its important role in making possible the production of effects, allows for much more than novel observations, affording us control over nature and also, over time, generating the kinds of autonomous structures Hacking identifies as experiments with lives of their own. There are now many good treatments of this mode of experimental practice. While I will not ignore this mode in what follows, I will be primarily concerned with the informative mode. My task will be to display the features of experimental systems that suit them to the regulation of the flow of information from the components of these systems to agents like us who know things about the systems, as well as to the material modes of storing and representing that knowledge.

I do not after all think that there is any sharp conceptual divide between these modes. Students of experiment have singled out the performative mode as a distinct class from the informative mode because of a frustration with standard accounts that take experiment as somehow the underlaborer of theory, rather than as a scientific actor in its own right. But such accounts are no longer standard, and the point is well taken that experiment does much more than confirm theory. It is at times autonomous from, at times anticipatory of, at times in service to, and at times coproductive with theory. That much is clear. Scientific activity as a whole, and experimental practice in particular, is (perhaps happily) seen as divided into the light-bearing and the fruit-bearing. Even so, neither of these is divorced from the generation of knowledge. Fruit-bearing inquiries of course are best understood as giving us a novel kind of control over nature, but it is still a knowledge of nature that supports counterfactual reasoning on the basis of these experimental activities.

In any case, in what follows my attention will be focused on experimental knowledge without any particular account of the connection between theory and experiment. That is, my concern always is to display experimental knowledge in its own right and to provide a conceptual analysis of that mode of knowing.

15.4. Data, Phenomena, and All That

What are the principle products of experimentation in the informative mode? In the crudest of terms we can say that experimentation produces noises, arrays of lights, inscriptions on pages, illuminated pixels on computer screens, etc. This way of speaking echoes the analysis of Latour and Woolgar ([1979] 1986), for example, and their attempt to give a sociological characterization of the happenings in contemporary research laboratories. That analysis is too abstracted from the practice of extracting information about the world from the things we do in (and out of) the laboratory. But we cannot simply answer that experimentation produces knowledge of distal systems based on the features of proximal systems. Of course it is my contention that they do that, but we need to explain how they do so, and so a prior account of the components of this knowledge must be given.

Experiments produce data, and the job of the experimentalist is to turn those data into generalized knowledge claims, in part in the form of scientific theories. This is not to assert that the business of experimentation is to either confirm or disconfirm scientific theories. However, much of what we know experimentally about the world is couched in a way that requires knowing it via the mechanism of some theoretical framework. My purpose now is simply to ask how we should think about the process of turning data into knowledge?

James Woodward (2010) tells us something useful about the connection between scientific theories and observations in his reconsideration of a long-standing theme he began investigating with James Bogen (Bogen and Woodward 1988). They asked how we can partition theories in a way that illuminates how they make contact with the world they are meant to describe. They reject (as does essentially every philosopher of science after the late middle of the twentieth century) the logical empiricist conception of scientific theories as a specific kind of structured language involving a calculus and correspondence rules that are supposed to connect purely theoretical talk with observational talk. In their rejection of this conception, Bogen and Woodward (1988) make a useful distinction between data and phenomena, and Woodward (2010) especially makes it clear how such a distinction illuminates the role that lists of measurement results play in inferring the presence of and features of various parts of the world. As Woodward points out, trying to use data as evidence is a perilous matter, even though that is its normal use.

Bogen and Woodward's takeaway message is one that I largely endorse: the business of scientific theories is explaining phenomena that are themselves inferred from data. These data are interpreted liberally and include whatever we may observe either with our senses or with instrumentation that is appropriately understood as extending those senses. A key point they urge is that while theories may predict phenomena, and so may preserve that aspect of the symmetry between prediction and explanation, they are generally powerless to predict data that are capricious, dispersed, and difficult to generate using the apparatuses we can construct.

Woodward next considers what he calls the *problematic* of using data as evidence for phenomena (2010, 792), which use is, he says, the normal reason for producing data in the first place (2010, 793). Suppose I want to know how some agent assigns monetary value to things in the world. The natural course is to take some data about how she values a variety of items, and then come up with her valuation structure. On Woodward's account, a data point d_i (perhaps the price we observe our subject to pay for some item of value) has the following functional structure: $d_i = f(P, u_i)$ where P is the phenomenon to which the data point pertains (in this case the valuation structure of our subject) and u_i captures the various interfering factors that intervene between the phenomenon and our observation of the data (perhaps the subject was confused, or misspoke, or we mistook the value of the bill, or it was counterfeit, or she quick-changed a cashier, etc.; 2010, 794). Once we have generated a set of data $\{d_i\}$ we want to use it to make inferences about P.

While this approach is basically sound, one might wonder why "phenomena" are the targets of our inferences from data, rather than, say, "properties" of objects or of systems. Taking "properties" as the target would leave the word "phenomena" free to designate the more traditional class(es) of thing that are the objects of our experience (or more broadly, the ways objects interact with our senses and with the portions of our apparatuses that are the analogues of our senses). There is in the philosophy of science a tendency to take the term "phenomena" as connected to theories rather than to our experience. The idea is that the phenomena are only the correlates for possible observables, within the models of the theory. For example, the continuous trajectories of Newtonian physics would be phenomena according to this way of thinking, even though nothing could ever be observed actually to be a continuous trajectory. I take Bogen and Woodward's account to be telling us, on the one hand, how specific classes of phenomena are predicted and explained by our theories and, on the other, how they are connected to happenings in the lab. My account will be largely in line with this approach, but I will focus more tightly on the role of information flow in connecting all of these stages. Generally, I think the right approach is to say that we have an observed feature of the world—some happening occurs in the lab or elsewhere (e.g., the counter attached to a scintillation detector advanced); a datum is extracted from this (the scintillation detector was triggered); multiple such happenings

provide data; these data support the inference to the presence of some phenomenon or other (a neutrino interacted with our vat of mineral oil, for example); then from multiple collections of these data (and perhaps from the
known neutrino interaction cross section with the target) we infer a class of
phenomena (neutrinos are passing through the earth at such and such a rate).

15.5. The Basics of Experiments

What is a scientific experiment? There's a pretty standard picture that identifies what experiments are good for and how they produce those goods: the
blinded, randomized control trial, or RCT, I have already discussed to some
extent in part I. What this kind of experiment is good for is giving us causal
knowledge of some features of the world. Those features may be the role that
certain drugs play in treating disease, or the fact that students' cheating accounted for part of the incredible improvement in the Chicago Public Schools
around the turn of the twenty-first century, or the field strengths involved in
interactions by which the Higgs mechanism generates mass itself. This kind
of knowledge is reliably produced in experiments, and the key question for
me is "How?" The RCT model allows us to give the following answer.

An experiment, on this model, is thought of as an intervention into nature, and that's not a bad place to start. The experimentalist has a question in
mind, let's say about the efficacy of a certain drug. The most naive intervention is simply to give the drug to a sick person and see whether she is cured.
But as we all know, there are many factors that confound us in such a case and
prevent us from learning whether the drug itself is the causal source of the
patient's improvement. Thus we introduce more subjects, as well as controls
(patients who do not receive the drug). Then, in addition to observing the
patients who did receive the drug, we also watch these control patients, and
we compare outcomes for the two groups. This procedure is much better. It
does, however, leave us exposed to a few more confounding factors. The first,
of course, is that we may well have chosen the control patients and the test
patients in a way that is systematically correlated with how sick they are. If
we give the drug only to relatively healthy patients and withhold it from the
relatively less healthy patients, we may overestimate the strength of the drug.
Likewise, if we give the drug only to those worse off, we may underestimate
it. There may also be some factor in some patients that makes them resistant
to the drug or very sensitive to it, and we may somehow have chosen our
samples to correlate this factor with our assignment of patients to test versus control group. Randomization—making the probability 0.5 that any one
patient is assigned to the control group versus the test group—mitigates this

possibility. Finally, it is necessary to hide from all participants and evaluators which patients are controls and which are test subjects. This need arises, on the one hand, from the placebo effect, in which patients who are convinced that they are receiving effective treatments respond medically as though they were given effective treatments, to some extent. On the other hand, the need arises because researchers, even without intending to, are prone to evaluate patients differently when they know whether the patients have been treated with the test drug. So when we are dealing with knowing subjects, we use a double-blind methodology in which neither the patients nor their evaluators can be allowed to know who are the test subjects and who are the control subjects. This of course introduces the necessity of giving placebos and otherwise making all the material circumstances of the test and control subjects as similar as possible.

Putting all these correctives in place gives us the RCT model; and now we have, in outline, the answer to how the double-blind RCT gives knowledge. It is an intervention into nature whose outcome is compared to the outcome in the absence of such intervention in such a way as to avoid confounding factors. The difference between the intervention case and the nonintervention case is the causal difference the intervention makes, and by seeing this difference we can come to know the causal power of the intervention. Again, this is just an outline; to fully support the causal knowledge we need to test many subjects in order to generate more robust statistics than I have suggested here, because without many test subjects we may be observing only fluctuations rather than a true difference between the cases. Indeed, there is a sense in which no finite number of test subjects can ever make us absolutely sure that we are not observing some wildly improbable fluctuation that will not manifest when we put the drugs on the pharmacy shelves. But we don't need such assurance. With proper study design we can be as sure of causal efficacy on the basis of double-blind RCTs as we are of any other general, empirical facts. Even if we are never sure, we can come to believe (or accept) these findings, and *in that good way* that results in knowledge.

I will, in what follows, take for granted that we can come to have knowledge on the basis of RCTs. While I have critical things to say about them, I will not be joining the voices that are calling into question the power and legitimacy of such experiments. Instead I will use my critical evaluations of them as probes to show that they do not function the way we have normally supposed. Further, once their function is clarified, we can see that other activities can give us experimental knowledge that is as robust as that produced by RCTs. Of course experimental knowledge can be produced by RCTs; but it can be produced by many other types of experiment as well.

Lest I give the wrong impression, I certainly do not think that even all standard examples of scientific experimentation could fall under the RCT model. Many experimental domains are simply not suited to that kind of thing. Experiments on electrons do not require that the electrons be blinded to whether they are being treated or not—that wouldn't even make sense. Electrons are not aware, so they cannot experience the placebo effect. Electrons are also noncomplicated, so it is immediately clear, at the location of the electron, whether or not some intervention has taken place. In addition, there are many types of activity that can secure all of the epistemic aims of the RCT model but that simply do not involve anyone providing controls. And, as I showed earlier, if one thinks of the RCT model as tied to human intervention and wants to subsume all experimentation under that model, one will miss out on important instances of experimentation—even in the case of drug trials, for example. Finally, drug type trials are not the only thing that get done in experimentation—there are other types, including chemistry and physics experiments. One could attempt to shoehorn these activities into that model, but it would be artificial and unilluminating, and ultimately done only to serve the model and not our understanding.

As a first pass I responded to the question "what is a scientific experiment?" by outlining the concept of the double-blind RCT. While RCTs can give us experimental knowledge, they are not coextensive with the class of things that can do so. Indeed, it is difficult to imagine any usefully succinct answer that picks out only the things that should count as experiments and rules out everything else. As I said above, I don't find Wootton's answer, "an artificial test designed to answer a question," very satisfactory, nor do I want to offer a swarm of bees as an answer to a question about what it is to be a bee. So instead of answering the question "what is an experiment?" I would like to ask instead a similar but contrastive question: What is special about experiment? That is, what, if anything, makes something count as an experiment? Even that is not quite right, for it leaves something out. Experiments are knowledge-generating activities, and to know whether something generates knowledge we are best served by focusing on the knowledge those activities are apt to generate.

So instead of asking about experiments full stop, we should be asking "What makes something count as an experiment for _____?" where the blank is to be replaced by the statement of a knowledge claim. The practices that are experimental practices are those that can generate experimental knowledge. Why this reformulation? Because of the nature of the project in which I am engaged. I am investigating the nature of experimental knowledge, and such investigations can seem very much like epistemology. However, what I am engaged in here is better understood as a conceptual analysis of the notion of experimental knowledge.

Conceptual analysis is related to, but different from, epistemology. Sometimes epistemic issues are distractions from conceptual issues, and sometimes intractable epistemic issues are clarified only in the light of proper conceptual analysis of the issue. I want to show that our understanding of experimental practice is in a state where conceptual issues are not clearly seen because of a misplaced focus on questions of justification and evidential considerations. Shifting our attention to clarifying the concept of experimental practice will itself shed useful light on the epistemic issues.

My analysis, in terms of information flow, begins by answering a more basic question: what makes activity count as an experiment for _____?

15.6. What Is Experimental Practice?

Experimental practices are those that are capable of generating experimental knowledge.[3] Elaboration of this answer is the main task of this book, but a few things should be clarified right now. First, I am taking a strong stand on the nature of experiment. When I say that the demarcation condition for activities to count as experimental is that they be able to generate experimental knowledge, it is because I think there can be no answer to this question in terms of any other criteria that will make sense and will cover all of the things that should count as experimentation. Why not? Here is the central reason.

Some kinds of activity count as experimental when some people do them, but not when others do, and those same activities can stop counting as experiments when they are done by the same person over and over. Let me return to a simple example, well known from discussions of the foundations of experimentation: Millikan's oil-drop experiment. Millikan examined samples of oil drops that had been forcefully ejected from an atomizer. When they were ejected, the droplets generally acquired some electric charge. These droplets were ejected into a region that was within the objective of a microscope and that was also between the plates of a capacitor that generated an electric field

3. Although experiments will be individuated by and typed according to their knowledge production capabilities, we will also see instances of experiments produced to persuade others of some result. This presents no difficulty, but it might sound odd when we see someone who already knows well about some causal connection or other described as performing an experiment establishing that connection. The claim there will be that the experimenter was already sufficiently well informed that no new knowledge was generated for her, but that the experiment, were it observed by (reported to, etc.) someone else, could produce new knowledge for that person.

which could be quickly switched on and off. The orientation of the plates was such that the electric field was parallel to the gravitational field. Thus, when the capacitor was energized the falling particles were subjected to a force opposed to gravity, and when the capacitor was not energized they were subject to the force of gravity alone. So the particles could be allowed to fall and then caused to rise, as the capacitor was turned off and then on. Millikan could observe a given droplet and control its motion up and down in his field of view, and record the time it took for the particle to move in each direction. After repeated observations of a given droplet, he would have a good observational basis for establishing how much charge was on the droplet, for by comparing the times of fall to the times of rise he could determine the force exerted by the electric field.[4] When, as an undergraduate, I first carried out Millikan's oil-drop experiment to determine the charge on the electron, I didn't know the answer I was looking for. We were given instructions on technique and on the analysis of our results, but we were not told what answers to expect. Also, I was that kind of undergraduate who doesn't believe the results found in instruction manuals anyway. So I had no belief about the value of the charge on the electron. Nor, to be honest, did I learn it from performing the experiment itself—my team's value was pretty close to one-third of the currently accepted value. However, by checking my results with other teams, and with the accepted values, I came to know the value of the electron's charge. I don't think what my group did counts as an experiment for *determining the charge on the electron*, though I may be wrong about that. Still, the activities I performed were very similar to the activities performed by students in other groups, and students in other colleges at various times, and by Millikan himself. So why shouldn't they count as performing the experiment on this occasion?

To begin with, similarity of activity is simply too weak a standard to apply. Some groups in my lab did even worse than mine, in part because they were careless in their work in addition to being unskilled as we all were. They were, in some respects, merely going through the motions. Do their activities still count as performing the experiment? Their activities were surely *similar* to

---

4. Actually, in both cases the droplets were subjected to a retarding force caused by the drag of the surrounding air particles. This feature is essential to understanding the significance of the observations themselves, because over the time and distance scales Millikan was using, the drag was such that the particles very quickly attained terminal velocity—both when falling under the influence of the gravitational field alone and when rising under the influence of the capacitor. This fact greatly facilitated the calculation of the force due to the presence of charges on the oil drops, which would have been very complicated if they were accelerating during most of the time of observation.

mine and to Millikan's. What about some group that would simply guess the time of suspension of the oil drops? or a group that would use a randomly shifting electrostatic field? All of these sequences of activity would appear very similar on a video recording. So at some point we will have to say that activities that are similar on the surface don't count as performing the same experiment, or even any experiment at all. But so what? Perhaps the right answer here is that the boundary between activities that do count as experimental and those that do not is vague. Probably that answer is right. But it is appropriate to patrol even vague borders in order to keep separate those regions on either side: however vague the boundary, there will be some way to determine whether something is clearly on one or the other side or is just too close to call—otherwise it isn't a border at all, but an undifferentiated field. The problem with the similarity-of-activities standard is not that it is vague and that it leaves vague the line between experiment and not experiment; the problem is that it doesn't really tell us anything of importance. Even if workable, it is parasitic on extant examples of clear experimental practice and thus gives no guidance either on which novel types of practice should count or on how to decide whether some practice is similar enough to another.

Surface similarity of activity is clearly insufficient as a guide to what is and is not an experiment. It simply doesn't get things right. But perhaps all it is missing as a standard is some way of grounding its similarity analysis in some other, more basic concept that will itself guide the application of similarity considerations. That is to say that perhaps similarity is the right analytical tool, but that it needs to be informed by a more detailed plan of application in order to do its job. No added formal requirement seems sufficient, because we'll end up back where we were. When are the forms the same, when are they different, and when they are sufficiently similar?

Perhaps outcomes are simply the wrong kind of thing to be considering here. Perhaps instead we should be thinking about methodology. After all, isn't that what we're often told is at the heart of contemporary science, a method, indeed the *scientific* method? But what is this method? How would we apply this method in determining whether or not some activity does or does not count as an experiment? In his *Design of Experiments*, Fisher ([1935] 1974) suggests that there are just two classes of criticism one might make against a purported experiment. Either there is something wrong with the interpretation of the experiment—that is, something to do with statistics—or there is something wrong with the design or execution of the experiment. Fisher's aim in his book is to develop appropriate methods. He puts the thing this way: "Statistical procedure and experimental design are only two different aspects of the same whole, and that whole comprises all the logical requirements of the complete process of adding to natural knowledge by experimentation" ([1935] 1974, 2).

Similarly, then, we can ask, "What demarcation criteria, arising from the application of scientific method, are appropriate to distinguish, on the one hand, experiment from other activities and, on the other hand, one type of experiment from another?" I submit that it is along the line of whether or not they give knowledge, and about what. A standard then naturally presents itself, and it is the one I will adopt: count practices as experimental when they are deployed in generating relevant experimental knowledge. This standard will, perhaps, itself be vague. Certainly it will not always be clear to us how to use the standard to decide whether something is or is not an experimental practice. However, it *does* tell us why some things count and other things do not; it is a *conceptually* useful separating condition. If the knowledge they generate is best counted as adding to our "natural knowledge," i.e., giving us knowledge of systems distinct from the apparatus itself, then that is what I am calling "experimental knowledge" proper. If, however, the knowledge is about the state of functioning of the device itself that is being used, then "calibration" is a more appropriate label. In each case the experimental aims of the researcher are relevant, as are changes in that researcher's state of knowledge. Similarly, activities that are best understood as giving students the skills necessary to engage in experimental practice will surely count as giving them knowledge, but knowledge *about* experimenting rather than experimental knowledge, and so should not count as experiments.

One might object to the theory at this point by noting that it seems to make any activity at all capable of generating experimental knowledge. If my definition of experimental knowledge is correct, then any activity that produces observational knowledge of some proximal system can be combined with background knowledge, for some observer, to generate experimental knowledge of some distal system. And that would meet my definition of experimental knowledge. While that is correct, it doesn't tell us anything that is interesting about scientific experimentation. My plan here will be to accept that virtually any activity *can* count as experimentation, but to point out that what we really want to know about experimentation is, experimentation for what? It is not enough, for example, to say of the Large Hadron Collider that some experiments are being done with it. One wants to know which experiments, and how those experiments are giving knowledge. But to say which experiments are being performed will be to say which systems we are generating knowledge about, and what the nature of that knowledge is.

What kinds of activity generally count as experimental? There is calibration of experimental tools (experiments on the tools), there is checking other experiments (experiments on those experiments using similar proximal systems), and there is generating novel experimental knowledge of nature at large. That seems to cover it. In each of these cases, to say which kind of experiment is being performed is impossible without assessing our state of knowledge of various systems before and after the activity. Before I begin calibrating my device, I don't know how it functions. I take some data, and compare those data with other, benchmark data. Then, if all is going well, I can say of my device that in the future it will generate data that are capable of supporting various kinds of inferences about other distal systems. But the distal system in the case of my calibration experiment is the collection of future states of the device I am checking (or past states, if I've already generated some data that I am unsure of). The same activities may count as checking up on the experiments of others. If I've already calibrated my device—if, that is, I already know that it generates data that are capable of supporting inferences about the systems that you claim to have experimental knowledge of—then when I take those data I may be generating experimental knowledge both about your experiments and about the systems you were studying. It may not count as generating such knowledge for everyone; some may already know about these systems and not require a check. But for those whose information store is not as rich, my activities can generate experimental knowledge of both types. Then, finally, I may be performing a novel experiment and generating novel experimental knowledge of nature at large. What is the difference between all of these activities? The state of knowledge of those who take note of the activities. In the first instance it's the knowledge state of those performing the activities, but presumably anyone else with the right access to what happened during these activities can get experimental knowledge of one or the other type.

Given that what counts as experimentation is demarcated by its connection to knowledge generation, why should we stop with saying that what counts as an experiment is whatever is capable of generating experimental knowledge, rather than saying that it is whatever does in fact give knowledge? That at least is a pretty clear demarcation and seems to respect the intuitions behind worries about whether undergraduate lab students are really doing experiments.

Recall my discussion of Todhunter from part I of this book. His idea was that it is only an experiment the first time around, and subsequent activities simply go through the motions. Todhunter seems to have a view of experiment that is motivated by concerns that are similar to mine. Isn't the point to experimentation, he wonders, to gather novel knowledge of nature? If so, then once we have that knowledge it seems no longer precisely an experiment even were we to repeat the same actions. For Todhunter, things are simple. Simply throw out as nonexperimental anything that happens *after* some fact is established (barring cases of ignorance on the part of the researcher, of course). There does appear to be a tension here for my view, though. I've been constructing an account that makes something not an experiment if I'm just displaying for my students how a machine works and not really seriously taking data, and also makes an "experiment" performed by accident not really an experiment. Indeed, that was the point of the story about my experience in replicating Millikan's experiment, and why some range of activities on one side of the activities I perform should not count as experiment while some range on the other should. Yet my account of the different activities of calibration, replication, and novel data generation seems to make all of that stuff experimental after all—and it appears to do so in a way that would in fact count Todhunter's schoolchild's observations of a teacher redoing an experiment as legitimately experimental activity.

The key to resolving the tension will be in my notion of experiment for _____. The calibration versus replication versus generation of novel results discussion shows the way. In each of those activities, the experiments were experiments for a certain end—either for checking the equipment or for checking other experiments or for generating novel data. But they could, given the right circumstances, count as experiments for any number of other things—should the state of knowledge of some observer be ripe for that. So I will treat claims about what experiment was performed as defeasible claims about the state of knowers who have access to the data about the apparatus and generated by the apparatus. Then we can say things like "this was an experiment for establishing the value of the charge on the electron" and be understood as saying that, as far as we can tell, the knowers who have access to the data do not know the value of that charge, and that the data bear appropriately on the question. The experiment may fail as an experiment for establishing the charge on the electron if, for example, the data do not really bear appropriately on the question, or if the epistemic state of those knowers doesn't change in response to those data, or for myriad other reasons. That is all consistent, however, with it succeeding as an experiment for any number of other things.

Now it will make sense, I think, to begin to conceive of experimentation as a way of manipulating the flow of information between the world and those who are observing the world. Because information flow is the way knowing subjects acquire new knowledge, we will need to pay attention to the way that knowers are hooked into the information transport structure of any experimental situation. Not only that; we will have to pay attention to how they are already situated with respect to the kind of knowledge the experiment is capable of generating. Do they already know these things? If so, then no information can flow into them about those things through this experiment. On the other hand, the exact same apparatus, treated in exactly the same way, but connected by observation to a different knowing subject may well allow information about those things to flow.

15.7. Characterizing Experimental Systems

Suppose we have been able to isolate a system on which to perform an experimental intervention. How do we determine the features possessed by the proximal system? Here's a rough guide. For conceptual clarity it may be best to begin with the degenerate case of learning about an object by observing it over some span of time. We consider some temporally distributed system, and we carve out of it some temporal section. (For example: my garden over the next week; this thermometer for two minutes; etc.) We can treat that as the proximal experimental system. Is the temporal section we choose arbitrary? Certainly some measure of choice enters in—we do, after all, elect to begin the experimental intervention at a particular time and to finish it at a particular time. I think it fair to say that the beginning time presents no special worry. Naturally we can, if we know something about the system in advance, choose judiciously in order to give a misleading result. That, however, is not an arbitrariness problem but a bad faith problem. The end time does present a worry, though—or rather, the duration of the experiment is something subject to troubling ambiguity.

This may seem complicated, but the point is simple: by selecting the wrong temporal interval we may fail to model the device properly at other times. Moreover, by performing the wrong tests, or not enough of the right tests, we may also fail to model the device properly at other times. And this of course is a persistent worry, one that is not ever truly overcome, but only addressed by appeal to our background practices and our present classification of the device as a whole. And our conviction as empiricists—or at least as empirically minded students of nature—is that by careful and persistent feedback of our results onto our methods, we come to have classifications of the proximal system that are adequate to the task of classifying the distal system. It is here that considerations of evidence gathering and analysis are appropriate. Here as well we apply the various epistemological strategies appropriate to the case. But once we have done all of that, and once we do have a temporal section of the device that is arbitrary—that does, by that token, bear information about the other temporal sections of the device—we can learn about it as an individual, generalize that knowledge by means of the way it bears information about its own future and past, and then infer properties about individual instances of its future and past states. Then we have the following picture:

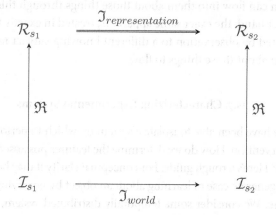

At the initial time of observation there is a worldly system, I, with some state or other. Our observations allow us to represent it according to its type, R. Our general knowledge of how objects behave, including the way things of that type evolve in time, allows us to evolve that representation in time. This trivial-seeming story undergirds all of our learning from observation that goes beyond the simple registration of spatial and temporal coincidences of qualities. This idea of registering these qualities sounds a bit like the old logical empiricist idea of protocol sentences. It also sounds a bit like Hume's (and other idealists') notion that all that is given in experience is the conjunction of various sensory ideas. That's no accident. Those do strike me as the fundamental epistemological building blocks of all science. What is novel here is the idea that, while the building blocks of Humean and, say, Carnapian accounts of science were sound, a house built on justification cannot stand— only information provides the necessary stable foundation. I will elaborate the information theoretic part of this shortly; for now it will suffice to reiterate that we know about the stability of objects over time when (and only when) our inductive inferences are information-conducting.

The degenerate case I just described—of inferring[5] that properties that are clustered together here and now (hardness, shape, color, etc.) will be clustered together elsewhere and elsewhen—itself constitutes a kind of experimental

---

5. Lest I be misunderstood, I will emphasize that I do not regard inference as essentially connected with explicit logical derivation. That is *one* type of inference, a particularly well-understood and useful type to be sure; but inference as I understand it is simply the process of coming to have new beliefs on the basis of other beliefs or experiences. The key for me is whether some inference is or is not information-bearing.

knowledge. It is implicit and little noticed, but it forms the basis for all of our further experimental inquiries. All of our experimental knowledge, from the most trivial to the most elaborate, arises from our acquiring information about some part of nature from our observation of nature's other parts. This works because these parts are similar in just the right way for features of the one to bear information about features of the other.

15.8. The Inference

And why do we think that some systems are similar enough to bear such information, and others are not? What is the "similar enough" marker we are tuning into when we make that determination? I am not asking a specifically test-based question here. It is not a question that is appropriately answered by an assertion to the effect that when we make such determinations we are correct (all, some, or part of the time). I am fully prepared to grant that. Indeed, I am presupposing all along that we really can and do get experimental knowledge, and that we do so by taking advantage of the fact that we are pretty good at picking out pairs of systems that are similar enough to each other that we can get knowledge of the one from knowledge of the other. Instead I am asking a question about how we do it. Is there a particular feature, _____, of natural systems that we are sensitive to that allows us to say, "Ah, these two systems are alike in feature _____, and so one can be used to produce experimental knowledge of the other"?

Even for repeated applications of experimental procedures to, say, particular items of experimental apparatus that are more or less the same over the course of these applications, we face the question of how to generalize our results. At each instant the experimental apparatus has some set of features. These features have a law-like connection to its features at subsequent times. But these features, the properties that we directly observe, are not token identical to the properties we observe at subsequent moments of time. Most obviously, the features may change. Some solid objects become liquid on being heated, a fragile object breaks on being dropped, etc. Less obvious, but conceptually quite important, is the fact that the properties we observe are fundamentally indirect even when we are measuring directly. That is to say, our measurements of such properties as weight are evidenced by the way an object interacts with a spring, volume by displacement of a fluid, color even by the way other objects react to the light given off by the object. Even in the best situation, we are faced with the fact that we are always looking at token instances of objects over some temporal duration, and we want first to use these instances to infer fundamental properties of the object, and then second to use those properties to infer the fundamental properties of yet other objects. That is the key step in experimentation. How then is it that we say that some result in the lab shows us that something or other will happen outside the lab? We must be relying on the entities outside the lab, whose behaviors we are predicting, being similar enough to the entities inside the lab, whose behaviors we have observed, to support these predictions. Do we think they are similar enough because we have classified them as the same in prior cases? But unless we have learned nothing new in this trial, the basis for the classification scheme may have shifted. That the entities have behaved the same in the past, in such a way that they can be classified together, does not bear *directly* on the question of whether their behaviors with respect to these new interventions will be the same. In the limiting case of identical particles and elemental substances, perhaps we can see more or less directly that our inferences hold from one sample to the next—that is, if we're confident that our notions of identical particles and elemental substances are realized in nature. But what of more complicated situations, like experiments that attempt to discover the properties of bridges, airplanes, volcanoes, cancer cells in humans, etc.? For such situations, to find out anything about them from the way other things are, we must rely on what we already know about their similarity to the things in the lab. What kind of similarity? That kind of similarity that allows the features of things in the lab to bear information about things outside the lab, and generally at arbitrarily large temporal and spatial remove from those things in the lab.

15.9. What Is an Experimental Claim?

I conclude this chapter by saying what it means to make an experimental claim: to say that experimentation on some proximal system p gives knowledge of some distal system d is to say that establishing that p is of type π bears the information that d is of type δ, and moreover that p really is of type π. The notion will be elaborated in subsequent chapters.

Experimental knowledge is indirect knowledge generated by witnessing the behavior of some system or other and then mapping the logical structure of that behavior (statics, kinematics, dynamics, what have you) onto a model of another system to give an appropriate model of that other system's behavior.

This may seem very abstract, and far removed from scientific practice. But to motivate things a little, let us consider a few examples from the previous part of the book. This sample, while small, is meant to capture a wide swath of the kind of thing that counts as experiment in the sciences. There will be many issues left unresolved, of course, and no one book could possibly cover the entirety of empirical study. However, the examples are chosen to be as broad as possible so that the insights gained may be as general as possible. Here I will briefly offer a broad sense of how things hold together.

Why, roughly, do we think that cathode rays are charged particles? We have observed that they are responsive to applied external magnetic fields, and we know that is characteristic of charges when they move through such fields; as we also saw, Thomson and Perrin caught those charges in a bucket (and so did Hertz . . .). But we haven't really observed *all* cathode rays. We have observed a very small sample of cathode rays, but we are convinced that the properties we have observed in those rays are properties that all tokens of the type *cathode ray* share. And why do we think that? Well, at some point we run into matters of definition, but what we mean by "cathode rays" is whatever it is that fluoresces the cathode ray tube when sufficient voltage is applied between the cathode and the anode. Then, in this particularly simple case, what we are doing is inferring from the observed properties of some of the rays that all the other rays have those same properties. That is, observing that these token cathode rays are of type *travel curved paths in magnetic fields* carries the information that all tokens of the type *cathode ray* are also of type *charged.*

Consider the more involved case of those claims that saccharin causes bladder cancer in humans. Why did anyone ever believe it? Well, the original claim derived from the belief that saccharin causes bladder cancer in rats. Why should one believe that? Well, because we gave large doses of saccharin to some rats, who developed bladder cancer at rates far beyond those of rats not given such high does of saccharin. We inferred from our observations of these rates of cancer that saccharin is causally relevant to the development of bladder cancer in these rats. We then used that to support the inference that saccharin causes bladder cancer in rats generally, and in turn used that inference to support the inference that saccharin causes bladder cancer in humans. Compressing all the intermediate steps, the fundamental experimental knowledge claim was that observing that these token rats are of the type *contract bladder cancer from saccharin* bears the information that humans generally are of the type *contract bladder cancer from saccharin.* The fundamental inference was from "neonatal rats get cancer from saccharin" to "humans get cancer from saccharin"; that inference fails because, in fact, neonatal rat properties just don't bear the appropriate information about humans.

Now a case from economics. Are people subject to the endowment effect? That is, do people place higher value on the things they have merely in virtue of having them? Many economists think so. Why? In large part it is because they have asked groups of people (principally their economics students) to evaluate the relative worths of relatively small-value objects (pencils, say, or coffee cups); they then give the people these objects and note that their evaluations change once they have them. That is, once having received something as a gift, these subjects will not trade that thing for another thing that they had earlier assigned the same value. On the basis of such observed behaviors, we conclude that the behavior is general. That is, we conclude that *they* assign higher value to all objects of value once *they* possess them. We then infer that humans in general are subject to the endowment effect. That is, the token observations that these humans are of type *subject to the endowment effect for small-value gift items* is taken to support the inference that all humans are of type *subject, generally, to the endowment effect*. There are other sources of this belief, but our strongest source is experimentation. We think we know that humans are subject to this effect because we think that students' economic behavior bears information about human economic behavior generally.

I won't focus now on the strength of these inferences, or whether they constitute particularly reliable experimental claims. I just want to use them to illustrate my basic account of what constitutes an experimental claim. Note a few key features. First, the proximal systems that are observed in the first instance do not have to be of the same basic kind as those distal systems about which the experimental claim is being made. Tokens of rat were used to conclude experimentally that saccharin causes cancer in humans. Note also that the types assigned to the proximal system need not be the types assigned to the distal system. In the case of cathode rays, we observed the paths of some rays curving (one property or type) and inferred that all cathode rays are charged (a related, but distinct property or type); further, we observed the endowment effect operating for very small-value gift items (one property) and inferred that the endowment effect applies generally (a related, but different property), and we saw the effect in college students but inferred that it is present in humans generally.

While I have chosen to characterize what constitutes an experimental claim in an unusual way, I do not think that the content of what I am calling an experimental claim is at all unusual. My hope is that the content is just what everyone already takes it to be, while the form will make it more apt for conceptual analysis.

15.10. Overview

Let's consider the various types of experimental hypothesis. Can they all, and should they all, be assimilated to claims of the form "that proximal system p is of type π supports the inference that distal system d is of type δ"? I think so.

Suppose I am seeking knowledge of how some natural system (or process) behaves, or about its properties more generally. How can I find out? It is a purely logical matter that I must either appeal to what I already know, even implicitly by drawing inferences about my overall knowledge, say; or I must gather new knowledge from the world in the form of reports, observations, or other data. To be an empiricist is to claim that all of our knowledge must find its ultimate source in the latter kind of finding out. While a contemporary empiricist may grant that much of our knowledge is implicit and reflects a great deal of conceptual structure that is rarely up for grabs, still the arbiter is the empirical world. But to extract knowledge of the world is, except in quite restricted cases, to rely on some kind of experiment. And the upshot is that many more activities count as experimental practice than we normally think. Of course, not all of these will be successful experimental practices in the sense that they really do license the fundamental inference. But they are of the type *experiment* in that to understand what is happening is to assimilate them to the practice of drawing that inference.

But prior to the question of arbitration is the question of source. Tracking, counterfactual supporting, information bearing, and so forth are all notions that tap into the basic idea that what we do, when we get knowledge of the world, is extract information about one part of it and try to use that as a nucleus around which to structure information about other parts.

Are there only two classes of activity that count as empirical investigation proper? We can count observation (including measurement) and experimentation. The former is a method for determining that some proximal system is of a given type: the rose is red, the triple-beam balance registers 17 grams, etc. The latter is a method for determining that the proximal system being of a certain type supports the inference that a distal system is of some (in general other) type: that these rats died from pancreatic cancer supports the inference that saccharin causes cancer in humans (i.e., rats are of type *get cancer from saccharin* so humans are of type *get cancer from saccharin*).

The totalizing claim here is that a great number of intellectual and scientific activities seem to share a conceptual profile: they're like experiments. So why think there is more to science and to getting around in the world than experiment? First, there's the banal point that we all seem to think there's more. The enterprises seem quite disparate. But drop intentions and worries about epistemology. Thinking carefully allows us to unify all these activities under the information theoretic account. And this will bear fruit when we apply the conception.

So I am making a purely conceptual point about what is (both descriptively and normatively) science, and that point transcends the descriptive versus prescriptive stuff. It just states what science really does when it gives us knowledge of the world. The place where we may find whatever remains of "theory-ladenness of observation" is perhaps in the application of local classifications. These classifications both express our interests and shape what we take to be the kinds of information appropriate to gaining knowledge of the systems.

The unification claim is very broad and seems outlandish. But it is not merely the claim that, essentially, all empirical practice is experimental practice (though it is that); it is further the provision of a way to classify these practices in a way that makes clear how they are and must be supported by other kinds of practice. Here is where epistemology comes in. Various experimental practices differ in the points at which epistemic concerns arise, and the way they are dealt with. How information flows and what kinds of backlogs in that flow are possible will depend on the details of the structure of the channels connecting the local logics, and this will be reflected in how the experimental type is classified. Things of that sort will certainly allow us to draw conceptual distinctions between activities that lie *within* the domain of experimentation.

What then does my own proposal amount to? First we require something that counts as an experimental observation. As a result of an intervention, for example, we may observe changes in some feature of the proximal system: that, say, one form of fructokinase causes obesity in mice. This gives experimental knowledge of a distal system if the inference therefrom—that the distal system has some feature or other—is sound. For example, if the mouse experiment bears the information that fructokinase causes obesity in humans, then humans are the distal system and we know experimentally that fructokinase causes obesity in humans.

Notice again that the proposal as framed is not in terms of experiments as such and in terms of the link some experiment establishes between some proximal and distal systems (or as is sometimes said, object and target systems); rather, it is in terms of the generation of experimental knowledge by means of experimental practices. So one and the same experimental activity may well give experimental knowledge of very different systems.

15.11. Looking Forward

I will next make more precise the idea of information flow in experimental systems. It may appear excessively elaborate to spend so much effort explicating what seems to be a rather easy idea: that when we perform experiments and generate experimental knowledge thereby, we do so by becoming informed about some system of interest. However, like many other apparently trivial points, there is more going on here than one initially suspects.

Information

16.1. Introduction

It is finally time for me to develop a general framework for tracking and answering foundational questions about experiment—a framework that begins not with general questions about the epistemology and ontology of experimental practice, but with what is perhaps the most crucial feature of experimental knowledge generation: the flow of information. By following the flow of information we get to see how and why it is that one system can be informative about the properties of another. On the surface the process of experimentation appears magical even for standard experiments. Remember that what appears to happen is that intervening on (or perhaps even simply attending to) some proximal system and noting its subsequent behavior causes information about some further, distal system to flow to the researcher. But where does this information come from? How, for example, do the sonic properties of microscopic samples of gas say anything about the fundamental structure of space-time, as they do in the case of gravitational analogues? Obviously there is no magic here, but keeping track of how the changing states of the experiment cause changes in the informational situation of the researcher, and making sure that the situation is exploited properly, is key to understanding how experiments generate knowledge.

In this chapter I will give a general overview of the concept of information as it is related to communication theory, and also of its role in generating knowledge. I will then connect that to a very general account of knowledge in terms of information flow., I will follow that by constructing the logical superstructure that connects the components of experimental systems, and explaining how the logical structures of those components fit together to allow information to flow in the experiment: from type-level facts about the systems that are the real targets of our experimental inquiry, through type-level facts about our devices and apparatuses, then into the tokens of them we deal with in our experimental practices, and finally to us to generate experimental knowledge. After that I will consider a generalization of that account adapted specifically to the case of extended systems, and adapt it to give an account of the elementary structure through which we gain experimental knowledge— a structure that I think is best understood as controlling information flow through various systems. The idea is that when we have constructed some experimental system we have also, sometimes explicitly but often only implicitly, developed characterizations of those systems that allow us to understand them as conduits for information flow. In fact we can view them this way at various levels of granularity, depending on how detailed an image we want of the stages of information flow in some experimental setup.

This approach will shed light on a question which may not seem puzzling at first, but which really is: how can interventions on a system local to an observer immediately give information about the state of distant systems, as they seem to do when we learn something general from a particular observation? Understanding experiments as manipulating the flow of information makes clear how that works.

An information flow–based account of experiment is also a useful tool for assessing the workings of individual experiments themselves. By making clear that certain structural barriers to information flow are built in, for example, it provides a novel way of seeing why certain types of experiment consistently give us less knowledge than we might otherwise expect. We will see this for the entire class of experiments on murine models of human disease, a class that is dense in failure and sparse in success. Familiar episodes in the history of science also look different from an information theoretic standpoint. My earlier discussion of Hertz's cathode ray experiments is best understood as showing that they were replicated in the sense that the information he was seeking about the rays in his own device flowed through both Thomson's and Perrin's devices. Indeed, it flowed through his as well, but as we saw, in Hertz's case the signal was not strong enough to cause the right belief.

I am by no means the first philosopher to think that information is an important tool for understanding our sciences or our scientific practice, and in fact I'll be drawing rather heavily on the work of Fred Dretske and of Jon Barwise and Jerry Seligman as I set up the account. What is novel here, in addition to the detailed account of experimentation as regulating information flow, is the use to which the account will be put: giving a unified account of the various things that fall under the term "experimentation." It will turn out that, surprisingly, even such esoteric practices as thought experimentation will come out as properly experimental.

16.2. Becoming Informed

Coming to know more than one knew before seems to consist in either gaining new beliefs or changing how those beliefs relate to something else. Most of epistemology centers on that latter bit—the thing that goes beyond the belief itself—and saying just what it is and how it functions and what is necessary in order that it transform mere belief into knowledge. Here I am going to adopt the position that the extra bit is that the belief is caused or causally sustained by information. This is Fred Dretske's proposal from the 1980s. His account has not gained the popularity it deserves, and yet it seems to me a pretty solid grounding of empirical knowledge. Before introducing my account of experimental knowledge, I will offer some reasons why this is the right way to go, why it's an appropriate ground for the analysis of experimentation, and I will defend it against some popular objections. Even before that, however, I will explore the idea of information itself: what it is, how it functions, etc. And I hope to make clear why information flow is a good candidate for explaining the nature of scientific experimentation by the time I move on to that explanation.

16.3. What Is Information?

Information has come to occupy an important position in the foundations of science. Interesting results from thermodynamics, relativity theory, and quantum mechanics appear to indicate that information is somehow the fundamental feature of the physical world and/or of our understanding of it through physical theory. And our current conception of evolutionary biology suggests that the regulation (storage and transmission) of information is the key role our genes play in constructing our bodies in ways that suit them to withstand the rigors of their environments and that give good odds that they will live long enough to generate viable offspring. Whether or not this appearance is correct, the importance of the role of information in physics (and other disciplines) to our understanding of the world is clear. Yet the nature of that importance, as well as the nature of information itself, remains unclear. Part of the reason for this lack of clarity is, I will argue, that information is not itself a good physical quantity. What, in this context, do I mean by a "good" physical quantity? For configurations of systems I mean in part that for a quantity to be good it should be a property of the physical state of the system and not be dependent on how that state is represented.

There is some controversy in the foundations of physics community about whether we should view information as physical and embedded in systems, or as a property of agents coming to know something. For those who take information to be straightforwardly physical, standard approaches to questions about how to understand the fundamental physical nature of information, both quantum and classical, begin with the presupposition that information exists: sometimes conceived substantially, and sometimes as a property of special kinds of or configurations of other systems. But even granting that information is physical (as I do) does not commit us to its substantiality, or to its status as system property. We need not view it as on a par with energy, say, or mass or other substantial quantities. The presupposition that information is substantial or a system property goes well beyond a commitment to its physicality.

I suggest that we can easily acknowledge the physicality of information, but deny that it is substantial. Unlike 4-momentum, say, or local stress-energy density, information is not a conserved quantity, and it is not a quantity that is independent of how we represent it. A given array of numbers, a given configuration of atoms, a given region of space-time carry no particular amount of information. Any scheme that assigns values of information to such systems does so with considerable arbitrariness. On the other hand, it is quite plausible that certain systems can carry information about other systems—that is, in a sense to be explained below, that information can flow from one system to another, and also that this flow may be quantified without arbitrariness. Thus I am making the claim that information is a bad physical quantity, but that information flow is a good physical quantity. That may seem paradoxical; however, there is good precedent in science for quantities of this sort, and the accepted facts about the nature of information support viewing things this way.

I am claiming that we should distinguish sharply information flow from information. The obvious analogy for this distinction is the difference between heat (the measure of which is not a good fundamental physical quantity) and heat flow (the transfer of molecular energy, which *is* a good fundamental physical quantity). The history of thermodynamics makes clear the value of distinguishing heat from heat flow in understanding thermodynamic behavior. Another analogy is to the total gravitational energy of an isolated gravitational source. If the space-time is asymptotically flat, then the total gravitational energy of the source is not well defined. On the other hand, the flow of gravitational energy across any surface *is* well defined. The latter is a good physical quantity and the former is not. These two examples are interesting in part because of the tight connection that has been drawn recently between information, thermodynamics, and gravitation. In addition, however, these two examples illustrate that there are widespread typical situations in physics at least where we do not have a well-defined quantity (because it is not uniquely defined, or because it is not a state property, or for whatever reason) but we do have a well-defined flow of that quantity. The case with information is the same: information is not a good physical quantity, but the flow of information is a good physical quantity, and so is the difference between the information states of various systems. While outside the mainstream in physics, this position is, I think, reasonably standard in philosophical discussions of information. It is not that physicists do not think of information flow as a good physical quantity—they do. So in that sense at least there is nothing funny going on here. However, discussions by physicists of, for example, the maximum amount of information that can occupy a region of space are predicated on the idea that there is some well-understood, well-defined, physical quantity that is the total amount of information in a given region. I see no reason to think that is the case, especially given the multiple incompatible measures of information that exist in the literature. Nothing in what follows hinges on my choice to treat information flow directly rather than grounding it in an ontology of information. However, those who (rightly, in my view) reject the possibility of an ontology of information can rest content, because my account is not tied to any such ontology, even though I treat information as physical.

Setting aside now the question of ontology, and taking it as reasonable to treat the basic object of inquiry as information flow rather than information itself, I turn to the concept of information as it has been developed for use in communication theory. That concept accords very well with the use of the term "information" in most areas of scientific inquiry. In outline the discussion in the next section will be as follows: I begin with the concept of information for use in communication theory as developed by Shannon and conveyed by Weaver. The notions of information source, channel, receiver, etc., are introduced, and this provides an account of the quantity of information. In chapter 17 I detail Dretske's attempt to apply and expand these notions as account of knowledge, along with some minor critiques that set the stage for subsequent development of these ideas by Barwise and Seligman, who attempt to understand information as flowing in distributed systems in virtue of the regularities in those systems. Their account of distributed systems, and the logical structures that govern them, follows. Theirs is a brilliant addition to the literature on information flow, and I will use their account as the basic story of experimental systems. The difficulty with the Barwise and Seligman view has to do with the concept of flow—and it is a difficulty that finds its source in the very beginning of discussions of information. The problem is that for certain kinds of coming to be informed about the world, there is yet no clear way to understand them as information flowing to us from the systems we are now informed about.

16.4. Communication Theory

Shannon[1] provides an engineering-oriented conception of information flow. The point to his exercise is to characterize the capacity of various media for carrying information from some source to some system capable of making use of that information. Shannon's approach is to think of this in terms of the possible contents that might exist at a source, and what it would take for messages keyed to those contents, but distinct from them in any material respect one might imagine, to reliably transmit the content from source to receiver. One might imagine that at the source there are some possible messages to be sent, and that at the receiver there are already copies of those messages; the trick is for someone at the source to get someone at the receiver to pick up and read the correct message.

For example: suppose that I want to communicate something to you. At the source I may have a copy of the complete works of Shakespeare, a love letter from me to my wife, a shopping list, and a recording of Tchaikovsky's *Pathétique*. And you, at the receiver, might have copies of those as well. What Shannon wants to know is what is necessary in order that when I choose one of these as the message, you at the receiver pick up the right document or recording. Rather, Shannon asks how capacious a channel of communication must be in order that I can let you know about my choice in the most efficient way possible. Here's a very inefficient way: I myself walk over from where I am, pick up the appropriate item, and hand it to you. A more efficient way (if you're far away) is that I photograph in detail the appropriate item, and have a carrier pigeon fly it over to you. More efficient still is that, since you and I have the same store of items, we agree in advance what short name we shall give each one, and I bellow that name as loudly as is necessary for you to hear me and then pick out the right item. Each of these methods is more efficient than the one before; but what method, if any, is the *most* efficient? That question Shannon answers by first asking, "how many possibilities are we talking about?" If we know that, then we know exactly how many choices you must rule out at the receiver for me to have succeeded in directing you to the appropriate item. With a given number of items, we can choose short names that are combinations of yes and no, or on and off, or 1 and 0. Each yes or no, or on or off, or 1 or 0

---

1. See in this connection the Shannon and Weaver ([1949] 1963) volume, from which the discussion below is largely abstracted.

allows you to cut in half the number of possibilities. For n items to choose from, then, I at the source can completely specify the appropriate item in $\log_2(n)$ such indications. (There is no necessity to choose the logarithm, since any monotonic function of the [finite] number of messages can be used. Shannon plumps for logarithm, following Hartley [1928], who pointed out its naturalness.)

Notice, though, that this description of what is going on requires that for each item, if that one turns out to be my choice at the sender, that fact remains ambiguous on your end at the receiver until you have answers to all n questions. We cannot, for example, agree in advance that the Shakespeare volume has the name "1," allowing it to be picked out after just one symbol has been transmitted. Well, we could of course do that, but then if it *weren't* the Shakespeare that I wanted you to pick up but instead the recording of the *Pathétique*, you might need more than the optimal *general* number of symbols. Notice as well that the whole discussion is based on the assumption that there is some fixed store of messages or a fixed number of items from which I want you to select. The larger that store or number, the longer the minimal message must be to let you know my decision.

With these facts as background, we can ask, "How much information was transmitted?" but we cannot really ask, "What information was transmitted?" The answer to the former will be, in the simple case of equal probabilities, that in order to select with certainty among n choices one must receive $\log_2(n)$ units of information. Those units are now called bits (binary digits), and they represent elements of a choice sequence for progressively dividing in half the store of messages or items in order to specify some unique message or item. This quantity, $\log_2(n)$, is the Shannon information contained in a signal that permits an unambiguous selection among n equally likely options. Conversely, the Shannon entropy is just a measure of how much information *would* be required in order to unambiguously select among n equally likely possibilities— clearly the same quantity.

Now let's back up a little and tie this more directly to Shannon's depiction of what is going on in communication. Once again, we have two systems, the transmitter and the receiver. These can be people, machines, or any other physical system. The transmitter by some process or other selects a message, encodes that message, and transmits a signal. On the way there can be some noise, which involves any manner in which the signal is degraded. Finally the signal is received, the message is decoded, and the information reaches its destination. Schematically we have the picture shown in figure 16.1.

An interesting and important feature of this conception of things is that we see that information doesn't have any substantive character. There are signals of various sorts, and there are messages of various sorts. The only clear

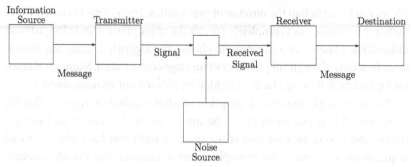

FIGURE 16.1 Shannon's schematic of a general messaging system

connection between the two is that signals are used to indicate features at the source, not to represent them.

This is not a book about the technical aspects of engineering communication channels capable of transmitting information-bearing signals, so I will only need a few of Shannon's technical results. One we have already seen: the most information about n equiprobable outcomes that a signal can reliably[2] carry is $\log_2(n)$. The logarithm was arrived at by assigning a single unit of information to each of two possibilities it eliminated. That by itself makes it a useful measure. But we can easily generalize to nonequiprobable circumstances using the same measure. If each of the n choices has a different probability of selection,[3] p_i, probabilities which are independent of each other, then H, the conventional symbol for both entropy and (rate of generation of) information by a source is given by

(16.1) $H = -\sum_{i=1}^{n} p_i \log_2 p_i$.

(Because probabilities are always less than or equal to 1, their logarithms will necessarily be negative, so we wipe that out with a leading minus sign.)

Another feature of this theory that will be important is the capacity of a communication channel. For now we will follow Shannon in thinking of capacity as best expressed in bits per time, but later when we come to generalize

2. The point is that, as I said, we could arrange that a single 1 symbol pick out uniquely one of, say, a million possibilities, thus making the overall naming convention less efficient. This would be silly for equiprobable messages, since it reduces the efficiency of the whole system, but it would sometimes give a big single information blast.

3. The probabilities here are a bit mysterious and must be left alone for now. The issue is that they need to be related both to the receiver of the information and to the actual situation at the source. We will revisit the issue when we turn to knowledge.

this story in a way suited to experimentation we may find other expressions more germane. The capacity of a channel then is

$$(16.2) \qquad C = \lim_{T \to \infty} \frac{\log_2(N(T))}{T}.$$

Here $N(T)$ is the number of different signals there are that take time T to send. Signals are understood as the allowed sequences of symbols, so any sequence of the basic symbols we can string together and transmit in less than the time T contributes to N. This expression is very general and allows for the possibility that some symbols take longer to send than others. When the speed of transmission of each symbol is some fixed number of symbols per unit of time, n/t, then $N(T)$ is just $N \times T$ and the expression reduces to $C = n/t$.

A nontrivial result of the theory is that, for a source with entropy H (bits per symbol) and a channel with capacity C (bits per second), there is always some scheme by which the source can be coded to allow an average transmission rate of $C/H - \varepsilon$ bits per second, and that no coding whatsoever will allow an average transmission rate greater than C/H bits per second (Shannon and Weaver [1949] 1963, 59).

But this value is an ideal. In reality outside factors can impact how much information a channel can carry; that is, there may be noise. Noise includes all kinds of things, from stray currents that disrupt the regular flow of voltage differences down the wire, to interfering electromagnetic fields that compete with the electromagnetic fields generated by a radio station, to the acoustical noise in a restaurant that interferes with the waitstaff announcing the daily specials, and so on. If symbols from a noise source get into the channel, then obviously there is less room for the symbols from our source. In order for our information-bearing signal to get through, the total capacity of the channel must be greater than the symbols needed to express our message (that is, to accommodate our signal) together with the symbols generated by the knowledge source. The channel capacity of a noisy channel is given by

$$(16.3) \qquad C = \max(H(x) - H_y(x)).$$

Here $H(x)$ is the entropy of the source that is best matched to the channel, and $H_y(x)$ is the entropy of the input when the output is known. Again, the measure here is in units of bits per second, but when it comes to experimental investigations where we want to determine features of various distal worldly systems by reference to properties of proximal systems, we will need to rethink this measure. But for now, this is all we will need from the general theory of communication.

Knowledge and the Flow of Information

17.1. Motivating Dretske's Account

I will now develop and defend Dretske's account of empirical knowledge in terms of information flow. It will not, however, be my goal to defend his account against all comers. For example, I will not be concerned to sort out the difficulties his account faces as a description of belief and belief formation processes, and I will not be offering a general analysis of the knowledge concept. Rather, I will be taking advantage of important insights offered by Dretske's model that will allow me to show that experimental knowledge results from control over the flow of information. Even so, I will offer a generally sympathetic account of Dretske's view, and in particular I will highlight what is sometimes taken to be a quite serious flaw in his account—the Xerox principle—and offer an argument that this principle does what it is supposed to do (reflect something like the stability of information channels over time) without, as is sometimes supposed, committing us to an impossibly high standard of knowledge acquisition. Another standard objection to the view, that it violates closure, I will show to be founded on a mistake: the view does not violate closure after all.

Dretske's (1981) *Knowledge and the Flow of Information* correctly connects the acquisition of knowledge with the transfer of information to the knower from the object of knowledge. The book begins as an account of the Shannon/Weaver model of communication; the idea is that by developing that model in certain ways it will suffice as a model of knowledge. In particular Dretske attempts to extract from the statistical account of information that underlies the Shannon/Weaver model a method for characterizing the information generated by individual events and then applying that characterization to what it is for us to have perceptual knowledge.

Starting from a description of the information generated at a source, Dretske follows Shannon and Weaver in taking the information to be related to the reduction of possibilities. The emphasis of his account is slightly different from theirs, because he does not directly confront the message in the source's store of messages as the target of our information acquisition; rather, he takes there to be a source of information, s, which is itself a process or mechanism constructed to transform n equally likely possibilities into a single one, giving us the quantity of information $I(s) = \log_2(n)$.[1] Thus, as Dretske says, the unit of information is the binary choice or decision.

He points out that events of probability 1 generate no information by their occurrence. Thus there is no information content generated by the fact that, for example, squares have four corners. (We will have to come back to this when information flow is relativized to states of knowledge.) He doesn't say so, but it does follow that in a deterministic universe there can be no information, as he is conceiving it, generated since the time of the big bang, because all probabilities are 0 or 1 in a deterministic universe: whatever situation is imagined in such a universe, its probability is fixed by the initial conditions. So we immediately find ourselves with a puzzle. Dretske wants probability to be objective, but he *also* wants there to be information generated by various happenings. This will be a recurring theme, because it is very difficult to reconcile these two demands. Dretske is a little vague about this, but my sense is that he is happy to drop determinism from fundamental physics when it is convenient. From my perspective, though, there is little to worry about, because I do not need to posit the ongoing generation of information itself (which I don't really believe in anyway). My only concern will be with what happens when people receive information-bearing signals. For that I will only need to account for information flow, which (*pace* Loewer) *can* be done with conditional probabilities. Moreover, as an early reader of this book pointed out, probability plays an important role in the theory of information at the heart of this story. So apparently we need not just a theory of probability, and a role for conditional probability, but also a role for *objective* probability, if information is to have anything to do with empirical knowledge. I will postpone discussion of this until the end of this chapter.

Turning to information received, $I(r)$, we are to understand things in the same way that we understood things at the source: at the location of the receiver, the signal serves as the process that reduces our open choices to a single one. The wrinkle, though, is finding out about the information transmitted from the source to the receiver. This requires knowing "how much of $I(r)$ is information about s." Generally the answer is $I_s(r) = I(r) - \text{noise} = I(s) - \text{equivocation}$, which is another way to express what we saw above.

1. Here I have, following Dretske (2006), shifted terminology from H to I.

Dretske offers a very nice discussion at this point (1981, 27ff) distinguishing causal signals from their information content. In a way this is the most crucial part of the opening of his book. The idea is this: While it is true that the physical symbols we use to make up our signals will themselves behave causally, that kind of behavior is not the only thing that is going on when information flows. We must, on this account of information and communication, employ a lot of counterfactual reasoning. Dretske claims that our use of this kind of reasoning is inevitable, but that we are not thereby committed to having a theory about how that reasoning works in order to employ it in our account of communication (and ultimately, knowledge). This seems right. One can easily imagine situations in which the very same information is conveyed by very different causal processes (seeing a murder and hearing about it will both inform you that it happened), while the very same causal process may convey different information (you holding four aces don't get the information that the up card isn't an ace from seeing that the up card is a king, whereas I do).

Dretske next distinguishes the formal notion of information from the notion of meaning. Having done that, he can attempt to show that communication theory tells us only about the former. To do so—that is, to tell us what the content of a signal is rather than merely its amount—he thinks we need to focus on individual signals rather than averages, even though it is only averages that are dealt with in Shannon and Weaver's account. While it may seem hopeless to specify fully the information in any one signal, we are saved in Dretske's view by seeing that we don't ever really need to know this in order to use communication theory as a foundation for an information theory that will underwrite a theory of knowledge. His principal point here is that his account of knowledge will not require that we know the exact content of information of signals. Rather, we will be content with a measure of the relative information. That is, he will only be concerned to answer questions of the form "did this signal bear to this agent the information that P?"

17.2. Chaining Signals

Along the way Dretske also introduces the very important Xerox principle.

> **Xerox principle:** If A carries the information that B, and B carries the information that C, then A carries the information that C.

The principle is intuitively clear, but it treats facts and their bearers as identical. That is to say, A carrying the information that B seems to indicate that some signal (sequence of symbols, perhaps just one) bears the information that the proposition B, or the fact that B, or something like that is true. Is the proposition that B, or the fact that B, itself a signal (a sequence of symbols, or perhaps just one)? It does seem that they are causally, even informationally, related—B the fact and B the signal—but they don't seem to be the same. Can this matter? For the metaphysically minded, perhaps; since I am not very metaphysically minded, I'll just ignore this.

More than merely a copy principle, the Xerox principle should be understood as a chaining principle for communication channels. Recall that on the Xerox principle if A bears the information that B and B bears the information that C, then A bears the information that C. Absent the Xerox principle, however, we do not have any instruction about how to chain channels together, because whether a signal bears information or not depends crucially on the recipient.

The Xerox principle should tell us that if the situation $\mathbb{A}$ being A bears the information that some source $\mathbb{B}$ is B, and that source being B bears the information that some further source $\mathbb{C}$ is C, then the situation $\mathbb{A}$ being A bears the information that $\mathbb{C}$ is C. It may be a little counterintuitive to think of these static situations as being signals in a communication channel, but that is only because we do not normally think of the stability (or lack thereof) of object properties as themselves being informative. But they tell us at least something about the past. So, for example, my being human now can let someone know that I was human some time ago; a copper statue being green can let someone know that it's been around a while oxidizing; etc. Generally, signals are simply specially arranged clusters of physical properties, so there is nothing surprising about static clusters playing a signaling role. To be sure, Shannon and Weaver discussed a much more narrow-seeming conception where signals were trains of symbols from some specified set. But their discussion makes it clear that we can, by liberalizing our understanding of what it is to be a coding scheme, expand what counts as a signal so that it runs the gamut from dashes and dots all the way to the Eiffel Tower, say. Then the Eiffel Tower being a certain way can be thought of as a signal that it is a certain way (i.e., it bears the information that it is a certain way). If it being that way (say, dark) itself bears the information that some other situation is a certain way (the sky is sunless), and the sky being sunless bears the information that it is night, then the darkness of the Eiffel Tower bears the information that it is night. That is the Xerox principle.

17.3. Semantics

Before any of this can play a part in Dretske's epistemology, he needs to provide a semantic theory of information. Here are its key features:

If some signal (a thermometer reading, an email, a dust cloud on the horizon) carries information that some situation or thing or process s is or has the property F, then it must be the case that

  A. The signal carries as much information about s as would be generated by
     s's being F;
  B. s is F; and something like
  C. The quantity of information that the signal carries about s is (or includes)
     that quantity generated by s's being F (not, say, by s's being G).

Since the Shannon and Weaver formulation does not address content, but only quantity, Dretske needs to say something about what it is for a signal to make some content available for the agents who receive it. Here we let K be an agent and k represent the agent's current background knowledge. Then

Informational Content: "A signal at r carries the information that s is F" =
"The conditional probability of s's being F, given r (and k), is 1 (but, given
k alone, less than 1)."

Nesting: "The information that t is G is nested in s's being F" = "s's being F
carries the information that t is G."

Dretske's relativization of the information carried by a signal shows again pretty clearly that on a view like this there can be no information *as such*, nor even a context-free fact of the matter about how much information a signal carries. It is not a stable or fixed quantity of the signal but a relation between the source and the receiver.[2]

17.4. Knowledge

Now we have the tools in place to introduce Dretske's account of knowledge: the Information Theoretic Account of Knowledge (ITAK). I'll present the account, show how it addresses some standard problems in the theory of knowledge, and defend it against a few of the most popular objections against it; then I will try to use it in my own account.

When there is a positive amount of information associated with s's being F, for example when F is not a logical truth, then

ITAK: "K knows that s is F" = "K's belief that s is F is caused by (or causally
sustained by) the information that s is F."

2. One may well be able to say that the information content of a signal is not, after all, a good physical quantity. However, one can place a limit on the quantity of information that any signal can carry. Perhaps the physicists identification of 1 bit of information with $k\log_2$ units of entropy really does do this. The resolution of this issue will not be pursued here.

Again, one might wonder how, on a view of knowledge of this sort, we are ever to get knowledge of necessary truths. An easy dodge is to say, "Well, this is a theory of empirical knowledge, so that's not necessary." I don't think that's either necessary or a good idea. Here is a better account of what we *can* say. We know that we can't get any new information about necessary truths once we have any at all. Naturally, because information transmission is dependent on the state of knowledge of the recipient, it is easy to see how certain redundant signals can convey information; as long as the recipient's beliefs (and hence knowledge) were not updated by an original signal, then that information is still available to be transmitted to the recipient. On the other hand, no one can get more information about necessary truths once they know anything at all because, according to the above account of knowledge, the conditional probabilities of necessary truths are 1 for any agent who has any knowledge at all. They may, however, still *learn* about these things by coming to believe on the basis of the information that they already have—information that was received the very first time they came to know anything at all. So knowledge of necessary truths is not a fundamental problem for the information theoretic view of knowledge.

An important first test of any theory of knowledge post-Gettier (1963) is whether or not it can be Gettiered.[3] An interesting test case is Zagzebski's (1994) suggestion that no theory that leaves any space between truth and justification can be immune to all Gettier cases. Her idea is this: whatever standards of justification one has, there must be some beliefs that are justified but not true. That is, there must be some gap between what counts as a justified belief and what is a true belief, otherwise justification is just a fifth wheel. All we need do, once "we find an example of a false belief which satisfies the justification and defeasibility conditions . . . [is to] make the belief true anyway due to features of the situation independent of those conditions" (1994, 71). Then we will have someone with justified belief that is only accidentally true. With ITAK, though, there is no straightforward notion of justification. Indeed, ITAK appears to be unique in its absolute externalism. Here knowledge is to be understood only as properly caused belief. It is a substantive and important question, on this view, what kinds of behavior or attitudes are necessary in order that some agent be thought of as a good epistemic agent. But it

---

3. That is, whether one can show that the theory classifies some claim as knowledge when intuitively it is clearly not so, as Gettier did for certain classes of justified true belief that should have counted as knowledge on the going theories of knowledge, but which clearly were not knowledge.

is not a question that pertains directly to the question of whether or not some belief counts as knowledge; rather, it is a question that focuses on the factors that influence whether or not some agent is habitually getting into position to have knowledge by being good at identifying and interacting with signals that bear information. We will return to this issue when we consider the debates having to do with closure of knowledge under known, drawn, logical inference, and their relation to the question of whether knowing something always puts one in a position to know that one knows it.

This account respects both the tight connection between information and causation, and the distinction we mentioned above between the two. This is a feature of the account, not a bug, despite Foley's (1987) objection. The objection goes like this: A spy wants to give me the message that she's standing outside my door. We have agreed on a particular code that she will use to send me that information. When things are going right, she knocks in a special way, I hear the knock pattern which I recognize as our code, and so I know it's her. Things go wrong, however: she knocks, and I don't hear her (perhaps I've been jackhammering in the backyard all day). On the other hand, the knock signal we agreed on is of just the right sort to cause something else (my wife being startled and whacking me on the head) that causes me to believe that the spy is outside. But it would be strange to think that such a causal chain could result in my having knowledge of the presence of the spy.

These and related objections have been clearly addressed by Adams (2005), but it's worth repeating here the main point. When the spy's knock indirectly causes the blow to my head, which in turn causes my belief that she's outside, information about the presence of the spy is not carried by the blow itself, and thus my belief is not caused by information. There are any number of situations that would have caused the same belief to be formed, but most of those do not have the feature that they raise to 1 the conditional probability of the proposition that the spy is outside. Only the event of my hearing the knock and recognizing it as the code for the spy's presence can do that in this circumstance. So while a causal chain is in place that is sufficient to make me believe the proposition in question, it does not bear the right information, and so cannot give me knowledge. Here intuition and theory are properly aligned.

This seems to be a general feature of arguments against ITAK that have to do with wayward causal chains and the like—they equivocate between beliefs being caused by information and being caused by things that are caused by other things that are themselves sometimes (even often) carrying information even though the proximal cause of the belief itself is not.

17.4.1. UNJUSTIFIED KNOWLEDGE

Let me pause here to say something about what we could mean by knowledge without justification, for to many that idea makes no sense.

Typically, when we discuss knowledge we are looking for a concept that includes true belief and justification, and then something else. Why something else? Because of all the various examples showing that even when we have justified true belief, we may still not have knowledge. Some of those are discussed below. But here I will just point out that, typically, what the justification part of the knowledge concept is doing is preventing folks from having knowledge by accident, so to speak, or by luck. A prolonged discussion of this notion is beyond my scope here (but see Pritchard 2005 for extensive discussion of this point). What I can say, however, is that views like Dretske's that connect knowledge directly to features of the knower's situation in the world do not have to add a clause ruling out accidents—that's baked into the connection. It is true that if one has the wrong view of what that connection is, it is possible to produce troubling counterexamples. What we have *here*, though, is an account that gets exactly right what features of the situation and the agent together are necessary to produce knowledge in a nonaccidental way.

Whatever justification was doing in older views of knowledge is done here. But what is missing from this view is why we should praise folks for their knowing if it has nothing to do with their good epistemic behavior. In fact, what could good epistemic behavior even be on such a view? Again, I can't address such questions at length here. But I can say this much: it is good to have knowledge. Knowledge guides action, and that by itself makes it worth having. But so does true belief. The difference in value between knowledge and true belief is not in any particular knowledge claim, however. Someone who reliably forms beliefs by dipping into streams of information-bearing signals is simply much better at getting true beliefs. She gets them by virtue of getting knowledge. One who gets many beliefs from streams that don't bear information may well have some true beliefs, but again only in an accidental way. That person then is not as good a knower, and that is going to be reflected in the fact that it's a crapshoot whether or not he does have knowledge.

Someone who thinks that each item of knowledge is somehow more valuable than the associated true belief *without* knowledge will not be satisfied with this little story. In my view, however, there is little to be gained in such a value assignment, so I find the cost of swapping out justification in favor of information-bearing signals to be very low.

17.4.2. FURTHER WORRIES

Two more standard issues are also straightforwardly resolved on this account. One is the lottery "paradox," and the other is the possibility of communicating knowledge and its possible degradation by retelling. Regarding the former, it seems that on any version of knowledge built around justified true belief, if we are dealing with probabilistic situations there will be a threshold of probability above which a belief is justified. Yet we will still not think an agent has knowledge. Thus in a fair lottery, whatever threshold for justification one has, there may be a large enough number of tickets sold that you are justified in believing that you will not win if you have only one ticket. Let us suppose that in such a case you believe you will not win, and in fact you do not win. Did you know it? It seems that most have the intuition that you did not have knowledge in this situation, but it is difficult to see why not on any justificationist model of knowledge. ITAK, however, with its demand that the conditional probability of the event be 1, has no trouble here. To be sure, it is an exacting standard, and we'll consider shortly whether that is itself a problem for the theory. But it does satisfy intuition in this case.

The possibility of knowledge degrading under repeated transmission is part of Dretske's reason for adopting such a high standard in the first place. His first restriction on the semantics of information, requiring the signal to carry as much information as would be generated by the fact it informs about, is meant to do just that. But the probability 1 condition is necessary to make that work. Suppose you learn something—that all red burgundy wines are made from pinot noir grapes, for example—and want to tell me about it, and then I want to tell someone else. On the information theoretic view, if you succeed in transmitting the information to me, then my conditional probability is 1, and if I succeed in transmitting it to someone else, their conditional probability is 1; and if in those cases the information causes our beliefs, we all have knowledge. All that clearly respects the thought that these signals should be carrying as much information as was generated by the thing you originally learned about burgundy wines. But on an information theoretic view *without* a probability 1 requirement, it seems that there would be pathological behavior. Suppose we choose any value other than 1 to put in our definition of information flow, say 0.85. Then you can send me a signal that increases my conditional probability from less than 0.85 to above it, thus giving me the information in question without making my conditional probability 1. For example, you may send me a slightly blurry photo of the finish of a race, and the probability that you won the race, conditional on what I know of the contestants, the conditions, etc., is 0.9. Even so, by assumption the signal bears as much information as is generated by your winning of the race. Then if I have the information that you won the race and I tell someone else that you did, they should be able to get that (and *that much*) information. Presumably I can produce a signal that bears this information that I now possess and send it to you. But suppose that I'm an occasional liar. Perhaps I lie 10 percent of the time. Since 90 percent of the time my signals are information-bearing, then if I know something and I tell someone then the probability that what I say is true, conditional on their background knowledge and my reliability, should (since my stipulation that I know what I am asserting entails that it is true) be 0.9, and that suffices for information transmission. So now that person has the information. That seems like a problem, however, because it would also appear that their conditional probability (given the dual uncertainty in my signal and in your original transmission) should be 0.81—less than what is required for information transmission.

Dretske's account can be troubling to some, because it appears to fly in the face of best practices for empiricists. Dretske's account tells us that we have knowledge only when the proposition in question has probability 1. However, insofar as we're empiricists we should never assign probability 1 to empirical claims. If we do, we will never be able to change our minds in the light of new evidence; but we can always come up with circumstances that will make us change our minds about any given empirical claim. So whether or not one finds the argument above persuasive grounds for adopting a probability 1 conception, there is still the question of whether it is too high a standard. If it *is* too high, then we have reason to be suspicious of the view overall. But I don't believe the standard really is too high. There are two considerations to keep in mind here that should work in tandem. The first is to accept that part of being an empiricist is to refrain from extreme knowledge claims about empirical matters. That may seem overly restrictive and fastidious, but it is also just another way of expressing the content of the worry that there are always imaginable circumstances (of greater than zero in the space of possibilities) that would make us change our minds. The other consideration is that the account of knowledge here is not an account of the *self-attribution* of knowledge; it is not a theory about when I can tell that it is OK for me to assert of myself that I have empirical knowledge. Rather, it is an account of the circumstances under which some agent really does have knowledge of some proposition.

At least as presented, this is a completely externalist account of knowledge, so it is irrelevant in assessing my knowledge state that I myself may not be able to properly assess it, and it is also irrelevant that you may not be able to assess it. What matters is that I believe the true claim, and that I do so based on the information, where doing so based on the information happens only

in the case where the probability of the claim given my background knowledge is less than 1, but is equal to 1 when an appropriate signal is received.
These requirements—that it be reasonable for me to assign a conditional probability 1 to the proposition in question and that the conditional probability
is 1—are sometimes conflated. But they must be kept distinct. If we assume
that I have some knowledge, then there will be plenty of times when my
background knowledge does not entail some proposition but some signals
in conjunction with that background knowledge do. For example, sometimes
my background knowledge does not entail that my wife is at home. I may
have been napping, or may have just arrived home myself. But when I hear
the kettle just starting to whistle, I do know that she is home. I myself would
never assign probability 1 to the proposition that she is home, because I can
imagine all sorts of other possibilities: someone broke in to the house and is
making tea; the teapot was left on the stove and a short circuit in the stove a
minute ago turned on the burner; space aliens are reenacting the scene from
The Prisoner where Number 6 is giving instructions on how to make tea; etc.
So as a good empiricist I am reluctant to assign probability 1 to this empirical proposition. But knowledge on ITAK doesn't have much to do with what
probability I myself would assign to a situation. If we stipulate that my wife
is home and is making tea, then in this situation my objective conditional
probability is less than 1 that she is home, based only on what I already knew,
and equals 1 in conjunction with the whistle from the kettle. The whistling of
the kettle is appropriately causally tied to the particular sound I hear, so that
sound can bear the information that tea is being made, and my background
knowledge about this circumstance does let me know that only my wife is the
tea maker. The other situations I can imagine don't bear on what information
is present in this situation.

    But how does that work exactly? What is it that prevents these other possi-
bilities from being relevant to whether or not some signal really does transmit
information to me? Is there a principled way to rule out the alien cosplayers
so that my awareness of various possibilities does not suffice to make them
relevantly possible for determining whether information is flowing here?

    To answer such questions we need to examine more closely the notion of
a communication channel.

17.5. Channel Conditions

To introduce this discussion about how, despite its being an empirical theory,
his account can function with the probability 1 demand, Dretske reminds readers of Unger's insight about absolute concepts: their application conditions may

be contextual, despite their absolute character. Concepts like flat, empty, and so forth are absolute all-or-nothing affairs; yet what *counts* as flat, for example, can vary from application to application of the concept. If we're building a road and we want to know whether we will need to bring our blasting equipment, an assertion that the terrain is flat is true or false with respect to whether or not we will need to flatten it to lay the roadbed that day. If we're hanging out with Euclid playing the circle and straight line game, that assertion will have different truth conditions: can we use the terrain to perform our geometry proofs or not? Knowledge, says Dretske, is absolute, like flatness. But also like flatness, what counts as falling under the concept depends on the situation.

In addition to clarifying the type of concept knowledge is, we need to clear up a possible confusion between:

1. The information (about a source) that a signal carries, and
2. The channel on which the delivery of this information depends.

The fact that some signal carries the information it does is dependent on many different factors. These factors are too disparate and extensive to even try to articulate. They include, for written messages, the existence of suitable surfaces for the inscription (malleable enough to take the inscription, durable enough to hold it long enough to be read), the practice of using such surfaces to convey information, the existence of language itself, the particular language being used in the message, the possibility of distinguishing messages in that language from those in others, etc. For scientific instruments there is also a huge variety of factors: the workings that implement various (approximate) laws of nature, the stability of matter, that these laws can be implemented in devices that are accessible to observation, that there are technicians capable of servicing the devices, and factories to build them, etc. Only when all the necessary background conditions are in place can information-bearing signals be implemented on those surfaces, or by means of those devices. So one might be tempted to think that the presence of information-bearing signals is *itself* an information-bearing signal that such conditions are in place. Naively, we might think that when I know something on the basis of a voltmeter, I then know that the voltmeter is working. After all, it is a necessary condition of knowing anything on the basis of any device that the device be properly functioning. But that can't be right; generally signals only bear information about their source, not their mode of transmission.

Dretske realizes that we need both to restrict the information that can be gleaned from an information-bearing signal, and to clarify when various signals can support the transition of conditional probabilities from less than 1 to 1. Those needs are addressed by conditions on the channels through which signals flow.

> **The Channel of Communication** = That set of existing conditions (on which the signal depends) that either (1) generate no (relevant) information, or (2) generate only redundant information (from the point of view of the receiver).

When appropriate channel conditions are in place, then, there will be no information transmitted by signals supported by that channel that are *about* the channel itself—neither its state of repair nor its capacity for supporting such signals. To get that information one would need an entirely new signal, flowing in its own channel, that bears information about the original channel. To put this more clearly in Shannon's picture, the set of possibilities that a signal is capable of ruling out does not include the possibility that it is capable of ruling out possibilities. If we're not sure whether a given signal can bear information, it cannot itself bear information that makes us sure.

A simple example captures much of the content of the above lesson. Suppose we have a working boiler gauge. In that case a technician taking a reading from the gauge can gain information about the state of the boiler. But suppose instead that the factory owner has installed a monitor of the gauge that is meant to give an alarm when the gauge malfunctions. What do we say about the case where that monitor is itself defective and sounds the alarm even though the gauge itself is working fine? Here the gauge still generates a signal that carries information about the state of the boiler, and the monitor does not generate any information. But can the technician get the information generated by the gauge, given that she hears the alarm going off and knows what it signifies? Here we simply appeal to the situation where the monitor is working well, and giving no signal. In that case, by stipulation the signal from the gauge bears information. But nothing about the good or poor repair of the monitor is relevant to that fact. So there is no reason to think that the information doesn't flow simply because the monitor is faulty. Our way of *checking* whether information is flowing is independent of the information flow itself. It may well be the case that the technician believes the report from the monitor (because she knows that type to be reliable) and thus does not believe the report of the gauge. In that instance she would have the information, but not the knowledge. But if she believed the gauge anyway (even *despite* her knowledge that that type of monitor is reliable, and not having any grounds for thinking this one defective) and that belief was caused by the signal's information content, she would have knowledge, faulty monitor or no. It's worth pausing a moment to reflect on how externalist this account is. It is (plausibly) irresponsible for the technician to ignore the monitor and believe the gauge, and yet she still has knowledge. The lesson is that information flow is user *relative* but not subjective, and knowledge is not grounded in justification.

A similar point is that announcing skeptical possibilities does not by itself make them relevant. If you tell me that a channel is broken, that doesn't somehow break it. The physical system that sustains the signal and connects it is what makes the channel function; only those features of the situation that are relevant to that connection determine its capacity to transmit information-bearing signals. It may not be known to us a priori (nor often a posteriori as well) what possibilities are relevant, or whether any given channel is functioning, but such knowledge is not one of the conditions of a properly functioning channel.

17.6. Channels and Counterfactuals

I am a bit allergic to counterfactuals whose explication relies on possible-world semantics. I find that such explication invariably relies on a strong, prior intuition about the truth conditions for the counterfactual itself. So I prefer to stick with that intuition, since that's ultimately what's doing the work. Not everyone takes that approach. The most interesting analysis in recent years of Dretske's channel conditions comes from Kipper (2016); Kipper's analysis simply equates the probabilities in Dretske's information flow condition with counterfactuals, and then evaluates those by reference to possible worlds. Kipper has a good reason for doing things this way: he is worried that as stated, the probability 1 condition will make it so that almost nobody ever knows anything. To stop that bad effect he interprets the channel conditions in such a way that they are merely the possible worlds where the channel gives information. On the surface that seems like an OK thing to do. The idea is to codify remarks like "necessarily, if you receive this signal with the background knowledge that you do have, then the probability of the event in question is 1" as being true only *relative* to the working of the communication channel. The advantage to this is that it makes the probability 1 requirement relative to the proper functioning of the channel. The disadvantage, from my perspective, is that by going beyond the theory as stated, Kipper has built the unclosedness of the theory in by hand. Then when he reinterprets the modal demands on the proper functioning of the channels to make them centered on our own world, he has in fact made the theory massively unclosed—to the extent of ruling out closure of knowledge even under simple conjunctions.

But we do not need to adopt Kipper's modification of the theory. Kipper's analysis in fact seems quite backward, because he overestimates the severity of the probability 1 condition. First let's see that his condition really is backward. He is motivated by a kind of generalized safety consideration: in the nearest possible worlds where I get the signal, it had better be the case that the event in question really does happen. Consider a case where you're the security guard in a thousand-story building. If I call you up and ask whether anything is going on on floor 743 and you look at the appropriate camera feed and see that all is clear, then when the camera is functioning correctly you have knowledge about that. And that knowledge is not undermined by the fact that the team that was going to rob that floor, and also put in a tape loop of the undisturbed scene, was put off by a freak accident with a probability of 11,000,000. Even though in all the nearest worlds where you look and tell me that all is clear, all is not clear, you still have knowledge that all is clear in this case. Despite the possibility that there is something about modality that is relevant to probability, it is simply a mistake to subsume the one under the other. The appropriate channel conditions in the Dretske theory are best understood causally. You might think that causation itself is best understood counterfactually, but it is not. Causal connections should support counterfactual reasoning, not the other way around.

17.7. Is This a Good Account of Knowledge? Of *Our* Knowledge?

The principle problems for any theory of knowledge are made clear in the *Theaetetus*, and they are two: the problem of how we come to know from a state of complete ignorance; and the problem of how our beliefs must be connected to the world in order that we know. The latter problem is often motivated by appealing to the intuition that mere lucky guessing cannot be the connection—or rather, that being correct in our beliefs only by luck does not suffice. Note my equivocation: what we want is to deny that luck, or the mere fact that I believe the truth, can suffice. But the second claim is that luck cannot suffice to give knowledge. Consider these two different but related scenarios: I generate, using a quantum random number generator, the winning lottery number; you are performing repairs in the ducts at mob headquarters and overhear the boss tell her henchperson to set the lottery number to such and such. On television we're both interviewed about our winnings and our methods and we both explain truthfully that we just lucked into these numbers. While that's true, there is a fundamental difference. Your luck is that you lucked into knowing, while I lucked into being correct (and didn't know at all). These are fundamentally different statuses and they serve to clarify how externalism solves the problem of lucky belief while still allowing that one can come to know by luck. (This is all discussed much more extensively, but with roughly the same sense, in Pritchard 2005.)

In the case of Dretske's ITAK one may come to know by luck, when one is connected in the right way by accident to a source of information and comes to believe through that connection. And while there is always some measure of good fortune allowed, Dretske's model also tells us why internalist considerations are relevant and important in the acquisition of knowledge: they are useful mechanisms for promoting behavior that is likely to get us into streams of information that will supply us with knowledge.

If I am going to base my own epistemology of experimental knowledge on Dretske's ITAK, however, I need to address a standard objection to the account: that the closure condition on knowledge fails. This means that on Dretske's account knowledge is not closed under the logical operations.[4] So S can know that P and S can know $P \rightarrow Q$, and still not know that Q even after drawing the inference to Q on the basis of the first two premises. How does this work? The idea goes like this (Dretske 2006): Suppose there are cookies in a jar. And suppose Nadja sees them. Then on the basis of this seeing, Nadja knows that there are cookies in the jar. But does Nadja know that they are not really papier-mâché? No, says Dretske, because the visual evidence she uses to find out that they are cookies is identical to the evidence the papier-mâché cookies would give. But note that the fact that there really are cookies in the jar entails that they are not papier-mâché. So closure would tell us that Nadja does know that they are not papier-mâché. If those who have the intuition that she does not know this are correct, then we must reject closure. Dretske is content with this, and his account was constructed to respect his view that closure does indeed fail in certain circumstances. Since most reject his view and take it that respecting closure is an inviolable condition on any adequate conception of knowledge, the fact that ITAK does not respect closure makes it a nonstarter for many. Inference from long, intricate chains of evidence seems central to any viable conception of experimental knowledge, and so especially here closure failure would be a disaster.

I will now argue, however, that Dretske's information theoretic account does not entail the failure of closure.[5] Rather, I think that Dretske's intuitions about knowledge tell him that in certain cases known implications of

4. There are many careful ways to phrase this, none of which is needed to get across the worry. The simplest is just to say that if I know P, and I know that P entails Q, and on the basis of those two pieces of knowledge I properly infer that Q is true and thereby come to believe it, then I should know that Q.

5. Compare this discussion to the interchange in *Erkenntnis* published after this section was written. There Dretske (2006), Baumann (2006), Jäger (2006), and Shackel (2006) consider the issue of closure. While the conclusion there is that Dretske's theory as stated does not, after all, violate closure, Dretske himself opts to change the theory in order to preserve closure violation.

known facts are not known. The account *itself*, however, does not have this consequence. Thus those who reject the account for that reason can happily accept it and enjoy all of its other virtues.

Let us consider again the case of Nadja and the cookies. What, on Dretske's account, lets us conclude that she knows there are cookies? The condition on knowledge is this: Nadja has knowledge that there are cookies in the jar = Nadja's belief that there are cookies in the jar is caused by (or sustained by) the information that there are cookies in the jar. What is it for information to cause or to sustain her belief? At least this: an information-bearing signal must have carried the information to her that there are cookies in the jar. What is the condition on a signal bearing information? Fix a channel C. Then, relative to C, to a person with prior knowledge k, r being F carries the information that s is G if and only if the conditional probability of s being G given that r is F is 1 (and less than 1 given k alone). Here r might be something like "Nadja's view of the inside of the jar," and its being F might be something like "looking cookie-like." Then s being G is just the fact that there are cookies in the jar.

Given that Nadja knows that there are cookies in the jar, does she know that they are not papier-mâché? Yes. The fact that she knows that there are cookies in the jar tells us right away that there really are cookies in the jar. That's no worry. But then it is also true that they are not papier-mâché. But isn't it possible that they were? In other words, isn't it possible that Nadja, despite drawing the known implication that "is a cookie" → "is not a papier-mâché cookie" and coming to believe that the cookies are not papier-mâché and basing all of this on the information that they are cookies, might fail to have the information that they are not papier-mâché? No. Why not? Well, suppose she does still fail to have that information, even in the light of all that. Then it must be because the information that they are not papier-mâché was not carried by the signal she saw. Then the conditional probability that they are not papier-mâché, given Nadja's background knowledge, must be less than 1, and thus the conditional probability that they are papier-mâché is greater than 0. But by stipulation the probability that they are cookies, given the signal, is 1. Thus if the probability, relative to this channel, is greater than 0 that they are papier-mâché, then necessarily the probability that they are real cookies, relative to this channel, is less than 1. Whatever we might *ourselves* believe, the condition on information flow according to the theory is that the conditional probability be 1. If the channel really is capable of transmitting the information that there are cookies in the jar, it also transmits the information that they are not papier-mâché.

The danger is to confuse the existence of the channel with our having knowledge of whether the channel really is possible. For example, I'm not sure, myself, that we do have any knowledge of the external world. This is because, due to a too heavy dose of Hume in my youth, I am not sure whether external world skepticism is false. But my personal surety is not relevant on this account of knowledge. I do believe in external world objects (in a narrowly restricted manner of speaking). If these objects are known to me via normal perceptual processes, then the channel that connects me to them is capable of transmitting information to that effect. And if that is true, then I do know that external world skepticism is false. (Or I would know it were I to believe on the basis of that information. For me that last step is the tricky one.)

Channel conditions on this view are relative to a context. We can never be sure that some channel or other is really capable of transmitting information. We make our decisions about which channels support information flow and which do not by appeal to intuition, to our past experience, to our analysis of the nature of the signals involved and their means of generation, and so on. All of this makes it very complicated and epistemically risky to assert of some channel that the context is right and that it is capable of transmitting some piece of information to some knower with some background knowledge. However, if the channel does in fact support the flow of this information, that is an absolute fact, and as such the channel supports the flow of other information that is law-like connected to this information whose flow it supports.

Part of what is going on here is that Dretske, in displaying his account, takes it to be a successor theory to an account on which knowledge is had only when the possessor of knowledge has conclusive reasons. Conclusive reasons are powerful knowledge generators, but Dretske and others are persuaded (correctly, I believe) that conclusive reasons do not generically transmit across logical connectives. But there is nothing like conclusive reasons on the information theoretic account. Reasons have been replaced here by conditional probabilities and channel conditions. These conditions are context dependent, but that doesn't stop them from generating the appropriate probabilities; nor, crucially, does it stop them from transmitting information by means of logical connectives.[6] Is it possible to secure such channel conditions? Again, I am not sure. However, if we are right that we have knowledge of everyday things like tables and chairs being tables and chairs in the external

---

6. Dretske (1981) does point out that information does not flow via mere extensional equivalence but requires law-like connections. That qualification is compatible with the argument I am making here.

world, then that tells us that the channel conditions can be met. At least that's the account as presented. Whether or not it fails for other reasons, the information theoretic view does not violate closure.

But now a natural worry arises about a kind of spoofing of knowledge claims due to Gettier. One important motivation for truth-tracking accounts of knowledge is that they block those Gettier cases on which what looks like a proper knowledge claim according to a given theory of knowledge is intuitively, clearly not knowledge. Perhaps it is the fact that truth-tracking accounts fail to preserve knowledge under closure that does the work in blocking those cases. Can we show that the information theoretic account, as I am presenting it, still blocks Gettier cases even though it preserves knowledge under closure? I think so. Consider the disjunctive case: Nadja sees her friend Ollie driving a new car, a Geo Metro. She knows that Ollie was planning to buy the car last week, and so she forms the belief that it is Ollie's car. As an exercise in epistemology, she then forms the belief that either Ollie owns a Geo or her other, extremely law-abiding friend, Otis, is in jail. We now turn Gettier's crank and suppose that Ollie was test-driving the car (for some reason they couldn't buy it last week) while Otis really is in jail (all due to some strange misunderstanding). Our intuitions say that Nadja has justified true belief that either Ollie has a new car or Otis is in jail, but most of us don't believe she knows it. What does the information theoretic account say?

For one to have knowledge of some proposition, one's belief in it must be caused or sustained by the information that the proposition holds. What causes Nadja's belief that either Ollie has a new car or Otis is in jail? It was first that she saw Ollie in the car. But seeing Ollie in the car cannot bear the information that they own the car (even given Nadja's background knowledge about their intentions), because they don't own it. Nadja has no information that Otis is in jail. Thus Nadja's belief that either Ollie has a new car or Otis is in jail is not sustained by the information that the proposition holds. Similar analyses can be performed for Gettier's other cases. Thus it is not the failure of closure that blocks these cases.

Elsewhere (Mattingly 2021) I consider the issue of closure for Dretske's theory at greater length. What I show there is that on some ways of extending the theory (in Kipper 2016, for example) it does continue to violate closure, but those are *extensions* of the theory. The theory as presented in Dretske's book, and outlined here and used here, does not.

Before moving on to the Barwise and Seligman account of the logic of distributed systems, and how information flow works there, I need to clarify one point about semantics.

17.8. Semantics for Information

We have seen the choice Dretske makes in his analysis of information, and his method of giving semantic content to a signal by stipulating that those propositions whose conditional probability for an agent moves from less than 1 to 1 are that content. But that seems to be giving too much semantic content to signals that themselves can, very often, only carry a much *smaller* quantity of information. A single bit, for example, can change all manner of contingent propositions from less than 1 to 1 given the right background knowledge, and thus it bears the information that the situations they pertain to have the properties they do. But that doesn't seem to square properly with the very small amount of information a single bit carries. We need to be much more precise here about how we're understanding the connection between content and quantity of information. I will begin by insisting that a signal simply cannot bear all of that content if it only conveys a single bit of information. Let's go back to Shannon's original discussion: A signal carries a single bit of information when it disambiguates between two possibilities of equal probability. If my background knowledge is connected structurally with some propositions so that, in conjunction with a single bit of information, I acquire information about many many new facts, then it must be that I already had almost all of the information associated with those facts as part of my background knowledge, and all the signal does is add the information "true" to one set and "false" to some other set. The *semantic content* of these propositions is not conveyed by the signal, and it is misleading to say that the signal bears the information, for example, that s is F, when what the signal really does is allows me simply to assign that proposition, instead of some other, the value "true."

In Shannon's theory of communication, there is a store of possible messages. These messages are sent from the source to the receiver with a specified probability distribution. It is the likelihood of a given message (that is, its probability) that determines its information content. For two messages with equal probability of being sent, the probability is 0.5 for each. The quantity of information contained in the message when either 1 is sent or 2 is sent is $\ln_2(2)$ bits (that is, 1 bit). As has been pointed out repeatedly (by Weaver, Shannon, Dretske, and others), this is not by itself a possible account of information content, since the two messages are presumably different and yet their quantity of information is the same. I think, however, that there may be an equivocation here. For when I send from my store of two messages either a love note or a shopping list, the $\ln_2(2)$ bits of information concern which of the two it is. But they do not take account of the decoding that is yet necessary for me to discover exactly what the content of the message is that I have received. The idealization involved in setting out the store of messages and the channel and so forth takes as given that I can transform what comes into the receiver into meaningful information without acknowledging that that itself is an information-processing and information-transmitting task. The note from you to me is fully delivered once it is in my hand, from the perspective of calculating how much information has been transmitted, but that calculation involves me already having the probability distribution governing your choice of message.

This is why, while it is easy to think that by transmitting a signal comprising a single symbol that carries only one bit of information (the spin state of an electron, for example) one could generate a variety of different contents, it is simply not true. The content of that is some version of yes or no, or true or false, or this or that *depending* on what convention we have in place for the way signals reduce uncertainty. But reduction in uncertainty is all they do. This is kind of a Fregean point. Suppose our convention is that an up bit corresponds to "by land" and a down bit to "by sea." It is natural to think that the content of the signal when I receive the up bit is that the British are coming by land. But it isn't that. Instead the content is the demonstrative "that proposition is true" where all the further content comes from parsing "that proposition." What the signal does in this case is allow a further flow of information—the information contained in that proposition. Consider: the combination to a safe does not bear the contents of the safe; rather, it allows access to that content. The useful thing about conventions and languages and so forth is that they allow a great deal to be accomplished by signals with the content of "yes" or "no."

While my overall account inherits its basic epistemological structure from that of ITAK, the principal tool I will use to model information flow in experimental systems is provided by Barwise and Seligman's (1997) *Information Flow: The Logic of Distributed Systems*. I model physical experimentation as controlling the flow of information between a variety of subsystems connected by information channels whose structure constrains and enables the flow of information throughout the extended system. I'll examine that account and use it to construct my own account of scientific experimentation in chapter 18.

17.9. The Problem of Objective Probabilities

Readers will no doubt have noticed, as did the early reader mentioned above, that probability concepts are part and parcel of ITAK and, by extension, of the story of experimental knowledge offered here. Moreover, these concepts are of an objective notion of probability according to which there is some fact of the matter, for a given event, constituting what the probability is of that event (at least in some cases). However, many contemporary thinkers take probability assignments *not* to be objective, but rather to reflect something like betting odds or a measure of degrees of belief. I do not share this view. Indeed, I take it that fundamentally all probabilities are either 1 or 0 or the objective chance of some quantum collapse event. But that's a contentious view, and one I do not need in order to make the account of knowledge and information flow work. To fully address the matter of probability in the sciences is well outside the scope of this analysis of experimental knowledge. But I do think I can offer some reasons to think that we can do everything we need with relatively uncontroversial assertions about the nature of probability. (For an overview, see as well Hájek 2019, esp. section 4 where aspects of the prospects for physical probability accounts are discussed.)

The main difficulty in developing an account of objective probability may have been best expressed by Eells (1983). He argues in favor of a pair of demands that must be satisfied by any acceptable theory of probability. On the one hand is the idealization/interpretation demand, and on the other the formalization demand. Respectively (and roughly), these are: That there be some good object/concept not involving probability which can be used to explain that notion, and which when idealized in an appropriate way results in entities that are apt for formalizing, after which the idealization makes appropriate contact with things in the world that we take to be probabilistic. And that the formalization of these idealized elements be of an appropriate sort—not necessarily Kolmogorov's axioms, for example, but something that gets right the bulk of our going conceptions of probability theory. Eells then examines the going contenders for interpretations of objective probability (two versions each of frequency accounts and dispositional accounts) and finds them lacking in a way that reveals that the two demands pull in opposite directions. That is to say, theories that do better on one demand do worse on the other.

I take it that the various challenges of constructing a theory of objective probability are so much standard wisdom on the matter (though folks like Maher [2010], who defends a logical conception of probability, continue to push back against this wisdom). Generally, I am sympathetic to attempts to meet some version of Eells's challenge (or the ur-version due to Salmon, for which see again Hájek 2019), but I will defer elaborating and defending a theory of objective probability to another occasion, in part because my story will itself be controversial, and so will not be as soothing as I would like in this context. Instead I will make clear how, appearances notwithstanding, ITAK does not really require the full resources of a theory of objective probability, and can make do with a much weaker but still objective notion.

While Shannon's account of communication theory is important for ITAK, it is hard to say whether it requires a theory of objective probability. Whatever type of theory it does require has to be about probability, though, because of the intimate connection between the measure of information received and the entropy of the source and the probability measure over the possible signals generated at the source. However, those measures are in no way involved in ITAK itself. Instead, in that theory we use only the distinction between the probability of some event being equal to 1 or less than 1. And that much we can develop an objective account of, for the question is always about the compatibility between our background knowledge and some event *not* happening (on the one hand), and between our background knowledge together with the signal and some event not happening (on the other). That is just a matter of logical consistency, and so there is no real difficulty in giving Dretske objective probabilities of the sort that are involved in the theory.

There is much more to say here, of course. We do want to be able to measure the flow of information in more detail, and for that we will need some story about how to ground that measure in objective features of the situation. As I said, I do think that is possible despite the general wisdom. But it is not at all *necessary* in order to carry out the conceptual analysis of experimental knowledge that I am carrying out here. In short: I believe that an account of objective probabilities is possible, and that all probabilities are either 1, 0, or those of objective quantum collapse.[7] At the same time I do not need to appeal to anything like that in order to make the account work. All I need for a fruitful analysis of scientific experiment are the objective conditional[8] probabilities involved in determining whether some signal has or has not carried information to some knower. I have those, and so I will now turn to that analysis.

7. Note that Bohmians do not have any of these worries because *all* of their probabilities are 0 or 1.

8. Incidentally, I am entirely persuaded by Hájek's (2003) argument that conditional probabilities should be taken as primitive and *unconditional* probabilities should be taken as derivative.

18

The Logic of Experimental Practice

An individual experiment is some particular situation in the world, but of a very special sort. As theorists we want to understand how it is possible to conclude, from the fact that that particular situation is of a certain type, that other particular situations are of either that or some other type. That is, we want to understand how we can conclude from the features of the experimental situation that the rest of the world or some part of it is itself properly classified in some way or other. How do we understand how a particular space-time region (the experiment over time) being a certain way can bear the information that the world is itself a certain way? Questions of this general form were addressed by Barwise and Seligman (1997), who proposed a framework for answering them. Their account comes with some heavy mathematical machinery. That machinery may be helpful in actual applications, but here I just want to introduce it as a potential new tool for the philosophical study of experiment. The account begins in the response to a common, and yet still significant puzzle: how does information about one system manifest itself in the features of another system?

Here is how their story works: Take some extended system made up of separated subsystems. Characterize these separate subsystems according to their own inferential structure—that is, classify them in a way that supports the basic kinds of inferences we can make about them—and characterize the total system in terms of how the local systems' inferential structures are related to each other. The relations between these structures, maps from the separate logics to each other, are the pathways along which information flows in the system. These maps are called "infomorphisms."

Consider this example as a way to motivate the ideas: one of the most important classifications for biomedical science is of murine mammals (rats, mice, gerbils), and there are huge databases of these classifications allowing researchers to select a variety of genomes, body types, immunologic responses, etc., and to make reliable inferences from the possessed properties of some strain to its other possessed properties. It is the use of such classifications of all kinds of systems, discovered by observation (and by experimental investigation), to infer classifications of other systems that generates experimental knowledge. Infomorphisms are the logical structures that support such inferences.

18.1. Infomorphisms: The Logic of Distributed Systems

An infomorphism is like a standard morphism of structures in model theory. Its main building block is the important notion of classification, presented this way by Barwise and Seligman (1997, 28):

> A classification: $\langle A, \Sigma_{\mathcal{A}}, \models_{\mathcal{A}} \rangle$ consists of a set A of objects to be classified, called *tokens* of $\mathcal{A}$, a set $\Sigma_{\mathcal{A}}$ of objects used to classify the tokens, called the *types* of $\mathcal{A}$, and a binary relation $\models_{\mathcal{A}}$ between A and $\Sigma_{\mathcal{A}}$ that tells one which tokens are classified as being of which types.

We can picture classifications like this.

$$\Sigma_{\mathcal{A}}$$
$$\Big| \models_{\mathcal{A}}$$
$$A$$

So far that's pretty standard. What they are offering is an account of how to classify the various parts of distributed systems so that they can be used to model information flow through these systems. Their idea is to figure out the logic governing some classification of a distributed system, and then to relate that to the logic governing some other classification (which may be a part of the first distributed system, or of the system of which they're both parts). Their initial model is of classifying a flashlight. I'll instead continue using systems of interest to biologists, which will be important later: model organisms in an experimental protocol. There are a number of different ways to classify such a system. On the one hand we have the observable features of each token animal: is it listless or active; does it have a good appetite or not; is it above, below, or at its baseline weight; at what level has it been intervened on; etc. This classification might allow us to relate various types (listless, or intervened on above x dosage, for example). Or we could classify these animals into types by means of each one's various organ system functions: how well is it processing sugar; how insulin-resistant is it; what is its viral load for some virus of interest. Additionally, we could be interested only in the livers of these animals; they would be the tokens, and their types might be has/does not have squamous cells; is/is not swollen; etc.

So our two classifications for rats might be $\langle R, \Sigma_{\mathcal{R}}, \vDash_{\mathcal{R}} \rangle$ and $\langle R', \Sigma_{\mathcal{R}'}, \vDash_{\mathcal{R}'} \rangle$.

Here R and R' are the same set, {rat 1, rat 2 . . . rat n}, while $\Sigma_{\mathcal{R}}$ = {listless, active, good appetite, bad appetite . . .} and $\Sigma_{\mathcal{R}'}$ = {insulin reactivity high, insulin reactivity low, etc.}.

And the classification of their livers might be $\langle L(r), \Sigma_{L(\nabla)}, \vDash_{L(\nabla)} \rangle$.

These tokens are $L(r)$ = {liver of rat 1, . . . , liver of rat n} and $\Sigma_L(r)$ = {is swollen, is not swollen, has squamous cells, does not have squamous cells, etc.}.

It is generally a matter of observation to determine what relations hold between the various types and tokens. However, not all types need be directly observable types; they can be of whatever sort we are interested in. As such, the types might well have interesting connections between them that we can represent with a generalization of the entailment relation from logic. If $\mathcal{A}$ is a classification and $\langle \Gamma, \Delta \rangle$ a pair of sets of types (here thought of as a sequent), then we can introduce the notions of *satisfaction*, *entailment*, and *constraint*:

- $a \in \mathcal{A}$ *satisfies* $\langle \Gamma, \Delta \rangle$ if when a is of type α for every $\alpha \in \Gamma$, it is also of type β for some $\beta \in \Delta$.
- Γ *entails* Δ in $\mathcal{A}$, $\Gamma \vdash_{\mathcal{A}} \Delta$ if every token a of $\mathcal{A}$ satisfies $\langle \Gamma, \Delta \rangle$.
- If $\Gamma \vdash_{\mathcal{A}} \Delta$, then $\langle \Gamma, \Delta \rangle$ is a *constraint* supported by the classification $\mathcal{A}$.

Thought of in this way, classifications are the main repository of the information we have about individual systems (qua collections of objects with their properties). As informative as these various classifications are, though, in isolation they may seem to have nothing much to do with each other. They would then not be very useful for drawing inferences about the type–token relations in one domain based on those in the other. What is missing so far is an account of the relations between the classifications that will allow us to draw inferences connecting the tokens of a given type in one classification with those in another. This relation is the infomorphism.

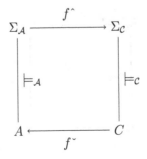

Barwise and Seligman put it this way: "If $\mathcal{A} = \langle A, \Sigma_\mathcal{A}, \vDash_\mathcal{A} \rangle$ and $\mathcal{C} = \langle C, \Sigma_\mathcal{C}, \vDash_\mathcal{C} \rangle$ are classifications then an *infomorphism* is a pair $f = \langle f^\wedge, f^\vee \rangle$ of functions satisfying the . . . biconditional: $f^\vee(c) \vDash_\mathcal{A} \alpha$ iff $c \vDash_\mathcal{C} f^\wedge(\alpha)$ for all tokens c of $\mathcal{C}$ and all types α of $\mathcal{A}$. Such [an] infomorphism is viewed as a morphism *from* $\mathcal{A}$ to $\mathcal{B}$. . . . It is the *fundamental property of infomorphisms*" (1997, 32).

An infomorphism between classifications is then just like a homomorphism between logical structures and their substructures, and classifications are called infomorphic just in case there is an infomorphism between them. We can use infomorphisms to make assertions in the classification of the subsystem $\mathcal{A}$ about the tokens of $\mathcal{A}$ that are themselves the representations in $\mathcal{A}$ of the tokens of the entire system $\mathcal{C}$, and turn these into assertions about the tokens of $\mathcal{C}$.

We can thus characterize information flow from one local classification to another through communication channels formed by maps between the classifications—the infomorphisms. These maps tell us what some token in one classification corresponds to in another, and further how the type structures of the two classifications connect (and hence what the composition of logics amounts to). Once we have all of this in place we can ask how the infomorphisms do or do not support inferences that are allowed in one classification when carried out in another classification. When we can answer these questions we can see how remarks about the tokens of one classification can be used to generate inferences about the relations between the tokens in another. We can then characterize the logic governing the various tokens making up the elements of the extended system as a whole. The local classifications I am interested in are about the component parts of an experimental setup (including the various theories of the devices and of the experiment as a whole), as well as about systems in the rest of the world at large, and the infomorphisms between these classifications provide the conceptual ground for experimental knowledge claims made about distal parts of the extended system based on observations of proximal parts of that system: facts about all rats based on facts about the rats in an experimental protocol.

For our rat example, if we want to characterize the relation between one of the rat classifications and the rat liver classification, we could represent it like this:

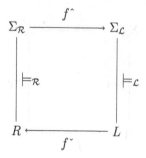

The infomorphism alone, however, doesn't take us quite as far as we need to go. We still need to account for the way infomorphisms can be strung together to allow us to model the kind of extended connections that exist between very localized lab situations. This will help to provide an understanding of the complex features that make up the connections between the localized situations, then of the entire class of system under study in the lab, and finally on out into the world at large. For that we will need Barwise and Seligman's (1997, 76) more general notion of information channel.

Definition: "A *channel* C is an indexed family $\{f_i : A_i \rightleftharpoons C\}_{i \in I}$ of infomorphisms with a common codomain C, called the *core* of C. The tokens of C are called *connections*; a connection c is said to *connect* the tokens $f_i(c)$ for $i \in I$. A channel with index set $\{0, \ldots, n-1\}$ is called an *n-ary* channel."[1]

The core of a channel is, in essence, the entire distributed system of interest—or at least it is infomorphic with all of the connected subsystems. When we come to engage with some system, we will become part of an even larger distributed system ourselves, and that new system will be the core of the information channel it comprises. As theorists, however, we can remove ourselves and simply ask of given researchers (or teams) that are connected to various distal systems through this channel how they bear information about them, by virtue of the regularities present in that distributed system. This way of talking is just what Barwise and Seligman were getting at in their main principle of information flow for distributed systems. They, however, were mixing in an engineering concern, by attempting to address irregularities and errors. I am offering with respect to experiments something more preliminary; I am asking only what happens when things go well, and what fails to happen when they do not. I am not, at this point, trying to model error correction.

When we have a channel in view, we can ask of various bits of it whether and in virtue of which regularities they bear information about other bits. That's the point to all of this machinery; it allows us to maintain a hold on the key question of how one (sub)system bears information about another (sub)system. It bears this information because its being a token of a certain type entails that some other token is of a certain other type, and it does so by being situated properly within our descriptive practice. The point to adopting the technology of information flow outlined by Barwise and Seligman is that it keeps this basic fact clearly in view. When one system's being of a certain type entails that some other system is of a certain type, it is because of their relationship within the information channel that connects them. While

---

1. Note that here they have adopted the convention of letting f_i stand in for $f_i^\wedge$ and $f_i^\vee$.

there are many ways to establish these channels (by convention in the case of mapmaking and word use; by observation and descriptive practice in the case of physical science; by proof in the case of some mathematical system), they all function to allow inferences in one classification to generate information about what is going on in another classification—that is, to allow information to flow through the various parts of these systems. This information flow has the same structure from the most banal situations to the most profound. For an example of the former: the fact that on the map of London that I am holding the icon for the Thames is between the icon for Russell Square and the icon for the road sign in front of me, can be coupled to my belief that I'm just around the corner from Russell Square—and when it is, it is that fact that bears the information that I am lost. For the latter: the fact that the sonic features of a leaking container of rubidium atoms are the way they are, when coupled to a particular reformulation of the Gross–Pitaewski equation, bears the information that pair production of scalar particles due to the expansion of the universe has the form that it does.

I will rehearse Barwise and Seligman's presentation of this for the practice of getting information about the world from maps,[2] so that the way it works is clear; then I'll apply it to the case of scientific experiments in part III of this book.

Barwise and Seligman speak in terms of people, maps, and locations in the world. They consider the first two as connected by the perceptual act of looking at a map; the second two are connected by the practice of making maps.[3] The perceptual acts connect the maps to the viewers of the map, and the mapmaking practice connects the maps to the world. Each set of connections is the core of an infomorphism of its own, and what we can do with them is join them together into a larger infomorphism that allows information flow between the world and map users. This is the infomorphic version of Dretske's Xerox principle.

So, let's think of maps, M, and users, U, as infomorphic to certain acts of map viewing, V, and maps and regions, R, of the world as infomorphic to certain acts of mapmaking, B. There will be functions f_{mv} ($= f^{\wedge}{}_{mv}$ and $f^{\vee}{}_{mv}$), f_{uv}, f_{rb}, and f_{mb} witnessing these infomorphisms. The elements of V and B are the connections between M and U, and of M and R, respectively. They are the cores of these two distinct channels.

2. See their discussion of this (1997, 90).

3. That practice is also connected to other groups of people, but we won't pay attention to that here. Naturally there is a useful classificatory structure there that we could be interested in analyzing if we are focused on the mapmaking practice itself. That's for another time.

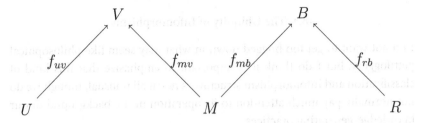

By making use of the fundamental property of infomorphisms, we can conceive of this as a single distributed system, with a new core, L, involving tokens of learning about regions, that connects V and B; we can then collapse the entire thing into a channel connecting U and R. This both exemplifies the Xerox principle and shows the power of that way of seeing the flow of information (and our knowledge as generated by it). However long and complicated the path from a source of information to a receiver, if it flows at each node it can flow through the entire distributed system. Thus the above diagram becomes

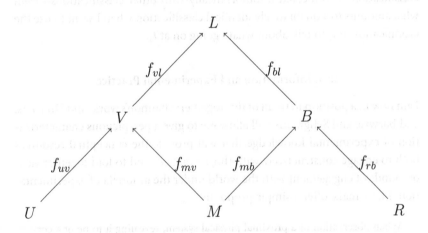

and that becomes

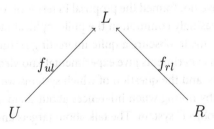

18.2. The Ubiquity of Infomorphisms

I do not want to get too bogged down in what may seem like philosophical pettifoggery, but I do think it is important to emphasize that this kind of classification and infomorphism structure is not at all unusual, though we do not normally pay much attention to its operation in the background of our knowledge-generating practices.

In fact it begins very early in the process of knowledge generation, when we take it that the way we classify some material object or system (a rock, say) is stable under the passage of time. That is, we generally take it that a rock at t_2 has the same properties as that rock at t_1, unless we are explicitly induced to acknowledge that something has changed (it was chipped, or shattered, or even melted). Indeed, the idea that over time we are dealing with *the same rock* arises in part from there *being* a stable classification scheme. I can use this scheme (even though implicit, these schemes will satisfy the basic properties identified above) to infer things about the rock when I know that that classification is connected infomorphically with other classifications—even what amounts to a qualitatively identical classification when I want to use the classification at t_1 to talk about what's going on at t_2.

18.3. Information and Experimental Practice

I am now in a position to tie all of this together: Shannon's work, and Dretske's, and Barwise and Seligman's will allow me to give a perspicuous characterization of experimental knowledge that will provide the conceptual resources both to engage constructively with those practices and to fold together various kinds of engagement with the world under the umbrella of experimentation. I now make offer a simple proposal:

> When observation of a proximal physical system, revealing it to be of a certain type, carries the information that some (generally other) distal physical system is of some (generally other) type, then inferring that conclusion on the basis of that information counts as experimental knowledge of the latter.

Notice that I have not framed the proposal in terms of "object" and "target" systems, as is increasingly common in the philosophical literature on experiment. Such talk seems to obscure a quite interesting feature of experimental practice: very often experiments give experimental knowledge about all kinds of different systems, and the question of which systems we do get knowledge about is answered by noting when inferences about them are licensed by experiments on the "object" system. The talk about targets and objects obscures the fact that our intentions with respect to these various systems are, in themselves, of no significance whatsoever. This is precisely what I was talking about when I focused on the problem of intentions carried along by the interventionist model of experimental inference (and causal explanation).

My proposal includes nothing about how to justify these inferences. For one thing, experimental practice is too complex for any general recipe of that sort, since for any given case we can classify the experimental system type by the various things we do to secure our trust in these inferences. For another, as we have now seen repeatedly, justification is simply the wrong model for what separates true experimental claims from experimental knowledge. Justification is just not an appropriate ingredient in the foundations of scientific *knowledge*, however useful it may be when we come to talk about the public (and institutional) uptake and promotion of that knowledge.

That, then, is the account of experiments as methods of controlling the flow of information from the systems in the world we're trying to find out about, to the systems that we actually deal with, and finally into us.

19

Moving On

The last few chapters have covered a lot of ground, and I'm afraid some of the thread may have been lost. So let me offer a brief recap. I believe the best way to understand what is going on in experiments is to think of them as controlling the flow of information through the systems that instantiate them. To make conceptually clear how that works, and to apply that idea to experimentation as a whole, requires having a clear idea of what information is, how it flows generally, how it gives knowledge, and then how the structural features of systems in the world (broadly construed to include theories and so forth) contribute to the flow of information in those special systems we use to experiment.

The best model for information I know of is Shannon's, and the best model for empirical knowledge I know of is Dretske's Information Theoretic Account of Knowledge (despite what some see as important limitations), and the best model for the structural constraints and enablers of information flow I know of is Barwise and Seligman's extension of Dretske's program. These three models allow me to offer a very simple model of experimental knowledge, which can be succinctly stated here: to say that experimentation on some proximal system P gives knowledge of some distal system D is to say that a signal bearing the information that P is of type Π bears the information that D is of type Δ.

Because experiments are used principally to produce knowledge, they should be judged in the first instance by how well they do so. Experimenting is the use of observations to assign properties (law-like relations, type designations, etc.) to systems that have not been directly observed: for example, concluding that effects observed in one system would be observed in another system if the latter were observed in the same manner. My task will be to display the features of experimental systems that suit them to the regulation of the flow of information from the components of these systems to agents like us who know things about the systems, as well as to the material modes of storing and representing that knowledge. To say that I learn something about distal systems by observing proximal systems is to say that I now have information about the distal system that I did not have before, and that that information came through my observation of the proximal system. The question is, "how did that happen?" The answer I offered is that the various systems we deal with in the world and about which we can get experimental knowledge are connected informationally. "Connected" was vague, and merely conveyed the idea that because it makes sense to talk about various systems in nature as instantiating properties that are common to natural kinds of which they are instances, or as being tokens of certain types, then it makes sense to say that by learning about these instances or tokens we are able to gather information about the kinds or types. We are able to do so because, by virtue of being of that kind or type, the systems' having certain properties bears the information that the kind or type is characterized by its instances or tokens having those properties. The law-like relations that connect the instances of a kind are, if nothing else, at least information channels connecting those instances together. All the work in this part has been meant to elucidate that idea.

Some may wonder why I attempt to characterize experiment as controlling information flow. What, for example, is wrong with a more straightforward sampling account? On such an account we would see the world as populated by numbers of token systems of various types. Our task would then be to select according to some randomization scheme exemplary tokens of these types. And should our randomization scheme be correct, we could draw reliable inferences about the more general populations to which these tokens belong. Presumably one could expand the usage to include constructed systems on which to intervene. The idea might be that we are constructing tokens that are appropriate exemplars of other tokens, that these tokens are produced by an appropriate randomization scheme, and therefore that by observing them we are in some sense sampling the population of possible tokens of this type. This may well be, in compressed form, a version of what we typically take to be the connection between tokens we observe and inferences to more general types. However, as the literature on the design of randomized trials makes clear, the conceptual foundations of this process are quite obscure. There appears to be agreement on whether certain experiments count as properly done and properly licensing inference to larger systems, but little agreement about why it is that things work this way.

My approach is to make considerations of experimental design subservient to accounting for the flow of information from these systems through the systems under study in the lab or in nature and then into the brains or notebooks of researchers. Approaching things this way was supposed to make clear how experimental design and analysis are related (Kempthorne 1992). It was also supposed to make clear the difference between assigning warrant for some experimental claim and saying what the experimental claim itself amounts to. I also hope to have made very natural the distinction between possible errors of observation and possible errors of inference.

My view of the characteristic feature of experimental knowledge—that it is the knowledge we get by finding that some distal system D has property Δ on the basis of some proximal system P having property Π—makes the class of things that count as sources of experimental knowledge much larger than normal, but it has the virtue of explaining how experiments give us knowledge, and making manifest how what seems to be experimental knowledge can fail to be such: practices that are apt for generating such knowledge are experimental, but do not always lead to experimental knowledge.

Consider, in outline, the process by which we "discovered" that saccharin causes bladder cancer in humans. Did we observe saccharin doing this in humans? No. Instead we observed some populations of rats and noted that very high doses of saccharin increased the likelihood of bladder cancer in this population (principally in the males in the population). On the basis of background beliefs about the similarities between rats and humans, researchers drew the conclusion that saccharin causes bladder cancer in humans. In my framework the proximal system is the rat population, the distal system is the human population, and the hoped for infomorphism was constituted by our background beliefs allowing knowledge about rats to be transformed into knowledge about humans. The observations of the rat populations uncovered a logical structure governing inferences about rat metabolisms that allows us to infer that increases in the rats' dietary saccharin increased their risk of cancer. We would then appeal to the supposed infomorphism between the logical structure governing rat metabolism and that governing human metabolism to transfer the inference about cancer risk to the human population.

Here we can see how the model illustrates the importance of information flow in understanding experiments and their hazards. Note that for a properly functioning experiment we need the observation of the proximal logic to be correct and generalizable (in this case, that the "metabolic logic" of the rat population we used was typical of rats); we need the infomorphism relating the proximal and distal metabolic logics to be finely grained enough to support transferring inferences from the proximal to the distal system (i.e., from rats to humans). In this case, it turns out that neither condition was met: these rats were not typical of the rat population; the mechanism used in rats to form the tumor-causing crystal deposits in their bladders is not present in humans (nor in mice, nor in monkeys, etc.; Health Canada 2007).

I asked "If to make an experimental claim is to assert that the observed fact that proximal token P is of type Π supports the inference that distal token D is of type Δ, what does that amount to information theoretically?" What we want is for an observation of some token or group of tokens to establish that those tokens are of some type. For example, we want to observe the behaviors of the rays in some given Crooke's tube or its equivalent and conclude for example that they bear some charge or other; or we want to conclude on the basis of observing a selection of rats, some of which are given more saccharin than others and some of which develop bladder cancer while others do not, that ingesting saccharin causes bladder cancer in them; or we want conclude from the actual behavior displayed by someone refusing an ultimatum game division that would have yielded a nonzero monetary reward that the person values fairness to some monetary degree. And then having observed these things, we want to conclude that we thereby have reason to infer that: all cathode rays are charged; rats get bladder cancer from saccharin; human agents can be said to value fairness. We want to do this by speaking about how information flows on the application of an experimental procedure. Here's a proposed formulation.

In the simplest instance, the distributed system is composed of some particular object's time instances.[1] Then the core of the channel will be the function encoding the temporal evolution of the object. For many properties of things we assume that the infomorphism connecting the temporal instances of these properties is the identity operator. We assume, for example, that when we observe the mass of the object, then that observation bears the information that its mass at some later time is that same value. Very often, of course, the function encoding some one of the object's properties will not be the identity. Its location, for example, may change over time. When this is so the function is more complicated, but the account is not. For the location of the object we simply take the core of the channel to be $x(t) = x_0 + \int_{t_0}^{t} v(t')\, dt'$.

---

1. I don't take any position here on the metaphysics of temporal identity. In particular, I don't choose between a 4-dimensionalist and a 3+1-dimensionalist view. In part this is because I think the debates around these issues are sterile and pointless—and clearly so in this case—and so cannot offer any insight into the conceptual issues that concern me here. In part it is because I do not think it makes any difference in this case.

Thus, given that the connection between the properties of the object at these times is what we think it is, its present property bears the information that its future properties have the values they do. We tend not to notice that we are reasoning at a distance in such cases, but that's what we're doing. And it's too bad we don't notice it, because appreciating information flow here is key to appreciating the way it works more generally, and in less obvious circumstances. Not seeing information flow here makes it difficult to see elsewhere, and not seeing it there obscures a key feature of the conceptual structure of experimental practice.

Consider something approximating a Hooke's law system, for example a spring not too greatly stretched. The equation governing the system is $F = -k(x - x_0)$.[2] This equation, while it can be used to produce the time-development equation for the position of the endpoint of the spring, also does something else. It tells us that one of the spring's properties is related to another. The equation has no time dependence in it, so it relates the force on the spring's endpoint to its location instantaneously. The force and the position are not, so to speak, spatiotemporally distinct—that is, they are both features of the tip of the spring at a given moment. It may therefore seem odd to call this a distributed system. However, that's just what we should call it. It is distributed not spatiotemporally but informationally—the different properties of interest here are not in a spatiotemporal relation, but they are related to each other via the structure of Hooke's law, and they are in that sense distributed within the classification we take to be governing them. In this case, of course, we can tell a nice, tidy, casual story that tracks very closely how information about the spring tip's location bears information about the force acting on the spring. But such a story is not always available, nor is it generally illuminating of the way information gets out of the device and into the heads of researchers. For that we are best served by noting that because we have reason to think that these properties are so related, we have reason to think that the location of the spring tip bears information about the force on the spring.

The above example is artificial, but was chosen to illustrate two important points. First, as I pointed out above, information and causation are often related, but they are not the same, and they may well have nothing to do with each other in certain cases: for example, two randomly chosen systems that happen to have the same dynamics will bear information about each other,

---

2. While this is true of a Hooke's law system, it is not true for any real spring, since the molecular forces that make up the spring and produce the restoring force do not all operate instantaneously, and changes in location of the spring's tip take time to propagate through the rest of the system. I don't believe anything I say relies crucially on this little fiction.

but will have no causal link (either overt or arising from a common cause). Second, distributed systems may well be distributed in unfamiliar and non-spatiotemporal ways. That a system is distributed has to do with our classificatory practices to some extent, and those practices need not line up with the properties of the systems that lie outside those practices.

Suppose that in a given experimental arrangement we have established by observation a local logic. For example, we may have established that for any given rat there is a given function connecting the marginal probability that it contracts bladder cancer with its saccharin intake. This is a constraint on the classification of the form $p(C_b) = p_0(C_b) + f(S)$ telling us that the probability has increased by a function of the saccharin in the rat's diet. Now we may wish to generalize things to have the results of one class of experiment provide the experimental evidence for another experimental claim. For example, we may wish to use one model organism as the source for a local logic that we project through an infomorphism onto another organism.

Isolating in the proximal system the mechanisms or processes of interest can have important benefits—it allows greater control over these processes, and it allows finer insight into their exact parameters and the inferential structure of the local logics they implement. But with these benefits come dangers as well—principally that the map between proximal and distal logics will fail. When the proximal system is coarsely characterized, then maps between it and distal systems are easy to obtain but consequently less informative. The more potentially informative the map, however, the greater risk that we are misled rather than informed.

Smith (1982) and Shapley (1958) focus on establishing that there is in fact an infomorphism between the entire experimental structure deployed to get experimental knowledge about one type of system and the experimental knowledge we want in a different system (either more comprehensive, or of another sort). Each one claims that we generally do so experimentally. I don't believe there's any difficulty of circularity here, but it is a kind of bootstrapping. We use the presence of another class of infomorphism to establish experimentally that the infomorphism at the heart of this experiment exists and can sustain the inferences we make with it. Thus, depending on what we are looking for and the kind of knowledge claims we are making, very similar processes may serve either as sources of local calibration or as experimental knowledge.

Smith appeals to what he calls "parallelism," the claim that "propositions about the behavior of individuals and the performance of institutions that have been tested in laboratory microeconomies apply also to nonlaboratory microeconomies where similar *ceteris paribus* conditions hold" (1982, 936).

He finds this concept in the context of Shapley's assertion that the evolutionary responses to similar evolutionary forces are similar. Whatever one thinks of that claim, Smith's application and reformulation from the notion that "as far as we can tell, the same physical laws apply everywhere" seems both unobjectionable and clearly intended to capture the central notion of what justifies claims to external validity in experiments (Shapley 1958, 43, cited in Smith 1982). What remains is to show just what these *ceteris paribus* conditions are and why we are justified in thinking that they hold. In other words, we need to understand what it is that enforces the parallelism between the laboratory context and the natural context in various cases.

I acknowledge here that the possible scope of the present work is pretty broad. My thesis is that my account of experimental knowledge applies to all fields of science, and indeed to all the activities by which we learn from experience, and this necessitates a large-scale broadening of the kinds of thing that fall under the concept of experiment. I claim that experimental knowledge can be generated by thought experiments, laboratory experiments, analogue experiments, natural experiments, and extended and mixed-mode experiments like those found in international collaborative drug trials, and they are all of the same type: experiment. But an experiment at the Large Hadron Collider consisting of miles of evacuated tunnel, vast computer arrays, countless engineered components of various scales and sophistication, and thousands of scientists engaged in controlling, tuning, troubleshooting, and analyzing all the various goings on for the purpose of finding the Higgs boson is vastly larger and more complex in every dimension than me, sitting alone in my office, coming to realize thought experimentally today that to get to campus it's better to take the bus than my bike. What I will show in part III of this book, however, is that from the point of view of characterizing how experimentation generates knowledge, both of these systems (and many more besides) function in just the same way. How? By controlling the flow of information from some distal system of interest, through a proximal system that we have access to, into our heads (or minds, or notebooks, or networks of communication, or whatever your preferred account of the location of knowledge). The rest of the book is an exercise in thinking about a variety of experiment types through the lens of this account and understanding how it is that they all are apt for the generation of experimental knowledge.

Ways of Experimenting

20

Introduction to Part III

There is a widespread conviction that certain of the activities that go under the name "experiment" deal with systems that are close reflections of the systems they are being used to study, while others are somehow more dissimilar. So I would like to begin this part of the book by addressing this conviction, which I take to be a persistent confusion that obscures the nature of experiment (and of modeling, simulation, analogical experimentation, thought experimentation, and other related activities). This confusion arises from the use of a folk category by those who discuss experimentation. There is not a specific term that indicates the use of this category, but its use is fairly easy to detect. It is the category I have been using all along: direct experiment. This category is supposed to comprise those experiments where the proximal system is identical to the distal system—where the system we experiment on is the system that the experiment is about. This folk category has little in the way of clear application conditions. Instead it is one of those concepts that is known by its negative instances, the experiments that are not direct: models, simulations, analogues, thought experiments, and so forth. These types of experiment constrain the class of direct experiments to whatever does not fall within their parameters. The problem, as we shall see, is that once we count up all the experiments that are not direct, there is nothing left over.

The distinction to which I am alluding is not new. Throughout the history of science we see attempts to exclude certain kinds of experiments as not giving the kind of direct access to features of the world that is taken to be the hallmark of the scientific method (the inappropriateness in the Renaissance of using tortoise lungs as clues to the behavior of lungs in other creatures,[1]

---

1. There is a nice discussion of this in Guerrini (2003).

the distinction between terrestrial bodies and superlunary bodies as reason
to exclude the latter as instances of gravitating systems, etc.). However, the
conceptual distinction between direct experimentation and the other, indirect types has simply not been drawn clearly enough to support attempts to
distinguish between actual instances of these types of systems. The reason
for that, I will argue, is that there is no such distinction. All experimentation
that yields knowledge of other than some instantaneous properties of some
immediately observed object will count as indirect experimentation—and all
production of experimental knowledge is directed to that end. My hope is
that when we realize that all experimental knowledge is analogical, our attention will turn toward characterizing experimentation by distinguishing
the type and strength and reach of the analogies being used to generate experimental knowledge. This conclusion should follow easily from the theory
of experiment outlined in part II, as well as the discussion in part I of the
nature of experimental knowledge. Experimental knowledge is just not the
kind of thing that is direct, relying as it does on the fundamental inference
using observations of the proximal system to generate knowledge of the distal
system.

I suppose one could think that certain kinds of observation generate
knowledge more directly than do others—for example, when the property
type observed in the proximal system is the same as the type inferred in the
distal system. Or, by a similar token, when the proximal system is the same
sort of thing as the distal system. Once properly calibrated information channels are in place, however, the similarity between the observed and inferred
properties of proximal and distal systems is not relevant to any conceptual
questions about experiment. Considerations of similarity, and of a kind of
remoteness, can be relevant to the pragmatic question of how easy it is to
calibrate information channels connecting the systems. We could call experiments exemplifying that remoteness "indirect," keeping in mind that direct-
ness is nonfundamental. But it is probably better just to realize that experimentation is an indirect kind of business.

The laboratory is an artificial environment, created to produce phenomena that are either not found outside the laboratory or not found in the quantity and/or quality and/or accessibility that may be found inside the laboratory. In this respect at least, what happens in the laboratory is removed from
the natural systems outside the laboratory, even for what seem to be the very
same natural systems we are trying to study by replicating them inside the
laboratory. This remove is temporal-spatial and material, or dynamical, or
conceptual, or what have you—the systems inside the lab are simply different
systems from those outside. So it is no wonder that we are inclined to try to

draw the indirect/direct distinction as outlined above. While that distinction doesn't seem to do any conceptual work, I will, to gain more insight into the nature of experimental knowledge, consider in what follows systems that are in one way or another quite remote from the systems they model. I do this to place the concept of experiment as generator of knowledge under stress, to examine it in extremes, and so to display in stark detail the ways that experiment can go wrong. I will then apply the insights so gained to the general account of experimental knowledge presented in part I to characterize what goes wrong in terms of the distinct ways that information flow can be restricted in these different cases. The experiments I have in mind for this are what are often called "analogical."[2]

In analogical experimentation one type of system is used to investigate the properties of another type of system, but a type that is radically unlike the first type in many respects. The experiments in this category are interventions into systems of the first type, observations of the results of those experiments for systems of that type, and then inferences about the observational properties of systems of the second type. The support for these inferences is the analogy between the two types of system. Even though they are radically different in some respects (as we will see, they can be as radically unlike as a small container of supercooled gas is to the universe at large), there are significant features that the types of systems share that make these inferences legitimate. In the case just mentioned, of the universe being experimented on via observation of supercooled gas (Bose–Einstein condensates), it's the shared dynamics resulting in a shared mechanism of particle production. In other cases it might be certain shared aspects of metabolism between very different kinds of living systems, or a similarity between the features of unfettered markets involving large multinational corporations and the "markets" set up in college classrooms by economics teachers. In all cases the strength of the analogy determines the strength of the results. And even though the experiments are only analogical, they still yield experimental knowledge of the same sort as any other experiments, or so I shall argue.

Next I will consider generally analogical experimentation and its relation to other forms of experimentation. We will see that a main experimental system used for biomedical research—animal models—is itself merely analogical. While I will point out some difficulties with these models as they have been used recently, I will still conclude that generally they can provide robust

---

2. Or "analog" or "analogue." I myself find congenial the convention that associates the first of these with the digital/analog divide and the second with the sense of analogy. But that convention seems nonuniform, and "analogical" works just as well.

experimental knowledge in just the way that other types of experimentation do. In addition, there is an important class of physics experiment that appeals to a tight infomorphism between certain terrestrial systems and the cosmos as a whole—the so-called analogue gravity models. And we will also find that economics experiments generally exploit a tight analogy between the behaviors of people in two very different kinds of system. The discussion will show that analogical experimentation is a wide and important class of experimentation. Indeed, we will see that analogical experimentation is a generic type that incorporates simulations and models.

Only after considering these types of system will I be prepared for the big reveal: that all scientific experimentation is analogical. The analogical features that are salient in any given case will differ, and these differences will give the various types of experimentation their individual flavors. However, the important place where they do not differ is in how they give experimental knowledge. They all give it the same way: by exploiting the infomorphic connection between proximal and distal systems, and allowing information to flow in the distributed system from the distal to the proximal system, and then into the minds of knowers.

Once we see that experimentation is generically analogical, we can see that models and simulations must count, generically, as experiments just in case they yield experimental knowledge. The way to check is to look at the associated knowledge claims and see how they function. There is a lot of great work on models and simulations, and I will only gesture to that work, my point being to show that there is no problem at all in taking these kinds of studies as providing experimental knowledge.

The final example is thought experimentation. There is, of course, a great deal of controversy over whether these are really experiments at all. Or I should say, there is very little controversy, with the strong consensus view being that they are not experiments at all. Even those who take them to be experiments do not take them to be experiments on the systems described in the articulation of the thought experiment. Rather, they take them to be experiments giving insight into the structure of the minds (and/or brains) that carry them out. As far as I can tell, I am the only philosopher who takes them to be full-fledged experiments on the systems described in their articulation. My aim is to establish through my discussion that I have the correct view, and that others should join me. Even failing that, I will have succeeded if I can convince my readers that essentially every case of so-called experimental knowledge we have is the result of a thought experiment. Indeed, on reflection, it is thought experimentation that provides the main *raison d'être* of the book. I will be concluding with the very strong claim that thought experiments are nothing other than full-fledged experiments. They can, naturally, fail to give knowledge, just as can any other form of experimental practice. Yet, when they give knowledge, it is experimental knowledge that they give.

First, though, I wish to consider the difference (if any) between laboratory experiments of familiar types and natural experiments.

Laboratory and Natural Experimentation

21.1. Introduction

Here the standard laboratory model of experimentation, familiar from science class and popular accounts, is brought under the information flow account of experimentation. This provides a kind of calibration of my own methods. There is reason at least to think that the information flow account is getting something right here. But we will see that the same analysis can be applied to the case of natural experimentation. These are experiments where we somehow find the data just lying around, and we also find that the data we find lying around are in fact information-bearing signals, yielded by a well-calibrated channel, that constitute observations of proximal systems that give us knowledge of the distal systems at the other end of that channel. Together these cases provide support both for the claim that my analysis of other types of experimentation is on the right track and for the unificationist claim that legitimate experimentation is a broader class than is normally thought.

21.2. The Laboratory

So what goes on in the laboratory? Much of what I have to say here is only a reminder of what we all already know about standard experiments. As I have said before, this is not a user manual for those wishing to generate experimental knowledge; it is instead a conceptual analysis of our practice of extracting knowledge from experiments. As we saw, a standard idea of experimentation is that I first gather a subpopulation of the population I wish to learn about, whether that is a population in a literal sense or simply a general class of entity whose properties I wish to know. We could be speaking here about cathode rays, or mice, or economic agents. I would then generate some cathode rays, or breed a typical selection of mice, or sequester some exemplars of *Homo economicus*. This selection procedure is performed carefully so that the subpopulation is a good reflection of the total population. At this point I carry out the important step of partitioning the subpopulation into smaller treatment groups. Then I perform whatever treatment I am testing (making trial of) on these treatment groups, according to a protocol that prevents any influence on the outcome other than the influence of the trial treatment itself. Then the expectation is that whatever I observe that is different between the various treatment and control groups is a causal consequence of the treatment at those levels, and I can know on that basis that typical members of the original population are causally impacted in that way by those treatments.

Two points should be kept in mind about this account. First, the typical double-blinding and so forth that are called for in standard accounts of a good experiment have little conceptually to do with what is happening in the experiment. Everything that needs to be said from that perspective has been said above. Yes, it is necessary to keep subjects ignorant of whether they are or are not in the control group. But that is because the experiment's outcome is influenced by subjects' knowing whether or not they are in that group. That is a fact about knowing subjects, not about experiment generally. In the same way, it is important not to hit control subjects with a baseball bat. The main reason for that is *not* that this will allow them (or someone else) to infer that they are the control subject, but rather that being hit with a baseball bat is very likely to have an important influence on the subsequent state of the subject. Second, it may seem unusual to continue to include such systems as cathode rays in the account, but a useful way to look at things is to see these rays as part of a continuing production of perfectly typical members of the class *cathode ray*, and partitioning as happening in the same act as treating them or not treating them (with a magnetic field, for example).

These various activities—of selecting the sample so that it is typical of the class, and preventing any extraneous influence, and observing the outcome—are naturally accounted for by the information flow–based account of experimental knowledge. In selecting the subpopulation so that its members are typical, what we have done is produce a proximal system that is in good infomorphic connection to a distal system. We have also ensured that the resulting infomorphism is manifest and apt for exploitation. Such infomorphisms, as we have seen, are the channels by which experimental information travels. Treating that subpopulation is a way to get signals to flow in the channel. It's a funny kind of signal, because it seems to exist only here at the near end of the channel, while it is supposed to be bearing information about the far end. That, of course, is the central mystery of experimental knowledge. For systems of this sort, though, where the proximal is a token instance of the type that is the distal system, the mystery is no greater than and also no different from the perennial philosophical puzzle of how we know types on the basis of acquaintance with their tokens. The resolution of that puzzle (or at least of our particular instance of it) is that the tokens themselves bear information about the types, and are themselves, in that sense, signals. Like other signals in a channel, the entity that bears them can pass them on to other entities along the way: from some orange and yellow flickers, to the child who shouts about it, to the runner who gasps about it, to the ringer of the bell, who signals about the fire to the whole town. The sound of the bell is at the very end of a communication channel that connects the initial spark to the bucket brigade. Here, in the laboratory, we have counterfactually rich signals, counterfactually rich bearers of information. When one is induced to respond to a causal influence, that signal reveals part of that information; when many are, sometimes they give us the whole causal story.

What now of all the blinding procedures and other steps to avoid confounding? Those are themselves measures both to keep the channels calibrated and to be sure that the observations we make are of the appropriate signals. It does no good to observe some system to see how it responds to a causal influence while at the same time subjecting it to some other causal influence over which we have no control and about the effects of which we have little knowledge. We know for example that humans (at least) are subject to the placebo effect. This is a genuine causal influence from our beliefs about what is happening to us to other biological effects over which we typically have no direct control. Biasing the beliefs of some group on which we are experimenting in a way relevant to the effect we are going to observe, then, is clearly a bad idea. I don't expect that anyone who has reflected on the nature of experimentation should be confused by this point, which is simply that there is nothing special about blinding or double-blinding experiments—it's just run-of-the-mill confound avoidance. When it is necessary, it is necessary because otherwise the influences of various factors on the outcome would be unknown, and our inferences would not be made on the basis of information, and we would thus not have knowledge. We would not have knowledge even if the beliefs we formed through such a process were correct.

What the information flow account identifies as the essential elements of experimental knowledge are infomorphisms connecting systems of interest to the systems with which we have to deal. I have said that selecting out subpopulations by random assignment is an appropriate method for this. It is pretty obvious, I think, how it works in this case, but let me just display it for clarity's sake.

Knowledge of, for example, the causal features of (the members of) some distal system will amount to knowledge of a variety of counterfactual statements about the system elements. To know that the tokens of the system are of a particular causal type requires knowing that normal tokens will, under the influence of certain causes, display certain effects. This extends as well to probabilistic cause–effect patterns. For example, should I know that a given level of dioxin exposure doubles the probability of liver cancer in rats, and then expose a rat to dioxins in that amount, the probability that it would develop liver cancer is twice what it would otherwise be. We are taking this hypothetical causal connection as a background classification of the rats, in order to analyze what it would take to have experimental knowledge of the claim. Recalling the definition of infomorphism from part II, there would be an appropriate infomorphism when: I select rats from the general population of rats so that the selected rats' cancer susceptibility bears information about the general population's susceptibility; my treatment protocol is as described in part I of the book, so that the cancer rates of the treated and untreated groups bear information about the pretreatment dispositions of those groups combined. Graphically, we have infomorphisms from rats at large to rats in the experiment prior to treatment to rats in the experiment after treatment.

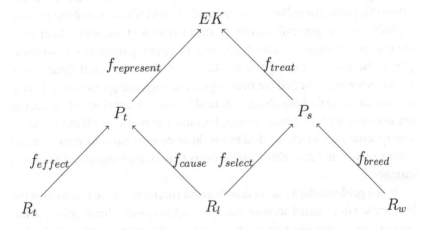

At this point, some may object that something fishy is going on with my account of experimental knowledge. If all knowledge arises from the flow of information, and information only flows when the probability of some event (given background knowledge and the signal) is 1, then how are we ever supposed to get knowledge of complicated causal relations, or knowledge of causal relations that requires complicated statistical reasoning? In those cases we will almost never have probabilities of 1, but rather a bunch of messy probability distributions over a whole range of samples and. To consider just the case used above in its easiest formulation: Suppose that rats in normal (laboratory) situations are at a 1 percent lifetime risk of liver cancer, and at a 2 percent lifetime risk when they are given appreciable amounts of dioxins in their diets. How do I find this out? Well, I gather a bunch of rats, and randomly assort them into two groups, and treat the groups the same, aside from giving one group dioxins at that rate. Then I wait and assess them for cancer at the end of their lives. But I'm doing this with only a sample of rats, so there is always a chance that these groups are atypical. Thus, even though I measured about 2 percent for the treated group and about 1 percent for the untreated group, it could be that these percentages are wildly off. Yes, we have good statistical methods for appraising how likely it is that we're off in our assessment of the entire population of rats based on the sample, but that doesn't help someone who needs probability 1 in order to have knowledge.

We have to keep in mind at this point the difference between my statistical *confidence* and the question of whether information flowed in a particular case. The fact that rats in my group got cancer at the rates they did may tell me about rats generally because these signals really do bear information, or it may not if they don't. Our methods for trying to find out about that are fraught with statistical risk and ambiguity, but the situation itself is not. When this particular collection of rats really bears information about the susceptibility of rats generally to cancer, then what I find out about them gives me that information, because the system comprising these rats is infomorphic to the system of rats generally. Our rules about statistical significance, for when we are "justified" in drawing conclusions, are proxies for this fact, and the more we find out about statistical features of samples with respect to populations, the better these proxies become. But they are still only proxies. The systems with which we deal either do or do not bear information about other systems, and thus either can or cannot give knowledge about them at a distance.

In the gold-standard case of double-blind control trials, we can see that the information flow–based account seems to track very well what is going on. But we can also see that the account does not favor this over any other method of generating experimental knowledge. What happens in this case is that we carve out of the world an infomorphic structure that serves our ends. The epistemic virtue of this method is not captured there. Instead, as we saw in part I of the book, intervening gives a ready choice of experiment for things we might wish to know about, because we can build these channels from scratch. The epistemology, however, is all in the channels themselves.

We might wonder whether other kinds of activity can result in experimental knowledge. For example, if there is no laboratory and no researcher standing by to set up the infomorphisms we will use to draw conclusions about distal systems, how will we get the information to flow? As a warm-up to considering a wider range of experimental practices that do generate experimental knowledge, I will begin with something close to the randomized control trial: natural experiments.

21.3. Nature's Laboratory

Can we have a natural experiment that allows for what normally counts as control? I showed in my examination of intervention in part I that, as a conceptual matter, the intentions of agents are inessential to the generation of experimental knowledge. But there the example was left in an awkward state. That's because the example I gave of a robot mouse lab, while supporting the claim that interventions were inessential from a conceptual point of view, was so implausible from a practical standpoint as to render it pragmatically impotent. Here I would like to go further and try to illustrate that intervention is not even practically necessary in all cases. I will do this by considering some examples of so-called natural experiments and showing that they meet all the requirements on (and so qualify as) double-blind, randomized control trials, which are the hallmark of experimental knowledge.

Normally, when we outline the requirements on a proper study design we tend to phrase things in terms of constraints on the behaviors of human agents who will construct and carry out these studies. And this introduces into the situation the specter of intention. Even when we note this possibility and discount it, we run the risk of being influenced by it in our appraisal of what can and cannot count as proper design and implementation of a randomized controlled study. One nice feature of analyzing experimental knowledge by way of the flow of information is that it makes clear which parts of our intuitive appraisals of situations really bear on the question of whether knowledge is generated in that situation.

Natural experiments are the found objects of scientific experimentation. They are such that all the conditions for the generation of experimental knowledge are lying readymade for our scientific consumption. Here I will proceed by stages, first looking at cases where one might argue that these experiments are natural in name only—cases where one might well suspect that intervention has been smuggled in the back door, so to speak. We will end up with perhaps the most extreme possible case: a randomized control trial in astronomy.

21.4. Snow's Pump-Handle Intervention

As an easy warm-up to natural experimentation, let's consider John Snow's recognition of the source of the cholera outbreak in London in 1854. Popular accounts of this episode give a picture of Snow, armed with the compelling visual evidence of his dot map, fending off a crowd of ignorant and disaffected townspeople who are trying to storm the Broad Street pump. These accounts also suggest that Snow discovered that the Broad Street pump was the source of a localized outbreak through the very construction of that dot map, detailing the distribution of cholera in the area surrounding the pump. Brody et al. (1999) decisively undermine that account. What they make clear is that Snow already had a theory that whatever caused cholera was ingested through the mouth and passed on via fecal matter. He had concluded that the likely mode of transmission at large distances was through contaminated drinking water. It is true that in *communicating* his view to the authorities, Snow made use of the map to indicate the way the cases were distributed in the area surrounding the pump. The map is a striking visual account of that distribution, but it is simply a myth that Snow used it for discovery purposes.

In the second edition of his *On the Mode of Communication of Cholera*, however, Snow (1849) outlines a beautiful natural experiment that was ripe for conducting. This is about all that is correct about the standard story. What Snow had noticed about the distribution of water in the Golden Square area of London is that the services of two distinct companies were thoroughly mixed. As Snow points out (1849, 74), the pipes from both companies are liberally interwoven and the choice of company is a result of decisions made long prior, by the then owner of each property. "In many cases a single house has a supply different from that on either side. Each Company supplies both rich and poor, both large houses and small; there is no difference either in the condition or occupation of the persons receiving the water of the different Companies" (1849, 75). Therefore, Snow continues, the only thing necessary to make the experiment work is to know the source of water for every residence where a case of cholera occurred. Snow was engaged in just that project during the cholera outbreak centered on the Broad Street pump. His idea was to show that the water supply of the Southwark and Vauxhall company was polluted, whereas the Lambeth company's water, drawn from a source upstream of the main sewer discharges, was not—and more importantly, that Southwark and Vauxhall's water was the source of disease.

Despite popular accounts that identify Snow's contribution to the Broad Street case as the result of a natural experiment, the natural experiment itself was not what prompted the removal of the Broad Street pump handle. The conclusions Snow drew that led to the handle's removal were prompted by interviews with residents of the areas surrounding the pump. Interestingly, the natural experiment Snow conducted would not even have been appropriate for identifying the source of the Golden Square outbreak, because (at least according to contemporary consensus) that outbreak derived *not* from the water source used by the Southwark and Vauxhall water company, but from a direct contamination of the pump well with runoff from the washing of a contaminated diaper worn by an early victim—a victim in fact unknown to Snow at the time he drew his conclusion.

Snow does not himself use the expression "experiment" or any cognate for his discovery of the source of the Golden Square outbreak. However, he does characterize his independent and ongoing investigation of the *general* difference in cholera outbreaks between Southwark and Vauxhall customers and those of the Lambeth water company as a "grand experiment" (1849, 75). And yet the experiment was never fully or satisfactorily carried out. Cholera was already on the wane, and sanitation was improving sufficiently that Snow would not have the data he needed to use that experiment to gather the crucial information. But it should be clear that Snow's use of "experiment" is both appropriate and in line with my own. The point is that the water lines from Southwark and Vauxhall were so deeply interwoven with those from Lambeth that use of water from either source was uncorrelated with anything else relevant to the contraction of cholera (personal hygiene, wealth, health, diet, etc.). Sifting the data that would be provided by water source and sufficiently extensive cholera incidence is all that would be needed to make trial of the hypothesis that cholera is caused by contaminated water.

21.5. Compulsory Education under Trial

Snow's case is nice because, even though not quite successful, it illustrates a researcher understanding explicitly that making trial requires no intervention. We don't need to rely on Snow's natural near-experiment, however, to see that the epistemic goods generated by double-blind randomized control trials can be had without explicit, planned intervention. We have plenty of successful experiments.

Laws enforcing certain social behaviors can interact in interesting ways with other social behaviors and yield real experimental knowledge. These laws can produce groupings of people that are arbitrary with respect to other facts about them. Consider laws regarding compulsory education. The laws in many states in the US, for example, mandate starting ages for schooling based on rigid birthday cutoffs. School-age children born after a given date begin school one year, those born before that date begin the prior year. However, those cutoffs do not apply to the age of majority; students who have attained that age with respect to schooling (whether that age is 16 or 18 or something else) can drop out of compulsory educational systems that same day. Noticing this, Angrist and Krueger were able to perform a randomized control trial examining the causal influence of schooling on future educational and financial attainment.

As the authors laconically observe: "Because one's birthday is unlikely to be correlated with personal attributes other than age at school entry, season of birth generates exogenous variation in education that can be used to estimate the impact of compulsory schooling on education and earnings" (Angrist and Krueger 1991, 980). That is all well and good, but does this estimate really amount to real experimental knowledge, and in particular to RCT experimental knowledge? I think it does.

If in fact birthday is uncorrelated with other factors involved in educational and financial attainment (intelligence, perseverance, etc.), then these compulsory attendance laws provide an entirely adequate selection of samples varying only in total days of such education. If, on the other hand, astrological sign is by itself a causal factor in such matters, this would confound our ability to gain experimental knowledge in this situation. It is false that astrological sign plays that causal role, but it is not in principle impossible for us to have found that it does. But that is no different from any other scheme we might use to prepare such sample populations. In *principle* the pseudo-random tables we use to divide samples could be interestingly correlated with the exact features we are trying to test. That would, in those cases, invalidate the study employing them. It would not, however, invalidate the idea of RCTs as providing experimental knowledge. Instead it would show that, in such a case, the appropriate samples were not treated with a potential causal agent in a way that could bear information. That reasoning cuts both ways, however. Given that we have such samples provided, so to speak, by nature itself, then observations of the result can bear information. Finally, even though they say they "estimate" the impact, there is no question that they find a causal impact; what they are estimating is its exact magnitude.

21.6. Astronomical Randomized Control Trials

Natural experimental systems, then, seem much like normal laboratory experiments. There is something that can be considered an intervention, even though it has nothing to do with the aims and desires of human agents. We have, in conjunction with that intervention, a "protocol" supplied by circumstances that separates the population into groups that are treated or not. These guarantee the infomorphic relation between the subsystems about which we can have observational information regarding their cause–effect dispositions and the larger systems of which they are parts. Nothing much has changed in these cases, other than that human agency has been removed from the picture. We have everything we had in the case of the RCT.

So if things go along much as before once we remove human agency, then our understanding of what kinds of system are available for randomized control experimental trials ought to expand to include a much wider class than we normally think. The following question then might arise to provide a test of our intuitions: Could planet-bound astronomers carry out randomized control trials on remote stellar systems?

Intuitions will diverge on this question. But I believe those intuitions that incline one toward a negative response are driven by an incorrect understanding of what is going on in an RCT. That understanding is deeply invested with commitments to the intentional and interventional character of randomizing and controlling. But as I rehearsed again above, and argued for at length in part I, those features play no *conceptual* part here. "Control" does not signal in any interesting way that somebody is *in* control, but instead signals that the unaffected system provides the background against which affected systems' changes appear. And randomizing as a process has no significance to the nature of the trial; instead it is a means to establish that that background appropriately reveals the changes in the affected systems.

The question of possibility is then a straightforward empirical one: are there sufficiently randomized groups of astronomical objects to allow for this? If the answer is yes, then astronomers can perform RCTs; if no, then they cannot. The answers here do not depend on our view of the power of astronomers to manipulate and intervene in the processes taking place in the heavens. Rather, we answer based on our judgment about the statistical distribution of stellar objects having or not having certain features. I do not know whether astronomers can in fact perform RCTs given the contingent facts about how things are in the heavens, but I do know that there is no conceptual impediment.

Suppose an astronomer comes across a store of images including spectroscopic records of solar systems that have planets. She examines these images for a while and then begins to classify them with respect to whether or not the planets are likely to have various compounds in their atmospheres: free oxygen, carbon dioxide, etc. This is a straightforward but not mechanical procedure, because the data are not so extensive for planets at such distances, but she seems to be reliable in this sense: when checked by her colleagues, her results and theirs generally coincide.

Suppose further that another astronomer has been cataloging these same solar systems, but with respect instead to their situations within the galaxy, and using different data. They consider such things as the age of the solar systems, the density of other stars in that region, their masses, and so forth, but they have no information about the internal characteristics of these solar systems. What this researcher discovers is that independent of all these other factors, there is a more or less uniform probability distribution of how strongly solar systems in this group were exposed to gamma ray burster–generated cosmic rays during planetary formation.

These two researchers meet at a conference during a session on extrasolar planets. When they share their data, they notice immediately the very strong correlation between the presence of free oxygen in the atmosphere of such planets and the intensity of their exposure to gamma ray burster cosmic rays. Have they together just performed an experiment establishing the causal efficacy of gamma ray burster cosmic rays to promote oxygen rich planetary atmospheres? Apparently they have.

What objections could be made against their conclusions? Is this not a blinded, randomized control trial of the hypothesis that cosmic rays promote free oxygen atmospheres?

There is a random control condition: exposure to cosmic rays. Even though there was no human intervention, there was an "intervention" in any respect that matters. Some solar systems were subject to these rays and others were not, and those that were range in the intensity of their exposure across an entire spectrum.

There is appropriate blinding. The researchers do not have relevant biases affecting their judgments, because by stipulation neither is aware of the other's data. The first researcher knows only about the outcomes of the trial, the second only about which systems were exposed to the trial condition. And clearly the solar systems themselves don't know in any relevant sense that they have been part of this study.

One could certainly object that there may have been confounding factors involved. Perhaps the distribution over the test condition was not truly random. The second researcher could well be mistaken about that, one might say. And indeed they could be. However, the difficulty of generating truly random assignments of test and control conditions to experimental subjects is not confined to this case. We are apparently never sure that any of our random number tables or generators give truly random numbers, for example. We hope they do, and we take careful measures to assure ourselves that they do. But in the end we may be wrong. Our success in performing randomized control trials is always against the background of that possibility. In that respect the astronomers here fare no differently than any other experimentalists. They take what measures they can to be sure that their trial really is governed by the appropriate probability distributions. The structural and conceptual situations these astronomers deal with are the same as those faced by interveners in the lab.

This example is in some respects more interesting and more plausible than my earlier fable of intervention. It is more interesting in the sense that there is really no human intervention at all. There are no robots (which are after all artifacts of human agents). There is no possibility at all that there could have been human involvement of any sort. The case is more plausible, though, because the observable universe is vast. Given the vastness of that scale, the likelihood that there is some structure of random distribution over "interventions" seems very high. In fact, it seems likely that there are a great many such situations waiting for us to develop the observational tools to exploit. Once we do so, then astronomy will become an experimental science. The real difficulty, then, is not that we cannot intervene into nature at that scale; rather, it is with finding the observational data that reflect the "interventions" that await us.

Analogical Experimenting

22.1. What Is an Analogical Experiment?

In this chapter I will give a general characterization of analogical experimentation. Then, to illustrate the use and robustness of analogical experimentation, I will consider three examples in this category: (1) experiments for cosmology; (2) experiments for human biology; and (3) experiments for economic behavior. All three have issues that make them difficult to see as proper generators of experimental knowledge about their intended systems, but I think they are all in principle apt for generating such knowledge.

For the cosmological case I will introduce an example of how to use a microscopic gas sample as a generator of experimental knowledge about the entire universe. In that class of experiment there are difficulties in establishing the reach of the identity between the dynamics of the proximal and distal systems. As we will see, the issue concerns not the strength of the conclusions we may draw, but the reach of our experiment for drawing conclusions about certain features of the cosmos. Once the domain is specified, we are in as good a position with respect to the claims as we are in any other case of not strictly identical proximal and distal systems. The use of Bose–Einstein condensates in experimental cosmology exemplifies all of the various interesting features of experimental practice, and so they are a good test bed for accounts of that practice. It also shows off in stark relief just how far removed systems can be from intuitive similarity and still be considered related as proximal to distal experimental system.

In the models of experimental economics we have not only the problem of reach, but also the problem of how to classify the proximal systems in the first place. A difficulty that has less to do with experiment than with the distal systems themselves is that we are trying to find out about the economic features of systems that are only partially economic. Generally, all systems are constrained from various directions, but in economics it is more than constraint—it is that the economy is not isolable in its outlines from other basic features of the same observable structures. In biological systems, for example, it is more or less possible to isolate effects we are interested in studying (e.g., drug metabolism) from those not of interest (e.g., intercapillary tidal forces caused by the moon). But to observe clearly that some property is the result of economic activity rather than of some other mass neurological effects of clustered human agents in a changing physical environment with their own changing interests and motivations seems intractably complicated in practice.

In the case of biology, and specifically animal models of disease, we suffer from (among other issues) the question of whether the type–token relations we can pick out in the proximal systems can be properly mapped onto those in the distal system. For example, it is difficult to be sure which of many different possible metabolic pathways and processes in model creatures are appropriately analogical to which pathways and processes in humans. For certain processes, humans may share all, some, or no metabolic pathways with their animal analogues, and may have some pathways their animal analogues do not.

Even so, it is my contention—and one shared by scientists at large—that these and other classes of system are apt for experimental knowledge generation. What makes an analogical experiment analogical is that we first do whatever is necessary to generate observational knowledge of one system, and then exploit the analogy in order to use that observational knowledge to generate experimental knowledge of the system to which it is analogical. Generally, analogical experiments will involve interventions on systems of one type to gather information about systems of another type. These type differences may be significant. Indeed, there may be no obvious similarity between the two systems. But there is one, by assumption, and that is the analogy that holds between them. This analogy supporting the experimental inference is generally a similarity of the relevant dynamics, and one that allows information about the dynamical state of the distal system to be carried by the dynamical state of the proximal system. Analogical experiments in that sense, then, are just the familiar experiments from earlier, but involving systems that are merely analogous to each other.

22.2. What Is an Analogue Experimental System?

If I want to know how billiard balls behave when struck with a cue, I strike some billiard balls with a cue and see what happens. But maybe I want to know about some type of system and there aren't any of those systems around; or there is only one and it's fragile and I don't want to break it; or I just don't know how to perform appropriate interventions on that type of system (because systems of that type are too big, or small, or remote, or require conditions that are not realizable with our current technology). Am I forced to conclude that such systems cannot be investigated experimentally? If so, then much of what we count as our current experimental knowledge will turn out to be illegitimate, and in particular much of our current medical knowledge about humans. For example, our experimental knowledge that cigarette smoking causes cancer in humans and that dioxins cause liver cancer in humans comes in large part from experiments on rats. Animal models of disease are one of the several types of system that can be used to construct analogical experimental systems. There are others, and their key feature is that by exploiting the right analogy we can see how the dynamics of one system are appropriately related to the dynamics of the other, so that finding out about one tells us something about the other. As Susan Sterrett (2017) makes abundantly clear, much of our experimental knowledge of geological processes and of fluid flow comes from various types of scale models, which are stereotypically analogical.

What I can do if I want to generate experimental knowledge of systems on which I cannot perform experiments is find systems that are *like* the systems I'm interested in, in just the right way, and perform experiments on them instead of on my inaccessible or fragile systems. The purpose of this discussion is to show that the experiments I perform in this way are fully realized experiments and not experiments only in some restricted sense. I will leave for later the question of how generic these experiments are.

Sterrett has also cleared up a number of confusions that arise in discussions of analogue models. She identifies three important misconceptions: that they are no longer used; that they can be dispensed with in favor of numerical simulations; and (from my perspective the most important) that they are only good for illustration or instruction. Instead, as she shows with vivid examples from physics and geophysics, analogue models provide robust *experimental* knowledge. She aptly characterizes analogue models as themselves physical systems being used for the purpose of modeling some other physical system. These are what I will take to be analogical experimental systems. "The modeling process for employing a physical object or setup as an analogue model includes the identification of a mapping that allows one to correlate something observed or measured in the analog model with something else (its correlative, such as a corresponding quantity) in the thing modeled. The modeling process also includes a justification of the mapping of some sort, usually invoking a principle or equation to establish the mapping" (2017, 858). I demur on her use of "justification," but I don't think it does any essential work in her analysis. Instead, we can see that what is required is awareness of the infomorphic relation between the analogue system and the system it is modeling. In order to get knowledge by means of this relation, we need be no more aware than is required to have our beliefs (or the things we accept) updated in the light of the information this relation conveys.

Sterrett quite rightly points out that there are all kinds of ways that something may be suited to function as an analogical experimental system. She mentions three in particular: there may be an appropriate equation in common that governs the behavior of both the proximal (analogue) and the distal system; there may be relevant features shared by the systems; or the systems may be physically similar. Something that should be emphasized, which Walter Warwick and I focused on a decade ago when we investigated the similarities between analogue models and simulations, is that the laws governing the behavior of the analogue system may have nothing obviously to do with the laws that govern the distal system it models. So, in a hydrodynamical model of electrical circuits, for example, it is not merely that flow of fluid is physically different from electrical conduction, even though both are (in some respects) governed by the same equations, but also that the fundamental equations of motion for the two systems are radically different and the systems are connected only by an equation that represents an artifact of one system and a feature of the other.[1] This is made clear in the case of Bose–Einstein condensates modeling pair production in the universe at large: the analogy comes at considerable remove from the system-level features of the condensates themselves.

Additionally, whichever of Sterrett's three conditions hold, the key feature will be in how the law-like behavior in one situation is a good proxy for law-like behavior in another. I would prefer, then, in the context of scientific experimentation, to say "ananomical" rather than "analogical." For here we are dealing with a special type of analogical relation. Systems that bear this relation to each other may, in any standard account, bear little resemblance: they do not share a logos, but they do instantiate a shared nomos. However, this species is very important in the context of scientific investigation, especially in coming to understand how experimentation serves to facilitate information flow between systems of interest and the minds of researchers (and hence to advance our general understanding of the world). The expression

---

1. I don't mean to suggest that Sterrett is unaware of this class of analogy.

is awkward, but it does convey an important truth about how we reason experimentally. While it is more common to think of certain experiments as analogical, the scope of analogy is generally unclear, and in cases where the inferences we make on the basis of analogy fail, it is not clear whether the failure results from the inadequacy of the analogy or merely reflects the fact that all analogies are limited in some respects. But ananomical relations are those that obtain when some law-like behaviors of one system are replicated in another system. The standard example here must be springs and pendula. When these systems are subjected to light force, they obey Hooke's law: $F = -kx$. An experiment on a pendulum is then as good as an experiment on a spring. Of course, in this case the experimental insight is rather slender. But the point is that the systems are experimentally equivalent in this domain, for this behavior, even though they are quite dissimilar in other obvious respects. For example, the pendulum operates through the action of gravitational force while the spring's force derives from the energy stored in the material of the spring itself. It is the law-like relation between proximal and distal systems that will bear the information necessary to generate experimental knowledge.

To bring the discussion into line with my previous account of experimentation, then, let us consider how to set up the appropriate infomorphisms. As always, the idea will be to classify the subsystems of some distributed system according to their local type–token relations and the local logics that govern those relations, and then to develop the appropriate infomorphisms that map the various subsystems onto each other. The trick with analogical experiments is that we now have to join together two different kinds of distributed system in a way that still allows information to flow through the new, larger system. We will still have the familiar task of isolating sufficiently the behaviors of interest in the two systems so that we can choose the most perspicuous classifications—classifications that display the similarity of the relevant dynamics. What are the relevant dynamics for the systems? As usual, they are precisely those that the experimenter is interested in learning about in the distal system by observation of the proximal system, but this time the distal system is at a further conceptual remove. Instances of this kind of thing abound. A typical case: I know the equations of motion governing some system, but I cannot solve these equations because they're intractable in some way. I also cannot observe all the behaviors in that system that I would need in order to fully understand its behavior. Often, however, I can find a system with the same equations of motion on which I can perform all of the experiments in which I have an interest. Once I know that the dynamics are the same, then it is obvious that whatever dynamical features I learn about the one system transfer immediately to knowledge of the other system.

In practice things are not usually so clear. And this lack of clarity is why analogical experiments seem generally less epistemologically secure than direct experiments. One way that things may be unclear is in how to isolate the appropriate dynamical behaviors in the two systems. One simple harmonic oscillator is the same as another, whether made out of springs, pendula, spinning disks, photons or whatever; so to perform an experiment on one simple harmonic oscillator is to perform it on them all. On the other hand, very complicated fluid dynamics systems are dynamically quite unlike even apparently very similar systems because of the crucial importance of scale, viscosity, etc. Consequently, finding systems that display the appropriate dynamics for mapping onto fluid dynamical systems is not easy. Another difficulty is that even when we have found such a system, the range of parameters of the proximal system over which it shares the dynamics of the distal system may be unclear or hard to isolate experimentally.

In the remainder of this chapter I will characterize analogical experiments according to the information theoretic account. In subsequent chapters I will turn to a consideration of the three cases mentioned above, using them to highlight some of the difficulties inherent in performing analogical experiments, and to show how these difficulties are overcome. These case studies will also point the way to a consideration of the connection between analogical experimental systems, thought experimental systems, and simulated experimental systems—and more generally of modeling as experimenting. Finally I will draw the general conclusion that analogical experiments are epistemologically as robust as their "direct" experiment cousins.

An interesting result will be that our intuitions about knowledge, bound up as they are with justificatory commitments, will not track very well the knowledge-generating capacities of various analogical experimental systems, while my model of experimental knowledge does much better. It is not easy to say why without going through the entire argument, but this should help make it seem plausible: much of what we do inferentially in general cases is to take advantage of the similarities we see between different things to learn about one on the basis of what we know about the other. Indeed, that's the vast majority of what we do. Fundamentally, justification has little to do with our actual epistemic practices, while information-conducting inferences from analogy are central to them.

22.3. Information and Analogical Systems

How are analogical systems to be folded into the account from part II? I think doing so is pretty straightforward. When I have access to some system for systematic observation, I have access as well to a number of different possible ways of classifying it. We saw this in my initial discussion of infomorphisms, where I characterized rats by turns according to their overt behavior, their viral loads, and so on. Generally, any property of a system can serve as the basis for a classification. The real question for the experimentalist is whether that classification can be made use of, possibly by being accessible via observation or via experimentation. For certain classifications of interest we may have only very limited knowledge of their type–token relation. Recall that a classification $\mathcal{A} = \langle A, \Sigma_{\mathcal{A}}, \vDash_{\mathcal{A}} \rangle$ is a collection of objects, a collection of types, and a relation between objects and their types, specifying which objects are of which type (have which properties). For analogical experiments, the choice of classification corresponds to a choice of which features of the proximal system and distal systems will be analogous (anamomous?) to each other.

Consider now as an easy warm-up the practice of making water-flow systems using such objects as pumps, pipes, flexible membranes, sand plugs, and water wheels. There is a parallel practice of constructing electrical flow systems out of wires, resistors, induction coils, and capacitors supporting electric currents. The classifications are easy to find and the infomorphism straightforward: starting with the concrete situation as pictured below, we see that when the pump is active, then if the wheel is in motion at its steady-state speed, the membrane is flat. But when the wheel is stopped the membrane expands.[2] We do not have, as part of the system pictured, any way of measuring the pressure at various places in the system, but values for the pressure exist and we can easily enough find out what they are—so they will form part of a fully informative classification. Likewise, fluid flow meters are not part of the system, but the flow at any point should be included in the classification.

At this point we can go a number of ways. Suppose we know already how hydrodynamical systems of this sort are nomically linked to electrical systems. That information suffices to give us knowledge of an infomorphism linking any such hydraulic arrangement to an analogous electrical arrangement. The arrangement is given by the infomorphism generated by the analogy itself: the one that maps pressure to voltage, fluid flow to current, resistors to sand plugs (resistance to product of sand density and length of pipe occupied), water wheels to inductors (mass of water wheel to inductance), and membranes to capacitors (membrane extension to charge on capacitor). With appropriate specification of units, we can translate configuration to configuration. Thus if what we are unsure about is how the hydraulic system behaves,

2. Harder (n.d.) offers a nice illustration of many aspects of the hydraulic analogy to electric circuits.

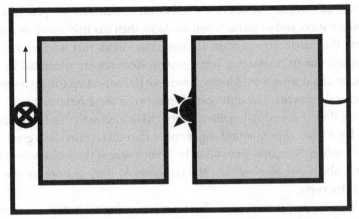

FIGURE 22.1 A hydrological analogy for an LC circuit (induction plus capacitance)

we can build the infomorphic electrical system and observe it in a state analogous to the one we are curious about.

We can easily see how to generate a classification of that diagram and construct the appropriate infomorphism; but we can also see how to extend it to a more general classification that includes possible elements not pictured,

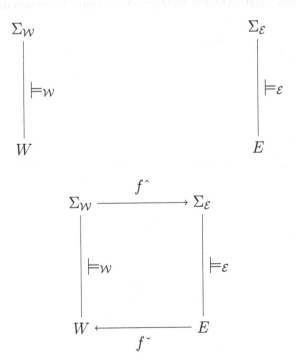

as well as ways to characterize whether those elements are part of a given configuration, and so forth. When we do so, then any inferences we can draw using the classification system for the water circuit that involve changes in pressure due to connecting water system elements are generated by information about what would happen were we to connect up the corresponding electrical elements. The infomorphism between these systems doesn't tell us much, if we're already competent in both basic electronics and hydrodynamics, but for someone ignorant about one or the other it provides a great deal of information. Someone ignorant of both but aware of the analogy could learn much about one by observing the responses to interventions on the other, and vice versa.

This is a pretty simple case. But for other analogical systems, even though only part of the one system may correspond analogically to the other system, and perhaps even only a part of it, we can still infomorphically identify those parts by means of the knowledge we have of the analogy, and derive experimental results about the one from observations of the other. Analogical experiments produce knowledge in the same way that so-called direct experiments do. Beginning with the classification of the proximal system, the experimental technique provides the observations, the analogy helps us pick out the appropriate infomorphism, and *that* allows experimental information about the distal system to flow through observational features of the proximal system.

23

Economic Analogues

Laboratory experiments in economics are strongly analogical in character. Stereotypically, the approach is to gather data about the behavior, preferences, and disposition of undergraduate students in economics classrooms in order to perform experiments on general markets. These general markets involve many layers of economic agency, from individual humans to families to microbusinesses to small, medium, large and multinational businesses, along with government agencies acting in and *on* the market with subsequent uncertainty about whether, what, and with what result that latter type of action will be. This is all folded up in a system with a vast amount of inertia, friction, outside forces, inherent sources of misinformation, and impediments to the flow of actual information.

It is not entirely clear why economics experiments are performed the way they are. In part it must be that there is general suspicion of any noninterventionist technique in experiment. There is also, as Plott and Smith (2008, ix–x) pointed out, a lack of natural experiments in economics that leaves us with

important questions that can only be answered by experiments. In the field it is difficult to study situations that have not occurred or institutions that do not exist because there is no natural experiment. For example, in the laboratory it is just as easy to study the effects of auction market rules that have never been observed in the economy as to study those that have. When left on her own, nature may never create a situation that clearly separates the predictions of competing models or may never create a situation that allows a clear view of the underlying principles at work. Indeed, much of the progress of experimental methods involves the posing of new questions or the posing of old questions in a way that experimental methods can be applied.

In economics we have a clear example where direct experimental knowl-edge of the systems of interest is practically impossible. Certainly we have a great deal of observational data about various economic systems. But we do not have any detailed picture of which antecedent activities leading up to those observations should be seen as causes and which parts of the observations should be seen as effects of some other, prior economic activity. It seems we can't really separate out the economic parts of these systems to study them directly. This, it is fair to say, is a decisive advantage of physics over econom-ics—in physics we have very good techniques for separating out the various causal components.

I will not undertake a detailed consideration of why and how that is. Clearly, subject matter differences are at the root of this somehow, but saying exactly how will have to wait for some other time. But we can note that the systems of interest are so complicated and so extensive that there is no real hope of isolating the effects of any particular intervention into the systems. The economy of the United States, for example, is made up of the economic activities of hundreds of millions of US residents; the economic activities of everyone else around the world as they buy things from and sell things to these residents; the many large and small corporations doing business in the US and around the world and the way they impact the flow of goods and money in the US; the actions of the various governmental agencies in the US; etc. To ask what might be the effect of raising the interest rate by some amount seems hopeless. Certainly one cannot do repeated controlled experiments on an en-tity as vast and complicated and unwieldy as the US economy, which is, in important respects, a new system each time new interventions are made on it. And more generally, it is difficult to isolate the influence of changes to real eco-nomic behavior without significantly altering the systems under investigation.

What we want in economics is to understand how changing conditions change the behaviors of economic agents. But to experiment on economic agents under controlled conditions is to experiment on them in situations where their normal motivational structures are disrupted—and those mo-tivational structures are key to understanding what they will do when con-fronted with the kinds of changing conditions we are hoping to model in the lab. What hope is there in all of this dis-analogy of finding any analogue of the one system in the other that will support the kind of information flow that is the hallmark of experimental knowledge? Given these complications and limitations, it is remarkable that we have any experimental knowledge at all in economics. But we seem to. So how does it work?

The entire structure of economic experimentation is fraught with real dif-ficulty, but economists are attempting to mitigate the concerns raised above.

In the next two sections I characterize the class of laboratory experiments in economics as analogical experiments and show how experimental claims in economics are structured. I follow with an overview of one important effect that has, apparently, been established experimentally and its significance for understanding what experiments can tell us in economics.

23.1. The Economics Laboratory

There is now a rich body of literature on how to perform economics experiments in the "laboratory," as it is called. This terminology makes a lot of sense in the light of the realization that even standard sorts of laboratory experiments are themselves fundamentally analogical. In the same way that physicists attempt to bring a segment of nature under their experimental control by building a smaller or simpler or in some way restricted version of that segment in the laboratory, economists are engaged in a process of world building in economics classrooms, which are their own laboratories. But just as physicists are unable to make versions of all of nature to study in their labs, the typical distal system in the case of economics is significantly restricted in scope and complexity. I have neither the space nor the expertise to attempt to characterize fully the status of current laboratory research in economics. But that is not my purpose. Instead I will focus on a single account of economics laboratory experimentation and then consider how to understand the experimental knowledge generated by such experimentation with respect to a specific experimental conclusion: humans are subject to the endowment (or entitlement) effect. The account of experimentation I consider is principally the result of work by Hurwicz, Smith, and Wilde; it is built around the notion of a "microeconomic environment."

Wilde (1981) clarifies and slightly modifies some precepts laid out by Smith in characterizing the nature of laboratory experiments for economics. "The basic idea behind a laboratory experiment in economics is to create a small-scale microeconomic environment in the laboratory where adequate control can be maintained and accurate measurement of relevant variables guaranteed" (1981, 138).

This small-scale microeconomic environment (MEE) requires two things: an *economic environment* and an *institutional setting*. These are defined as follows:

> The *economic environment* is composed of a list of agents $\{1, \ldots, n\}$ and commodities $\{1, \ldots, \ell\}$. The agents are described by preference relations α_i (generally represented by some utility function u_i), a technology T^i (represented by the

set of possibilities the agent has for turning initial endowments into subsequent goods), and an endowment vector ω^i. So the ith agent is described by $e^i = (\alpha_i, T^i, \omega^i)$. The commodity space is $\mathbb{R}^\ell$. Since there is a fixed list of commodities, the environment, e, is just given by $e = (e^1, \dots, e^n)$.

For the *institutional setting*, we assume that agent i can make a decision, $d_i \in D_i$ in order to maximize u_i. The institution is then given by the decision vector $D = D_1 \times \cdots \times D_n$ and a mapping $I:D \to \mathbb{R}^{\ell n}$ that takes decisions to final allocations.

The total MEE then is described by $E = (e, I)$. Wilde identifies the following features as essential:

1. It is composed of agents and an institutional setting.
2. Individuals are characterized by two assumptions: (a) consistent preferences; (b) decisions that maximize well-being.
3. Decisions act through the institutional setting in order to determine final outcomes.

The purpose of the MEE "is to uncover systematic relationships between individual preferences, institutional parameters, and outcomes" (Wilde 1981, 139). At least, he says, this is Plott's view. As I see it, we will have direct observations of some given MEE. And then we will wish to assert that our observations of this MEE give us experimental knowledge first of *it*, as is the case for any other laboratory system we experiment with, and then of some other, distal, real-world economic system. How that happens is the big question for the model. "To realize this purpose," says Wilde, "the experimenter must have control over both the preferences of the individuals participating in the experiment and the institutional parameters which govern final allocations" (1981, 139).

These small-scale environments are intended to meet a fundamental requirement on economic experiments: that they are themselves genuine economics systems. As Smith, whose requirement this is, puts it, these systems are "richer, behaviorally, than the systems parameterized in our theories" (1982, 924), and thus can give us at least some informational advantage over them. But can they give us experimental knowledge of other economic systems? Wilde takes it that the task of the theorist of such systems is to generate a list of sufficient conditions to guarantee the existence of a genuine MEE and the necessary control to yield genuine experimental knowledge of other systems. To that end, a number of precepts must be satisfied. Smith identifies three: *nonsatiation, complexity,* and *parallelism* (1980, 346, 348, 349):

Nonsatiation: Given a costless choice between two alternatives which differ only in that the first yields more of the reward medium (e.g., currency) than the second, the first will always be chosen (preferred) over the second by an autonomous individual, i.e., utility U(M) is a monotone increasing function of the reward medium.

Complexity: In general individual decision makers must be assumed to have multidimensional values which attach nonmonetary subjective cost or value to (1) the process of making and executing individual or group decisions, (2) the end result of such decisions, and (3) the reward (and perhaps behavior) of other individuals involved in the decision process.

Parallelism: Propositions about the behavior of individuals and of markets and other resource allocation mechanisms that have been tested in laboratory environments apply also to nonlaboratory environments where similar *ceteris paribus* conditions prevail.

As Wilde notes, however, these precepts do not suffice as stated, because for the system to count as an MEE, it is necessary "that individual decisions act through an institutional setting in order to determine final allocations which, in turn, determine rewards" (1981, 140). Smith's precepts do not guarantee this, so Wilde adds the requirement of saliency. Together with nonsatiation, it guarantees that the system is a MEE (1981, 140):

Saliency: The reward earned by an individual is tied to decisions made by that individual.

Once we have all of this in place, what can we do with it? How much experimental knowledge of economic systems can be generated with such a model of experimentation? As things stand here, it might seem like very little. The clearest legitimate uses of such MEEs seem to be the head-to-head testing of competing theories, and the degenerate case of that: determining whether some theory has zero hope of being correct. The idea is that if one of two theories does much better at predicting what happens in the laboratory system, then in the much more complicated nonlaboratory economic situations it should also do better. And if a theory cannot even get right the restricted laboratory situation, there is no chance that it will do better in nonlaboratory economic situations.

In order to understand how MEEs give experimental knowledge of real-world economic systems, rather than about the comparison of two theories,[1] more will have to be said about what Smith calls parallelism.

---

1. Yes, this comparison of theories can give some knowledge of the distal systems to which they are supposed to apply. But not much.

23.2. Parallelism

What Smith (1982), there following Shapley (1958), calls parallelism is essentially the assumption that MEEs are properly infomorphic to at least some nonlaboratory economic systems. As stated, that given the right *ceteris paribus* conditions the experimental propositions derived from observation of MEEs apply to nonlaboratory systems, parallelism is a trivial claim. It tells us only that, given the right classification, an infomorphism will exist between any two systems. But so what? This is no more illuminating and no more probative than the old antifunctionalist canard that any sufficiently complex physical system can, under the right description, be interpreted as executing any function one cares to name. Of course *that* is true, but it is very different from specifying in advance the interpretation function and then declaring correctly what function is being executed by it. And in the experimental knowledge case, what we really want to know is, given a classification of a distal system, whether there is a proximal system and an appropriate classification *of it* such that observations of the latter bear information about the former under that given classification.

So merely noting that experiments in economics will have external validity does not suffice to show that these experiments give any clear information about the behavior of any particular external (distal) system. What is needed is a more detailed expression of the way the logic of the lab system maps onto the logic of some real-world system. In economics, those conditions will be in place only when, for example, we compare the laboratory microeconomic system to some narrowly isolated portion of the total economy. A favorite example is that of government-controlled auctions—of treasury bonds, for example. What we have in that case is a clear display of system equivalence, for a sharply curtailed classification of the distal system.

The equivalence between systems in economics is not to be confused with material equivalence between systems. But it does nicely show how material equivalence can be misleading. Students in the classroom (a standard economics laboratory) and traders on Wall Street are clearly made of the same stuff. Their bids are made using the same media, and their rewards in monetary units are the same kind of stuff as well. Yet this does not suffice for establishing their equivalence. What is needed is to show that they are in an environment with an institution that shares enough structure that the dynamics of the one are informative about the dynamics of the other. Only then can the systems be said to be equivalent. But here "environment" embeds information about all the other agents in the system, so it is very unlikely that we will have strict environmental equivalence. And the fact that for small numbers of agents in the lab, smaller changes in those numbers can have large effects on such issues as whether the market clears, whether there is stable pricing, whether the market is efficient, etc., shows that shared materiality simply does not suffice, even barring absurd notions of shared materiality like "all natural systems are made up of quarks and leptons" and focusing on the material identity of units at the relevant scale. Of course, an even more restrictive notion of material equivalence would rule out that equivalence because of the divergence in the number of agents, their endowments, and so forth. That's not a very useful notion, since it would designate essentially every system in the world as not materially equivalent to all the others.

Here's one way to see what is going on here. In many cases of interest to theorists, physical systems may be very similar, and their behaviors may be very similar from one perspective but radically different from another, and those differences tell us what classifications are appropriate. For example, if a doctor tells a patient to kick out her leg, and she does, then the leg straightens and the foot rises. Similarly, if the doctor hits her patellar tendon with a rubber hammer, the leg straightens and the foot rises. In both cases the physics of leg raising, grounded in musculature and bone mechanics, all of which is ultimately grounded in electrodynamics (as is essentially all of the physics of everyday life), is the same. And yet the causal story appropriate to the case is very different. The proper way to characterize these causes, to classify the system of interest, is by reference to what we are trying to understand when we ask why the leg moved.

An instructive example is Thaler's (1980) treatment of the rationality of billiard players of differing skill levels. In any given configuration of the table, many different shots are possible. Of those possible shots, there is likely to be one that is the best to make for improving the score and leaving the table in a useful state for the next shot. An unreflective appraisal might indicate that the most rational shot to take is that one. But despite the absence of any overt material differences between an expert and a novice billiard player (one can imagine identical twins, one of whom has studied billiards and the other of whom has studied darts), they are very different systems and are not at all equivalent from the point of view of assessing the most rational course of action. Thaler points out that the best shot may be impossible for the novice, and the novice may well not even be able to see it as a possibility, while for the expert it may be a very routine shot with a high probability of success. The point for the experimentalist is that here we have nearly physically identical systems, and yet as billiard-playing systems they may require radically different classifications, leading to radically different possibilities for the experimenter (if, say, we wished to perform training experiments on them).

In any case, in the example of bond auctions mentioned above, we are engaged with a kind of MEE already, because the agents, their endowments and technologies, and the resources making up the environment are fixed. So there is little difficulty in showing that parallelism is satisfied between that and some given laboratory MEE, even without getting bogged down in questions concerning system equivalence. Of course, the ease of finding the analogy seems to vary inversely with its informativeness.

I am not suggesting that there is any confusion about this matter on the part of those (such as Smith and Wilde) who advocate for an important role for laboratory experiments in economics. Rather, the point is that the applicability of these experiments remains restricted. Even so, it should be emphasized that they do not differ conceptually from any other experiments in this respect. The important difference in the case of economics experimentation (as in social science experimentation generally) is that the scale we are most interested in learning about experimentally is one where is it very difficult to classify distal systems in a way that meets the *ceteris paribus* conditions required for experimental knowledge.

This may appear to be a rather negative assessment. But it should be seen more as an acknowledgment that, in economics, if we are looking for very general results we may find them of limited predictive power; conversely, powerfully predictive results are likely to be of very limited applicability. That is just a feature of the very complicated dynamics of economic systems that comprise many different forces. These different forces happen to all be relevant at the levels of description we find interesting. Parallelism, the assertion that there is a useful infomorphism between the classification applied to the proximal laboratory MEE and a distal system of interest, is then the crucial informational bottleneck for economics experiments. To gain access to such an infomorphism requires a careful isolation of a distal subsystem.

23.3. The Endowment Effect

I will now illustrate some of this by looking at one class of experimental result and its interpretation and significance for our understanding of economics generally. What Thaler called the "endowment effect" is the increased preference people have for things they already possess over that same thing prior to possessing it. Typically, if this effect really is present, people will be willing to spend less to acquire an item of value (a pair of sunglasses, say) than they would charge to give it up if they already possessed it. The body of experimental data seems to begin with some hypothetical tests of willingness to pay for versus willingness to accept payment in lieu of a given good—in this case, access to wetlands for hunting, where mean willingness to pay for access was $247 versus a mean willingness to accept loss of access of $1044.[2] Thaler (1980) gathers up some data and produces some of his own to further support the existence of this phenomenon. He asks the following questions (1980, 43–44):

a. Assume you have been exposed to a disease which if contracted leads to a quick and painless death within a week. The probability you have the disease is 0.001. What is the maximum you would be willing to pay for a cure?

b. Suppose volunteers were needed for research on the above disease. All that would be required is that you expose yourself to a 0.001 chance of contracting the disease. What is the minimum payment you would require to volunteer for this program? (You would not be allowed to purchase the cure.)

He reports, "Many people respond to questions (a) and (b) with answers which differ by an order of magnitude or more! (A typical response is $200 and $10,000)" (1980, 44).

All of this seems to contradict a strongly held principle of economics, the Coase "theorem," which asserts that the final allocation of resources will be independent of the initial distribution of rights if trades are without cost. However, these survey methods are subject to a great deal of ambiguity, so Kahneman, Knetsch, and Thaler tested the effect in the lab. Experiments of this sort can be performed with quite simple MEEs. As I understand it, Kahneman et al. constructed partially new MEEs for each experiment. That is, they begin each experiment with a list of agents (the students in a particular course, in attendance on the day of the experiment), and then select a list of commodities for each new experiment. The experiments generally comprise several repeated markets in a variety of commodities, iterated over multiple MEEs involving the same agents. For example, in their experiment 1 (1991) there were three different markets: an induced value market, a market in coffee mugs, and a market in ballpoint pens. Each MEE had the same list of agents: forty-four undergraduate students in an advanced law and economics class at Cornell University.

In the first markets some students (called sellers) were given tokens and others (called buyers) were not. Different students were given different valuations of the tokens involved: those valuations were what the experimenter would be willing to pay that student for the token. Then, for a series of prices, sellers were asked whether they would prefer to sell their token for that price or trade it in for the set valuation, and buyers were asked whether they would

2. Hammack and Brown (1974), also cited in Kahneman, Knetsch, and Thaler (1991, 2008).

prefer to purchase a token at that price and cash it in for the set valuation or not purchase a token. Once these questions were answered, the market clearing price was calculated and three buyers and three sellers were selected at random to be paid off according to their stated preferences. Three such markets were held, with different buyers and sellers and different valuation set points. The MEEs for these three markets had the same commodities, but different assignments of endowments and different preference functions controlled by the valuation set point.

The second series of four markets used Cornell coffee mugs ($6) as commodities, and the third series of four markets used ballpoint pens ($3.98, with visible labels). In each of these series some students were made buyers and some sellers, and those assignments were maintained for that series. There were two explicit differences between the goods markets and the induced value markets: (1) only one of the markets in each series would have its trades executed, and that was chosen at random from the four markets after all had been conducted; and (2) in that market, every trade was executed instead, of just a sample as was true for the induced value markets. The trading instructions given were similar in the commodity and induced value markets. In the commodity markets, though, value was not induced, so the experimenter did not have control over it. Instead, each agent had his or her own valuation of either mug or pen. So while we have MEEs for these markets, it is the valuation that we are seeking to determine. The question is whether supply and demand curves will be mirror images of each other, as is assumed on the basis of Coase's and others' arguments.

What Kahneman et al. find is that in the induced value markets, the expectations of Coase's "theorem" are met: they expect about half the valuations to be above market price and half below, and that's what they find. There are twelve, eleven, and ten trades, respectively, in the three markets, and the market clearing price is what they expect in the light of their valuation assignments. But in the mug and pen markets, trades are well below half of what is expected, and selling reserve price is much greater than the buying reserve price. The measurement of undertrading is calculated by dividing the actual number of trades (V) by the expected number of trades (V^*). In the first series, the average of V/V^* is 1, while in the mug market it is 0.2 and in the pen market it is 0.41. This is a pretty strong indication that students who understand the basic principles of markets and who trade in accordance with standard economics expectations for tokens—items that have value only in terms of what they can be traded for—do not trade in accordance with those expectations for items having a value that is not induced but arises out of their appraisal under the agent's own valuational system. Moreover, the way in which these students trade against economics expectation is to assign higher value to things once they own them, and lower value to things before they own them.

I think results of this sort are interesting in their own right, showing at least that more psychological features of agents than merely standard rational choice considerations are in play when those agents enter the market. That agents on average are susceptible to endowment effects is, however, not accepted by all on the basis of these experiments. Kuval and Smith (2008) discuss some other experiments showing that in certain circumstances, divergences between willingness to pay and willingness to accept can be reduced if close substitutes exist for the commodities in play. Even so, significant undertrading is observed in many such cases. So there are at least two things to say about these alternative experiments. First, Kahneman, Knetsch, and Thaler are clearly aware that the lack of availability of close substitutes is part of what activates the endowment effect (1991, 1344), but that does nothing to undermine our knowledge of its existence. Second, even were the endowment effect as such not the right explanation for what is going on here, it remains clear that something has gone wrong with the hypothesis that $V/V^* = 1$. Thus some story about the impact of human psychology must be told in order to make sense of these observations.

But whatever precise assessment is made of the psychological operation at play in these experiments, the dispute is over which psychological features we think are relevant to the decisions of individual agents themselves. In addition to the question about whether individuals on average are susceptible to endowment effects, we can ask another, more far-reaching question: what experimental knowledge of markets as a whole do these observations of individual agents give us? In a subsequent publication, Kahneman, Knetsch, and Thaler (2008) suggest that the potential significance of the effect is profound, affecting our entire understanding of how economic systems function. That leads them to ask whether there are more than merely aggregative effects of this feature of human agents on the marketplace that must be taken into account. One interesting aggregate feature they focus on is the total volume of trade that can be expected in markets. If it turns out that endowment effects are generic features of many market situations, and (as the authors suggested; Kahneman, Knetsch, and Thaler 1991, 1344) there are many cases in which close substitutes for commodities are lacking, then one should expect that total trading volume will be less than otherwise predicted. One consequence would be that gains from trade are less than would otherwise be the case (Kahneman, Knetsch, and Thaler 1991, 946–47).

Coase's (1960) arguments are all directed toward the hypothesis that it does not matter how one begins a trading situation, at least as far as final price and volume are concerned. And these arguments are predicated on the idea that "the question is commonly thought of as one in which A inflicts harm on B and what has to be decided is: how should we restrain A? But this is wrong. We are dealing with a problem of a reciprocal nature. To avoid the harm to B would inflict harm on A. The real question that has to be decided is: should A be allowed to harm B or should B be allowed to harm A?" (1960, 2). Not only does Coase view this matter as reciprocal; he assumes in example after example that cost to discontinue a practice or to remove an obstruction or to abandon a field is precisely the same as the value to start the practice or to raise the obstruction or to cultivate the field.

The Coase "theorem" has been used as a reason to think that the entire system of compensation for victims of harm can be structured around the idea that valuation curves of willingness to pay are mirror images of valuation curves of willingness to accept. Whether or not they *ought* to be, they are not for the kind of agent that we are, and these experiments and others like them, though merely analogical, give us the knowledge that we should restructure that compensation system.

23.4. Conclusion

This chapter has examined briefly the nature of laboratory experiments in economics, illustrating their analogical character and hinting that in general social science experiments are of this same analogical sort. It has also, I hope, made clear that while analogical experiments in economics, as described, have their own characteristic informational bottlenecks, these bottlenecks are such as to make a pragmatic but not conceptual distinction between them and any other sort of experiment. When appropriate classifications are in place, and when the right mapping between these classifications is found, observations of economic laboratories give experimental knowledge of worldly economic systems.

24

Cosmological Analogues

I discussed in the context of natural experiments the possibility that there might be cosmological and astronomical RCTs. These would allow us to draw conclusions about the universe at large on the basis of sufficiently randomly partitioned classes of astronomical events. I don't think the prospects for such cases are particularly good in the foreseeable future. Another way to do experimental cosmology is to begin building pocket universes, for example by mimicking the conditions at the big bang inside the Large Hadron Collider. The problem with this approach is that the cosmoi so constructed will be causally inaccessible, and so not particularly fruitful as experimental apparatuses. There is also the worry that we might destroy ourselves by constructing a world-consuming black hole during one of those experiments. Neither of these suggestions then is particularly fruitful.

On the other hand, there are things that can be done closer to home that will also give experimental knowledge, if they can be set up properly. There are many possibilities, but the cases I will be most interested in are the gravitational analogues. These are a species of analogical experimentation where the analogy is between systems that are surprisingly different in almost every respect. These analogue models are almost as old as general relativity itself, and they seem to have begun with Gordon (1923). But Gordon, as I understand him, was using the analogy in the other direction: to use what we know about gravity to learn something experimentally about solid state physics, rather than the other way around. Today, the field of analogue gravity is almost entirely directed toward learning experimentally about gravitation by means of earthbound systems whose salient observable behavior is governed by a dynamics very different from gravity itself. These models have been drawing a lot of attention recently, in part because we now have technologies that will allow the proximal analogues to be built and observed in ways that were not possible even a few years ago.

These analogical experiments are an especially clear illustration of how what really matters to the production of experimental knowledge is not the intuitive likeness of the analogically connected systems, but rather the strength of the law-like connection between the behaviors of the tokens observed in the experiment, whatever they may be, and those about which we are seeking information, whatever *they* may be.

Analogue gravity is an enormous and growing class of techniques both for attempting to discover a quantum theory of gravity and for investigating what are expected to be generic features of any such theory. But the topic is too large to address fully in this space.[1] So I will give a brief overview of a few kinds of analogical experiments that have been proposed for cosmology, and how they would give experimental knowledge were they to be realized in the lab. First is an analysis of an old chestnut from Bill Unruh: attempting to realize the properties of black holes in the terrestrial flow of liquid down a drain. I follow this with a very brief account of a new, exciting possibility for realizing a quantum gravitational system. Finally, I give a slightly more detailed account of one of the most extraordinary experiments I know about: the use of Bose–Einstein condensates to study the features of the expanding universe.

The radical character of this chapter's claim should not be overlooked. I am arguing not that we should apply the name "experiment" to certain of the practices in cosmology; rather, I am claiming that one kind of knowledge-generating activity underway now in cosmology goes beyond the practice of observational theory checking. For example, the gravitational lensing observations, observations of the red shift of receding galaxies, and the observation of the precise precession in the perihelion of Mercury are all used to test the predictions of general relativity. This kind of test in cosmology is useful, important, and powerful. It is not, however, experimental. On the other hand, analogical models of gravitation theory allow for genuine experimental cosmological knowledge.

24.1. Dumb Holes

In an early contribution to the literature on analogues for gravitating systems, Bill Unruh (1995) showed that so-called dumb holes were usefully analogous in important respects to black holes. A dumb hole is an analogue to a black hole, but instead of its gravitational field being so strong that it prevents any

1. See the useful overview in Liberati (2017) covering analogical gravity generally.

normal matter from escaping, including light rays, as is the case with black holes, the dumb hole is a flow of fluid where the speed of flow of the fluid increases in the direction of the flow. The consequence of this increasing speed is that, at some point along the flow, the speed of the fluid exceeds the speed of sound in the fluid. This point then corresponds to the event horizon of a black hole, the point beyond which no signals can emerge from the direction of the hole.

An interesting and by now well-known feature of black holes is that they radiate: even though nothing can travel out from beyond the event horizon, still the semiclassical theory of gravity seems to predict that they will continually radiate, losing mass and energy, until eventually they disappear. Semiclassical gravity is a pretty good phenomenological theory of the universe at large, but all physicists seem to think that it's completely hopeless as a fundamental description of nature. One main motivation for Unruh's analysis is that, because of the presumed nonfundamental character of semiclassical gravitation theory, it is unclear whether the calculations of how black holes radiate that had been performed as of 1995, using that theory, were trustworthy. As Unruh says, the calculations are performed using the theory in a domain where it is pretty clear that it breaks down as an adequate model of gravity. So if we are to trust the results of these calculations, we will have to find some independent method of checking them.

Unruh puts it like this: "if one could show that for dumb holes the existence of the changes in the theory at short wavelengths did not destroy the existence of thermal radiation from a dumb hole, one would have far more faith that whatever changes in the theory quantum gravity created, whatever nonlinearities quantum gravity introduced into the theory, the prediction of the thermal radiation from black holes was robust" (2008, 2908). His idea here is that the molecular nature of fluids is well known, so we can use that to discover what happens with the dumb hole. And then the strong claim is that the way we expect things to change when we go from a continuum to an atomic model in the case of the presumed quantum theory of gravity can be derived from the way we actually find things to change in the known case of fluid dynamics.

But what supports this argument? How robust is it? Can it really be so simple as to see that the breakdown of the one theory is similar in form to the breakdown of the other, and then to use the one as a proxy for the other? This is what he is exploring—the "possibility of using [dumb holes] to understand where and how the particles in the black hole evaporation process are created" (2008, 2909). That is, Unruh supposes that the process in one analogue can be used to gain insight into the analogous process in the other. Again we must ask how plausible this suggestion is. Is it enough that there be a shared dynamics in the theories outside of their failure regime? Do we also require a shared *substrate* for that dynamics? Or is the idea perhaps that dynamics is all there really is to such processes?

Unruh, however, has rather more modest goals for the analogy than fully understanding the exact character of black hole radiation. His view seems to be that we can come at some important features of black holes obliquely. We don't need an exact nomological connection between the two systems—the dumb hole and the black hole—to ask certain classes of experimental questions. For example, we can ask whether it is *necessary*, in order that a system like a black hole may radiate, for there to be incredibly high frequencies of radiated energy. The problem is this: When we do calculations for black hole radiation using the naive semiclassical theory, our calculations make apparently essential use of all ranges of radiation frequencies, and we know that the very high frequencies are not well modeled by the semiclassical theory. So if those high frequency modes make an essential contribution to the radiation profile given by the model, we may be suspicious of predictions of black hole radiation.

Fluid models are the same in that respect. The fluids we have are not continuous, but the theory of fluid dynamics is a model built for continuous fluids. So when we consider what is going on at the scale where our fluids are no longer continuous, we won't find waves at the frequency corresponding to the lengths between atoms—those distances are too small to be generated by the fluid character of that atomic medium. Yet the radiation we get for dumb holes *is* independent of the details of what happens at these scales. Here is the important insight: for a wide range of dispersion relations in various fluids, the same behavior for radiation is seen. A combination of numerical calculation and analytic equation solving shows that the Hawking radiation for dumb holes and the expected horizon "temperature" are independent of what goes on in the physics at very short length scales. A number of other such results regarding the point at which "particles" are created outside the dumb hole makes it clear that for the phenomenology outside the black hole, the dumb hole is appropriately infomorphic (2008, 2909–10).

Direct observation of what is happening with black holes in the universe at large is very difficult because of the incredibly tiny energies radiated by even the most energetic of gravitating systems. This difficulty is also present in the case of their fluid analogues, but the prospects there are much better. In any case, given the infomorphic relation between fluid flow models of dumb hole radiation and black hole radiation, we can expect the former to give us experimental knowledge of the latter.

24.2. Gauge/Gravity Duality

As with any other analogical experiment, cosmological experiments begin with a proximal system that is classified in an appropriate way. What does "appropriate" mean in this context? It means that the classification makes clear to a user of the system a type–token structure whose logic (dynamics) can be fruitfully mapped onto that of a distal system of interest. The trick as always will be to find a way of describing the proximal system, or an isolable part of that system, so that it becomes the range of an infomorphic function between that system and the distal system, which may itself be an isolated part of some larger system.

For cosmology our distal systems can range from the entire wave function of the universe itself, to the process of galaxy formation in the early universe, to fundamental space-time processes as they manifest for example in features of the expansion of the universe or in the process by which black holes radiate. An analogical experiment in cosmology relevant to the structure of the universe as a whole is a possible analogical model where very cold atoms realize a so-called Sachdev-Ye-Kitaev model. This model could, if implemented physically, provide an analogical model of a quantum gravitational system on which we can do cosmological experiments (Danshita et al. 2017). That is, we could do those experiments if it turns out that a speculative claim from string theory, the gauge/gravity duality principle, is true. This conjectured principle tells us, among other things, that for any quantum gravitational theory on a space-time region of $n+1$ dimensions, there is an equivalent quantum theory without gravity, but one which is determined by only the states of a many-body system on an n-dimensional boundary of that region. This principle is the essential feature of the universe that would allow for an infomorphism between itself and a small number of particles. That infomorphism would be a combination of two other infomorphisms: the one that maps the many-body system onto an equivalent quantum gravitational system, and the one that maps a generic quantum gravitational system onto the particular one that is our universe.

We could not find out much about the *specific* configuration of our universe as a whole from such an experiment. These experiments are much more general, telling us features about the law-like behavior and causal structure of the universe in which we find ourselves. The problem, of course, is that we only have one universe to look at, and we don't have the capacity to intervene on it in a way that would make it reveal counterfactual features about the whole thing. Nor can we expect the right observational features that would suffice for a natural experiment to simply appear. Analogical cosmological experiments allow us first to gain experimental knowledge of many instances of specific cosmoi from which we can infer features that are common to all those that are globally like our own. At least that is the promise. We don't know that the gauge/gravity duality is correct, but if it is and we can build and manipulate Sachdev-Ye-Kitaev models, then we can do experimental cosmology with them.

While physical realizations of the Sachdev-Ye-Kitaev models cannot be realized at the moment, the above discussion should make clear how the general idea of analogical cosmological experimentation should go. We have a physical system that we know a lot about theoretically, even if we cannot calculate its properties in any great detail. We do, however, know enough to characterize its dynamics in a variety of ways, and in particular we can produce a map from its dynamics to a classification scheme that is appropriate to the dynamics of some large-scale cosmological system. The output of that mapping is a kind of system about which we can have experimental knowledge. If a further feature holds—that the output system is *itself* of the sort that the universe is—then we can read off features of the universe from the observational features of that first system. In other words, we will have experimental knowledge of the universe at large.

24.3. Bose–Einstein Condensates[2]

In addition to these speculative experimental arrangements, cosmological experiments are under way on microscopic volumes of gas—Bose–Einstein condensates (BECs). The idea is to collect 10^8 rubidium atoms, supercool them, coalesce about 10^4 of them into a condensed quantum state, and start scattering sound waves through them. The obvious problem with this approach, as with all analogical experiments, is how different the distal system is from the proximal. It is difficult to imagine two systems that have less to do with each other than the entire cosmos, which is filled with structures on every length scale and every kind of matter, and a microscopic sample of gas in a single coherent quantum state. But these supercooled BECs are being used now to perform cosmological experiments—at least they are being constructed and observed, and soon they will give us experimental knowledge of the cosmos.

The universe at large appears to be well modeled by general relativity, but there is reason to doubt that that is the final story. There is now near consensus that something funny happens near the Planck length that deviates from

2. Much of this section has been drawn from joint work done with Walter Warwick.

the basic story given by general relativity, but there is little consensus about what that something is, except that probably the continuum model of space-time breaks down there and is replaced by something else. Nor is there much evidence to appeal to in forming a view. There is, however, good reason to think that the phenomenology of the universe at many observable scales is the phenomenology of a classical general-relativistic space-time coupled to c-number expectation values of quantum fields—i.e., semiclassical general relativity. Here then is something we wish we knew about the cosmos: what basic (or more basic) physics underlies this observationally good, but theoretically unlikely model? Physicists over the last several decades have made some progress in illuminating the issues, but without guidance and feedback from experiment there appears to be an impasse.

Two generic features of space-times roughly like ours with its quantum matter do present themselves as good sources of information, if only we could understand them better experimentally: the final consequences of Hawking radiation from black holes, and cosmological particle pair production caused by the expansion of the universe. These features of the universe are strongly influenced by the physics at or beyond the length scale where both gravity and quantum effects are very strong. They are known as trans-Planckian modes because that scale is the Planck scale, at about 10^{-33} centimeters. Trans-Planckian modes are promising as sources of information about the breakdown of our theoretical account of the universe at large because they are the result of transitions between the physics generating the phenomenology of the semiclassical theory and the physics on (so to speak) the other side of the Planck-length barrier. The difficulty with the trans-Planckian modes is to observe them at all, much less to observe them in the kind of detail that would give insight into their fundamental characteristics. And certainly experimental access to these modes by astronomical observations seems impossible to obtain, because we have no prospects at all for intervening either in the expansion of the universe or in the evaporation of black holes.

But there is reason to be hopeful. Matter is discrete, and yet much of physics is predicated on it being well described by continuum models, models that we know break down once we probe more closely. For systems made out of normal matter, standard experiments are possible that probe the scales at which the emergent continuum characteristics of our basic phenomenology give way to the characteristics of their discrete constituents. Many solid, liquid, and gaseous systems are suited to this kind of investigation. This much is not new. BECs, however, may allow much more than this: they may provide just the access we need to see what happens in the specific case of semiclassical cosmology, because they can be constructed to obey phenomenologically, in the continuum region, the same dynamics that the universe does in its continuum region. In particular, BECs can be made to produce phonon pairs using mechanisms whose dynamics appear to be of the same form as the source of cosmological particle pair production in the expanding universe.

Since the first production of a BEC in the gaseous state in 1995 by Cornell and Wiemann, many physicists have become interested in these systems as possible experimental test beds for studying quantum cosmology. This is extraordinary on its face—that fluctuations in the phase velocity of sound propagating through a quantum gas should give information about the large-scale behavior of the universe. And yet one can find analogous behaviors in these systems that make the one an appropriate experimental system for probing features of the other. Barceló et al. (2003) have shown how to manipulate a BEC in such a way that it will mimic certain features of an expanding universe exhibiting semiclassical particle production. That is, they show how to mimic in a BEC a semiclassical scalar field propagating in space-time that produces particle pairs as the universe expands.

Here's how they work. The theory of quantum mechanics, and especially the spin-statistics theorem, implies that quantum particles of integral spin will, under the right conditions, cluster together and condense into a single, coherent quantum state. This state is a BEC. These systems display many interesting behaviors and are ripe for philosophical analysis. I will consider here only their acoustical properties. Because fluids can flow, in certain circumstances, faster than the speed of sound in the fluid, sonic horizons may form. These are analogous to the event horizon surrounding a black hole: around black holes, light (and everything else) is trapped behind the event horizon by the powerful gravitational field, while in fluids sound waves are trapped behind the sonic horizon because they are carried by the fluid flow back from that horizon faster than they propagate toward it. Similar behaviors can be found in BECs with the right dispersion relations for their sound waves. So in a very coarse way we can say that these gases are analogous to singular space-times. While interesting and fruitful for our understanding of the physics of black holes, by itself this would not make the BECs particularly fruitful as experimental test beds of quantum cosmology. In fact the analogy can be made much more fine.

Barceló, Liberati, and Visser (2003) have shown that it is possible to manipulate a BEC so that it will mimic certain features of an expanding universe that is exhibiting semiclassical particle production. That is, they show how to mimic in a BEC a semiclassical scalar field propagating in space-time that produces particle pairs as the universe expands. Curiously, the analogue system has very little qualitatively in common with the target system. For there is no actual expansion of the gas in the model they consider. Rather, the interaction length of the particles comprising the gas can be changed by the application of an external potential. So, as Visser and Weinfurtner (2007) later argue, we have a clear example in which the underlying system differs from the system as described at the level of our experience—and markedly so in even those respects we take to be fundamental from the point of view of our coarse-grained physics. And yet the relevant dynamics is just that of a Lorentzian space-time undergoing expansion and exhibiting quantum pair production.

Jacobson (2000) makes it clear that one of the key features of experimental systems of this kind is that they allow us to probe certain behaviors of the cosmos in ways that give us information that constrains the kind of underlying dynamics that could be the source of the phenomenology we have in the present state of observational cosmology. That is, features that we discover through investigating these gases can provide us with experimental knowledge of the cosmos at large despite the roughly 32 orders of magnitude that separate their respective length scales. What, if anything, does this tell us about how to think about experimental practice?

It is well known to theorists of BECs that all of their important features can be captured in the Gross–Pitaewskii equation:

$$(24.1) \qquad i\hbar \frac{\partial}{\partial t} \psi(t, \mathbf{x}) = \left(-\frac{\hbar^2}{2m} \nabla^2 + V_{\text{ext}}(\mathbf{x}) + \lambda |\psi(t, \mathbf{x})|^2 \right) \psi(t, \mathbf{x}).$$

This is a nonlinear approximation to the Schrödinger equation with the self-interaction term given by a function of the modulo square of the wave function. In their proposed setup, Barceló et al. propose a series of generalizations to this equation. By allowing arbitrary orders of the modulo square of the wave function, by allowing the nonlinearity to be space- and time-dependent, by allowing the mass to be a tensor of third rank, by allowing that to be space- and time-dependent as well, and finally by allowing the external potential to be time-dependent, they arrive at a new Schrödinger equation:

$$(24.2) \qquad i\hbar \frac{\partial}{\partial t} \psi(t, \mathbf{x}) = -\frac{\hbar^2}{2\mu} \Delta_{\hbar} \psi(t, \mathbf{x}) - \frac{\xi \hbar^2_{(3)}}{2\mu} R(h) \psi(t, \mathbf{x}) + V_{\text{ext}}(t, \mathbf{x}) + \pi'(\psi^* \psi) \psi(t, \mathbf{x}).$$

And this equation has characteristics that allow it to be cast in the form that describes perturbations in the wave function propagating through an effective, dynamical Lorentzian metric. With a suitable form for the potentials, one can use this equation to replicate a general-relativistic space-time geometry.

It is also possible to show that, in the regimes of the experimental setup they identify, the BEC mimics very well the behavior as a whole of the expanding universe, and especially the behavior of scalar fields propagating in that universe. As the interaction between the components of the condensate is modified, the effective scattering length changes, and these changes are equivalent in their effect to the expansion of the universe. Under that "expansion" these scalar fields will exhibit pair production. And Barceló et al. give good reason to suppose that actual experimental tests can be conducted, in the near future, in these regimes. Thus the BECs are appropriate analogue models for the experimental study of important aspects of semiclassical cosmology. We can therefore use the condensate to probe the details of cosmological features of the universe, even though the analogue system has very little qualitative similarity to the universe as a whole. (For example, as I said, the condensate isn't really expanding.)

Look again at the BEC. That analogue system is a strange one. The target of the simulation is a continuous classical space-time metric that is coupled to the expectation value of a quantum field. This is being simulated by the analogue system of a single, unified quantum state supporting classical sound waves. As we saw, Barceló et al. generalize the governing equations for the BEC by proposing new potentials, and then show that the new system is governed by equations of motion that are essentially nonrelativistic but encode a Lorentzian space-time geometry. Their formal analysis allows one to consider the metric encoded by these equations of motion to be dynamical.

However, the metric of the actual space they consider is nondynamical across the span of the system. The processes operative there are radically, qualitatively unlike those of the semiclassical Einstein equation. Instead the "metric" is really a feature of the tensorial mass distribution. So the similarity is neither by approximation nor by suppression of parameters; instead it is something other. This is more like simulating ideal particle mechanics using standing waves in a river, rather than billiard balls. And we could see "particle" creation in such a simulation too—some dip and some hump might emerge from the same location and move off scene. Here the connection between the simulation and its target is as indirect as that of the leg-kicking case. The behavior is being caused in the one case by peculiar features of the condensate, and in the other by the interaction of the space-time geometry with a quantum field. We have a system with new "physical laws" that are merely artifacts of the analogue system. And it is those artifactual laws that we hope will shed light on the target system, the universe as a whole.

To emphasize the point: even the descriptive terminology we apply to the BEC is merely artifactual. We have a "mass" term, and we talk about "particles" and the "metric," but these are no more than names that we apply to certain projectible features of the condensate to indicate the analogical role they will play in our later analysis. The key work is in finding the stable features of the condensate, identifying them with features of interest in the cosmos, and showing that the subsystem in which these features are the operant causes is independent of the vagaries of the supersystem.

Before closing this section I must point out one further, important conclusion to be drawn about this class of experiment: the members of this class are much closer to what is normally called a simulation than they are to more common model systems. That is, we try to mimic the behavior of the target system by features of the simulation that arise from qualitatively dissimilar causes. I'll take this up shortly when I turn to simulations more generally.

We do not yet have fine enough technological control over these systems to allow for the detailed measurements that would be necessary to fruitfully investigate the mechanisms of the trans-Planckian modes in space-time. However, the rate at which such control is developing makes the odds very good that we will have that ability soon (see, e.g., Barceló, Liberati, and Visser 2011).

24.4. Conclusion

Extracting experimental knowledge of the cosmos from a volume of gas the size of a human red blood cell is complicated in the extreme. Relating the data models built out of the actual experimental results when they come to what we then infer are models of what the data would be in the case of the universe at large presents serious conceptual challenges. It is not even clear whether constructing such models is the right way to proceed. Additionally, concluding from the fact that BECs behave in certain ways that the universe is a certain way is not at all straightforward. It is not merely a matter of epistemic risk, but rather a question first of where the risk lies and what techniques are available to ameliorate it, and second of where the epistemic reward lies and what techniques are available to exploit that. All of the standard experimental worries are in play as well—questions of repeatability, effect isolation, and so on. These all concern what Franklin characterizes as "strategies that . . . provide grounds for rational belief in experimental results" (1986, 190). But rational belief in the phenomena is not all there is to the conceptual appraisal of experiments. We also wish to understand, *given* that it is rational to believe that the phenomena are as we take them to be, how one system can possibly tell us about the features of another, perhaps strikingly different system. This is the fundamental question of experimental knowledge, and it is a matter of understanding—and only partly a matter of epistemology. For in general we can know that something is true without really knowing why it is true. Indeed, much of our basic scientific knowledge of the world has something of this flavor. Remember that the theory of knowledge I am using has to do with information, not justification. For example, we know that cigarette smoking causes cancer, that the universe is expanding, and that collateralized debt obligations lead to economic disaster. However, we know these things not because we deeply understand why they are true, but because we defer to the epistemic authority of those from whom we have learned them. While there are many items in our scientific worldview that we understand well enough to offer detailed explanations for, the vast majority must remain known by authority. There is nothing wrong with such a state of affairs—indeed, the deferral to epistemic authority is a tremendously powerful tool by means of which we vastly increase our possible knowledge base. Still, the basic process by which experimental knowledge is generated can be understood, and that understanding comes by following the flow of information.

Turning again to the BECs, we can see how their dynamics being of a certain type bears information about the universe at large being of a certain type. The local logic used to classify the dynamics of the BEC is connected to the local logic used to classify the dynamics of the space-time as a whole by the fact that the Gross–Pitaewskii equation (and its generalizations), the equation that best characterizes the dynamics of the BEC, can be analyzed into a form that expresses the form of perturbations in a wave function propagating through an effective, dynamical Lorentzian metric. Once we have classified the BEC by finding a suitable form for the potentials, we can use the Gross–Pitaewski equation as the classification of a general-relativistic space-time geometry. And this then connects the dynamics of the two systems. The transformation between the dynamical equations is the infomorphism we need. Further, the cause of phonon production in the BEC system can be identified as the element of the classification corresponding to expansion of the space-time. Thus we have ground to apply what we learn about this phonon production (the details of its generation and development) to the production of particles in our actual space-time.

These are clearly experiments. However, the experimental domain is different from the target domain about which the experimenter is attempting to extract information. Having noted that, though, we must be careful to distinguish between factual differences (as in the difference between some token of a situation and some other token of the same type of situation about which we have done an actual experiment—a difference that includes mere temporal difference), and structural differences, and material differences. Is there ground to regard these as experiments in or on or about the target domain? Yes, but of course one must be careful. This care is necessary not only in the case of analogue and thought experiments, but also in the case of more standard laboratory experiments. We might well wonder about such an experiment whether it can be used to generate knowledge of any system other than itself. Our conclusion was that it can, but understanding how required some work.

Is there a principle of information flow for analogical systems that amounts to the fact that critical phenomena that are parameterized the same are in fact all the same? Is that even a fact? If so, we have an answer to the question of what kind of control we can have over the strength of analogical reasoning (and hence experimenting) across changes in scale that are shared among our systems of interest. If we are looking, for example, at critical phenomena, then the fact that some critical phenomena are characterized by the same set of parameters gives us good inductive evidence that the analogies that hold on one side of the phase transition also hold on the other side. Part of the strength of this induction comes from the way we derive predictions of critical behavior generally: such derivations usually proceed by taking limits of the appropriate sorts.

This is messy, but the core insight is perhaps right: that knowing that two systems share some class of phase transition supports the claim that their dynamical behavior near the phase transition will be similar. As always, such similar dynamics will not entail similar fundamental ontology, but that's not relevant to the kind of experimental knowledge we are trying to generate by the analogical experimentation under discussion.

Speaking now about the cosmos, we can abstract away from our own cosmos and evaluate experimentally what happens in cosmoi of certain types. The possibility seems strong in that case that we can learn about differences between the cosmoi that would be detectable based on this analogical experimentally generated evidence. The difference between that prediction and theoretical predictions generally is that the knowledge is generated by experimentally investigating to find characteristics—which may or may not be theoretically derivable—that distinguish various systems from each other, and then seeing which of these characteristics is actually in play in our universe.

Here there are two very broad ways to characterize what is going on experimentally. First we might say that these experiments are probing the character of systems obeying a Gross–Pitaevskii equation. When we know that these BECs are governed by a G–P equation, then what we learn about them experimentally constitutes experimental knowledge of G–P systems. Then there is what we learn about a particular class of system that we know is governed by a G–P equation. Here we have general knowledge not of the broad class, but of the more narrowly constituted class that we can appropriately capture by analogy to the actual experimental system.

On the other hand, there is a further interesting question: what can we learn about the general breakdown in the appropriateness of the G–P equation to those systems that are in some regimes well modeled by the equation as we experimentally probe the edges of the regime where the BECs are well modeled by G–P equations? Are there, in other words, universal failure classes for the applicability of a particular system of equations to various systems? Less strongly: are there classes of system that fail in characteristic ways to be modeled by certain classes of equation? I do not think we have answers yet to these questions, but they are all, it seems to me, targets of experimental investigation.

25

Nonhuman Animal Analogues

Over the last several hundred years, and starting I suppose with Morgan's fly lab, we have bred ever more uniform, tailored, nonhuman animal models to serve as proximal systems for experimental investigation. Sometimes the distal systems in these experiments are other nonhuman animals, but in many cases the distal systems are human. We use nonhuman animal models in an effort to learn about the basic biology of humans, as well as to determine what the outcomes will be of interventions into human biological systems. Indeed, there is a very strong presupposition that before novel drug therapies and other medical innovations are to be evaluated in human trials, they will be tested in nonhuman animal models. This basic protocol has two supporting pillars: first, that it is ethically less problematic to subject nonhuman animals to invasive and potentially dangerous testing than to subject humans to the same testing; and second, that these nonhuman animal trials give us useful information about the dangers and benefits to humans of the drugs and therapies under test.

These nonhuman animal trials are, in some measure, the foundation of a great deal of our knowledge of the risks and dangers of various toxic substances, the hazards and potential benefits of new drugs, the safety of cosmetics, and so forth. But they are also pretty clearly analogical in character. The nonhuman animals used as proximal experimental systems in such trials range across all manner of mammal species. If nothing else, this tells us that the medical and biology communities at large acknowledge the power of analogical experimentation to generate experimental knowledge. The thrust of this chapter will diverge somewhat from that of the previous few, however. Here I will be questioning the aptness for generating experimental knowledge of nonhuman animal trials. There are very good reasons to be suspicious, and

I will lay some of those out. Also, it isn't clear exactly what is going wrong in such trials when they fail, and I will try to spell out how this all fits into the picture of experimental knowledge I have been painting throughout this book. I will conclude the chapter with a reflection on the prospect for improving the experimental knowledge generating power of nonhuman analogical trials. This is an urgent task given the criticality of the questions these trials are attempting to answer and the fact that nonhuman animals will continue to be sacrificed in great numbers in service of these trials, and it is our minimal responsibility to ensure that such sacrifice is not in vain.

25.1. Inflammatory Problems

Of the many nonhuman animal models of human disease, perhaps none is as important as the mouse. They breed well and have very short lifetimes, making them good experimental subjects for multigenerational studies. However, it is not clear that our reliance on this model is a good idea. Does the biology of these animals really bear useful information about human biology?

In an influential paper, Seok et al. (2013a) asked how well mice modeled the way humans respond to inflammation. They measured correlations in the gene expressions across human populations for a variety of inflammatory events (burns, trauma, endotoxemia, acute respiratory distress, sepsis, acute infection), and in the gene expressions for models of those events in mice, and between the gene expressions in humans and in mice. They considered gene changes, temporal characteristics, and regulated pathways. What they found, broadly, is that for humans there is a strong correlation between their genomic responses to this wide variety of inflammation types, but that there is very little correlation either between murine and human responses for a given inflammatory stimulus or between the murine responses to different types of stimuli. Their conclusion was that mice are simply a poor model for humans in the case of inflammation. In short, they show that the biology of inflammation in mice is not information-bearing about the biology of inflammation in humans.

This is not what I would call a direct replication of any trials that had been done before. Rather, it is a kind of study that is becoming more and more common in the case of animal models. The study shows that a standard appeal to parallelism between animal and human doesn't work because there is no parallelism of the appropriate kind. In other words, this kind of study is not externally valid; or in the language of this book, the conclusion is that humans and mice (and murines, and mammals, and other animals, and so on) are not infomorphic[1] for biological types relevant to medicine. The social scientists would refer to this kind of study as a "conceptual" replication, where what is shown is that the general claims made in another study don't hold, but the prior study's methods were not used in the new study. I don't think that is quite the right way to put it. What is happening here is that a technique used widely in the biomedical sciences is itself tested, and the studies show that it cannot be calibrated in such a way that it can produce knowledge of humans. There seems little question that (by and large) biomedical studies produce valid results; they just don't produce experimental knowledge of humans, which is their ostensible purpose.

Seok et al.'s conclusions have not been without criticism.[2] Takao and Miyakawa (2015) offered some indication that there are (limited) shared inflammatory pathways between humans and mice. As Seok et al. point out in reply, however, the most that can be claimed is 12 percent overlap in the genes activated in humans. That is not at all predictive and gives little confidence in mice as a proximal system that bears useful information about humans. Also, Osterburg et al. (2013) questioned the choice of strain, the time-course methodology, and other features of the study by Seok et al. My sense of the consensus in the field, though, is that at least for inflammation, mice are pretty clearly a poor model for learning experimentally about humans. In particular, regarding the choice of mouse strain, since Seok et al. were using a strain that has been in widespread use for a long time and is unlikely to be retired soon, at the least their study undermines a large swath of results in the field. Moreover, given that other strains of mice haven't been subjected to the kind of modeling study Seok et al. undertook, it's cold comfort to say that those other strains have not yet been undermined.

Another line of critique comes from Cauwels et al. (2013), who suggest that, despite some possible difficulties with the mouse model, still "in the field of inflammation, valuable human therapies have been derived from mouse studies (e.g., anti-TNF treatment for rheumatoid arthritis and inflammatory bowel disease). In the field of sepsis, failure of the phase III clinical trial inhibiting NO synthases, which had to be terminated because of excess mortality, was actually predicted in assorted murine shock models" (2013, E3150). But as Seok et al. indicate in their reply, there are two problems with this line

1. At least we are unaware of what infomorphisms link them.

2. Much of this criticism is of legitimate scientific interest. But I do recall some discussion at online medical fora around the time of their study to the effect that, yes, animal models are essentially useless, but they are all we have, so we need to keep using them.

of attack: First, the anti-TNF arthritis and inflammatory bowel treatment was developed for use in mouse *sepsis*, and failed in subsequent human trials for sepsis. Its use for arthritis and inflammatory bowel disease was a later seren-dipitous discovery. Second, their claim is not that whatever happens in mouse models does not happen in humans, whether that something is the expres-sion of a gene, or the uptake of some drug by an organ, or something else. Instead their claim is that "the many trials for sepsis in which drugs protected in mice but failed in humans suggest that ability of mouse efficacy models to predict human inflammatory diseases is close to random, and therefore it should not be surprising that occasionally there is a correlation. However, for such a model to be helpful, it needs to prospectively predict the human condition. In this case, mouse models appear to perform very poorly indeed" (Seok et al. 2013b, E3151).

Seok et al. are quite right about this. The real issue that is at stake here is whether these models, and other nonhuman models, can be used for pre-diction. Of course there will be many cases where some feature possessed by humans will also be possessed by a nonhuman animal. But for genuine experimental knowledge, that is not good enough. We need a map between the inferential structures of the relevant type–token structures in the proxi-mal system that informs us about the type–token structures of the distal sys-tem. That sometimes humans react the same way to analogical causes tells us nothing beyond that very fact. Without an available map between classifica-tions of humans and nonhuman animals, information about the former will not flow to us from studying the latter.

25.2. Dioxins

What do we know about the carcinogenic effects of dioxins in humans? Very little, it seems. And yet, in the "Report on Carcinogens" prepared by the Na-tional Toxicology Program, the various substances listed as either known hu-man carcinogens or reasonably anticipated human carcinogens are evaluated in several ways. On the basis of multiple test methods, substances are either not listed, listed as known carcinogens, listed as reasonably anticipated car-cinogens, or delisted following petition and evaluation. Among these meth-ods is an evaluation of the effect of the substances in question on various ani-mals, ranging from mice to rats to guinea pigs to monkeys and so on, as well as on isolated cells grown in vitro. Here is how a typical claim involving test-ing on animals is presented (in this case the substance is phenolphthalein). Notice the standard of evidence for when we can expect that mechanisms of carcinogenesis will be present: when there is no evidence available to call into question the claim that the mechanisms in experimental animals are not present in humans (National Toxicology Program 2011, 342):

Studies on Mechanisms of Carcinogenesis

Phenolphthalein caused genetic damage in several in vitro and in vivo mammalian test systems. It caused hprt gene mutations, chromosomal aberrations, and morphological transformation in Syrian hamster embryo cells with or without mammalian metabolic activation, and it caused chromosomal aberrations in Chinese hamster ovary cells with metabolic activation. In vivo, phenolphthalein caused micronucleus formation in mouse erythrocytes after repeated, but not single, exposure by gavage or in the diet, and dietary administration for 13 weeks caused abnormal sperm in male mice (NTP 1999, IARC 2000). Dietary administration of phenolphthalein to female heterozygous p53-deficient transgenic mice for 26 weeks caused micronucleus formation and malignant thymic lymphoma. In the tumors, the normal allele of the p53 tumor-suppressor gene had been lost, suggesting the involvement of a mutagenic mechanism in tumor induction and/or progression (Dunnick et al. 1997).

Phenolphthalein is absorbed from the gastrointestinal tract and undergoes extensive first-pass metabolism in the intestinal epithelium and liver, resulting in almost complete conversion to its glucuronide, which is eliminated in the bile (NTP 1999). Phenolphthalein enhances the production of oxygen radicals in in vitro systems (IARC 2000). In vivo, reduction of phenoxyl radicals could allow re-formation of phenolphthalein, establishing a futile cycle of oxidation and reduction, thereby generating more free-radical species. Thus, phenolphthalein may be a significant source of oxidative stress in physiological systems (Sipe et al. 1997).

No evidence is available to suggest that mechanisms by which phenolphthalein causes tumors in experimental animals would not also operate in humans. In rodents, phenolphthalein caused oxidative stress and altered tumor-suppressor gene pathways, both of which are mechanisms believed to be involved in human cancer.

Since phenolphthalein was listed in the Ninth Report on Carcinogens, an additional study relevant to mechanisms of carcinogenesis has been identified. Dietary administration of phenolphthalein to transgenic mice with the human c-Ha-ras proto-oncogene promoted the development of lung cancer (adenocarcinoma) induced by a single intraperitoneal injection of em N-ethyl-N-nitrosourea (N-nitroso-N-ethylurea, which is listed in the Report on Carcinogens as reasonably anticipated to be a human carcinogen) (Imaoka et al. 2002).

This quotation includes a great deal that is of interest. What forms the basis of the view that dioxins cause cancer in humans? There are three paragraphs of data derived from rodents, and no positive reason at all to think that those studies translate to humans. Consider: the gene *pathways* by which mice respond to inflammation are also present in humans; it's just that they're not activated the same way, so their presence doesn't tell us much. We're in a similar position in this case. There are of course shared metabolic systems, systems that in some cases operate the same way in humans and rodents and in other cases do not. The standard of inference here is that *there is no reason to think not*. But that by itself is a very bad reason to accept the positive claim.

So the principal ground for thinking that dioxins cause liver cancer and various other types of liver toxicity in humans is that it causes liver cancer and various other types of liver toxicity in rats. While for many potentially dangerous substances we do not have good evidence, we do have some direct evidence about dioxins. Yet there is essentially no evidence that dioxins cause cancers in humans similar to those addressed in the Report on Carcinogens. While there is some evidence of increased rates of cancer based on retrospective studies of exposed chemical workers (Fingerhut et al. 1991), this evidence is quite weak, and in fact shows no increase in liver cancers at all—Fingerhut et al.'s principal finding was a statistically significant increase in *soft-tissue sarcoma*. In the Seveso accident in 1976, a great deal of TCDD escaped in a cloud that covered about 2.8 square kilometers. There were many deaths of animals, and inhabitants were evacuated. We have some data on the health effects of this cloud on humans (Pohjanvirta and Tuomisto 1994). Bertazzi et al. (1993) find an increase in soft-tissue sarcoma in residents of areas affected by the Seveso incident, but these cancers have nothing to do with the liver.

Pohjanvirta and Tuomisto (1994, 486) make the following two points:

Toxicologically, TCDD possesses several properties that make it unique among all xenobiotics. One of the most intriguing and perplexing features is a specific intracellular receptor (see section II.F) which occurs in a variety of species, but the physiological role, if any, is still a mystery. This receptor mediates at least part of the actions of TCDD, the best known of which at the molecular level is induction of cytochromes CYP1A1 and CYP1A2 in the liver. With regard to the induction ability of these monooxygenases, TCDD is the most potent compound known.

Finally, a plethora of health effects have been associated with exposure to TCDD in humans, ranging from mood alterations (Levy, 1988) to diabetes (Wolfe et al., 1992) and cancer (Zober et al., 1990; Fingerhut et al., 1991; Bertazzi et al., 1993). Yet, compelling evidence exists almost solely for TCDD-induced chloracne (May, 1973, 1982; Oliver, 1975; Pocchiari et al., 1979; Assennato et al., 1989).

So perhaps dioxins are quite harmful to humans. We still don't know, though, because none of the observations that we have are suited to bear that information. Indeed, in the Kociba study I discussed earlier in this book, the medical community swings back and forth over whether those *rat liver tumors* are cancerous. So we lack both the strong observational data of rats (our proximal system) and the reasonable maps from those observations to human disease states (our infomorphisms) we would need to derive any experimental knowledge about whether dioxins cause liver cancer in humans.

25.3. Other Murine Problems

There are a number of other places where questions about our reliance on animal models appear: Ostrand-Rosenberg (2004), von Herrath and Nepom (2005), Woodcock and Woosley (2008), Davis (2008), Hayday and Peakman (2008), to name just a few. Davis in particular notes that "mice are lousy models for clinical studies" (2008, 835), due to the many successful therapies developed for mice that fail in humans, and the generally very low success rate for translating murine therapies into human therapies.[3]

In any case, it is not a single inflammation study that is causing biologist to worry more and more about whether mouse models are doing the work they are supposed to do. There is a real failure of evidence here that mice are usefully infomorphic to humans with respect to medical classifications.

Mouse models are important, but the scope of their relevance is limited. I turn now to consider the *general* question of whether nonhuman animals provide experimental knowledge of humans. My conclusion will be that, as far as anyone knows, they do not.

25.4. Concordance Is Not Predictive

There is little in the way of systematic analysis of whether nonhuman models have contributed much to the development of human biology and medicine. That is surprising, given the very strong conviction in the scientific community regarding its efficacy. One would suppose, naively, that any such strongly held and widespread conviction would be well supported by the evidence. Instead, the picture that emerges from the history of the use of animals for drug trials especially is one of haphazard development of a nonhuman animal testing standard based on very little evidence. More than anything, apparently,

---

3. I don't want to give the wrong impression; Davis does continue to believe that they give us good insight into the basic biology of human immunology.

what solidified the preclinical animal testing practice in the United States was the thalidomide disaster of 1961 (Junod 2013).

Shanks et al. (2009) consider the available data on animal models of disease generally. Their question is a straightforward one: are the data about outcomes of various interventions in animal models predictive for humans? They focus on that narrow question (though they do attempt to draw broader lessons, to which we shall return). But first let's consider their answer to that narrow question. The data we now have about animals seem inadequate to support predictive inferences about humans. If true, that conclusion is quite surprising, given how much we have relied on these models over the years.

After their analysis of various data, Shanks et al. turn to the Olson report (Olson et al. 2000). That report attempted to find out what the connection is between human and nonhuman animal toxicity for drugs tested by pharmaceutical companies. "This report summarizes the results of a multinational pharmaceutical company survey and the outcome of an International Life Sciences Institute (ILSI) Workshop (April 1999), which served to better understand concordance of the toxicity of pharmaceuticals observed in humans with that observed in experimental animals" (Olson et al. 2000, 56). Their findings indicate a very high concordance for these trials, so one might be tempted to think that these trials are giving us useful information about humans based on nonhuman animal testing. Olson et al.'s report at least attempts to directly answer the predictive question, and it does appear at first blush to conclude that animal models do give us good predictive outcomes for humans. So perhaps Shanks et al. are too hasty in their appraisal of the data.

But Shanks et al. point out that the Olson report is not really about results with predictive power. Rather, that report deals with the notion of "concordance." This notion is never clearly defined, but we can determine from the data analysis Olson et al. provide that it is a measure much closer to statistical sensitivity than to something that would support counterfactuals, as would be necessary for true predictive power. So in fact it doesn't really matter what correlations are observed between nonhuman and human drug interactions. Sensitivity is simply the wrong measure. Things are slightly worse than this, in fact; drug companies were told only to report a subset of their data. If they noted an adverse effect both in humans and in a nonhuman animal, then that animal was reported; if no adverse effects were noted, then the companies were simply to report all animals tested. Thus we don't have a good sense for the statistics of positive correlations with respect to the total class of animals that may have been tested with harmful drugs. In short, the Olson report, apparently one of the strongest collections of data on the question, gives us very little reason to think that nonhuman animals are predictive for human reactions.

Shanks, Greek, and Greek analyze the data that now exist on the effects of thalidomide on nonhuman animals. This event, where many children were born with severely atrophied limbs as the result of thalidomide being taken by their pregnant mothers, is one of the main impetuses for our current requirements on drug testing. But despite the coincidence in timing between the onset of requirements for preclinical animal testing and the thalidomide disaster itself, there seems to be no reason to think that our current protocols would have predicted thalidomide's negative effects. While it is correct that these animal models would have included some that showed teratomies, there is also, according to Shanks et al., good reason to think that these models would not have been seen as predictive in a robust way, because there are very few that show poor outcomes and no reason is given to think that these would have been picked up on by the medical community. And so it goes for the general animal testing protocol, as they see it. Here they reiterate what is essentially Seok et al.'s point, but applied to all nonhuman animal models (Shanks et al. 2009, 10):

> Even if we retrospectively picked all the animals that reacted to thalidomide as humans did, we still could not say these animals predicted human response as their history of agreeing with human response to other drugs varied considerably. Prediction vis-à-vis drug testing and disease research implies a track record. Single correct guesses are not predictions.

25.5. No Analogical Animal Testing?

I think Shanks, Greek, and Greek do a very good job establishing their case. Nothing I have seen in the literature does a better job of indicating the positive predictive value of nonhuman animal models than Olson et al., and that study does a terrible job. Shanks et al. try to go further, though. They question the very idea that we can use animals in a predictive way to find out about human outcomes. Their argument turns essentially on the observation that all animals are the result of complex evolutionary processes, processes that have resulted in complex systems whose basic causal mechanisms are extremely difficult to track, control, and understand. But this is hasty. In fact the results they point to can be used in two important ways for advancing medical science. The first is to use the fact of divergent outcomes (when we do find them) as further specifying the facts about the human systems the animal models are meant to be targeting. For example, we may think that making predictions that fail using animal models is a reason to abandon them. Instead we might well use these failures to select among the various models, of the various situations, those that are better predictors of human outcomes. Shanks et al. are quite right to focus on the question of whether we have the data we need to support our predictive inferences. But we should focus on that not merely because we want to know *whether* animal models are predictive as a large class, but also because we want to know *which* animal models are predictive, in which contexts they are predictive, and how predictive they are. It is one thing to note that as a class our models are nonpredictive, and quite another to conclude on that basis that the possibility of predictions from animal models is somehow chimerical and that we should move away from the use of animals in human disease study.

An interesting counterpoint to the discussion comes from Schnabel (2008). While he is as critical of mouse models as many of his colleagues, Schnabel's reasons are somewhat different. He raises the possibility that the real problem in mouse studies is with the observation of proper type designations in the proximal (murine) system, rather than with the infomorphism itself.

Shanks et al. will claim that there is very little reason to think that we can have the right kind of infomorphism between animals and humans—in large part because of the evolutionary distance between the complex evolved systems of animals and humans. There is some sense to be made of this claim even a priori, given that we create strains of animals to be particularly susceptible to the disease analogues we're investigating. But their evidence is not clear in the light of Schnabel's investigation, which shows that the proximal typing studies are radically underpowered and the culture of generating these results is so strongly entrenched. The way many of the problems with the murine models arise appears to be in how the models fail to express the genes they are specifically bred to display strongly so that the analogue effect will manifest. The suggestion is that we have learned not that the course of drug interaction is interestingly, markedly different between murine models and humans, but rather that the drugs just don't do what they're said to be doing in the murine models in the first place.

In addition to this lack of power, Schnabel points out that many experiments involve pretreatment of conditions—a program that is completely impossible with human patients. Schnabel also argues that there is not a good mouse model of human Alzheimer's disease. Even the SOD1 mice, which were taken to be very good models of the end state of neuronal destruction in sporadic ALS, are starting to seem much less accurate. Indeed, he wonders, is it even plausible that there could be a full-course analogue of a human aging-related disease? By itself this string of worries is not a good reason to be hopeful about the prospects for useful analogical experiments on humans using mice or other animals. But it does point out that we really don't have a good handle on what the data as a whole are telling us.

Is the problem here that we know much less than we think about the proper classification of animals as possible disease models for humans? Is it that the dynamics of the nonhuman systems are more complicated than we're allowing? Rather than concluding with Shanks et al. that the prospects for analogical experiments on humans using nonhuman animals are hopeless, I would suggest that we need to try instead to understand better both the proximal systems comprising them and the potential infomorphisms between them and humans.

Shanks et al. seem to me to be making a very bad argument for a true conclusion. Their argument that is we're evolutionarily too far from other animals for them to provide good analogies for our metabolic processes; therefore animal models don't work. They're right that animal models in general don't work at the moment. But there are fruitful analogies between certain animals' metabolic processes and our own. Drug kinetics in dogs, for example, give very good information about drug kinetics in humans. The real problem is not that we have reason to think that animal models can't work, but that we just don't have guidance from evolutionary similarity as to when such analogies are available.

25.6. How Can We Fix Animal Modeling?

A great deal can be done to improve the way we generate experimental knowledge using nonhuman animal models. The most important thing, perhaps, is to realize that within the vast data sets that have not proven adequate to generate knowledge of human diseases there resides a wealth of information. Large-scale data analysis may provide important Baconian insights here; it may allow us to infer common cause without a causal model, in various aspects of the metabolisms of humans and various animals. That information can be extracted with the appropriate tools (Bayes nets, causal algorithms, etc.), using the kinds of causal inference software and methods developed by, for example, the Center for Causal Discovery at the University of Pittsburgh. We do not have to endorse all of their causal models to see that the huge data sets generated by biomedical studies of human diseases on animal analogues are likely to be storing a great deal of information about human diseases *and* about the right infomorphisms to use when conducting such studies in the future. I do not perform any such analyses here, but that is one of my future research projects, inspired directly by thinking of experimentation as regulating information flow.

Drug trial data sets (including trials halted before human testing) could provide a great deal of evidence about the nature and strength of analogies between animal models of disease and human disease modes. Many such sets are available from publicly funded trials. The main point is that surface probability assignments say that there is no information, but creative data mining can change that. The information may be there, but be unable to flow because the channels have not been sufficiently mapped.

I have been speaking as though the lack of analogy between certain species prevents proper experimentation, and have proposed in preliminary fashion possible ways of overcoming that difficulty. Here I just want to remind the reader that the problem arises in principle and in practice for intraspecies comparison as well—indeed, for human–human analogical experimentation. But we are familiar with the latter worry, and we mitigate it using our experimental strategies. That is just what we should do in the case of interspecies analogies.

26

Analogical Experimentation Is Generic

In chapter 22 I contended that many analogical experiments are just as robust in their generation of experimental knowledge as so-called direct experiments. My stronger contention here will be that in fact all experimental systems are of the analogical sort. However, our long practice with identifying and exploiting useful infomorphisms completely obscures the fact that we are making use of them at all in a wide variety of cases. We simply see right through the part of the exercise where we exploit the analogical character of the two systems. That is fine, as an experimental practice, *because* we are so good at it. But it should not remain hidden from our philosophical analysis of experimentation, because it gives the wrong idea about how other practices than these, whose analogies are so clear and familiar, can count as robustly experimental, and it gives the wrong idea about what is at stake in experimental knowledge claims generally.

In this chapter I will show (by reminding us of what we already know and pointing out its obvious implications) that even standard laboratory experiments must be understood as analogical in that there is no strict identity between the dynamics of the proximal and distal systems, or the appropriate classification of one or the other is unclear or contested, or establishing that some tokens are of some type is problematic. There simply are no experimental systems that don't suffer from at least some of these issues; when all such issues are absent, what we have is no longer an experimental system, but an observational system in the sense that the observations involved cannot be extended beyond the spatiotemporal instant at which they are made. Rather than a negative conclusion about the existence of direct experiments, though, this is a positive conclusion about analogical experiments: direct and analogical experiments differ from each other only in degree and not in kind, so we should accept the latter as robustly knowledge-generating experiments.

26.1. A Folk Category

There is a folk category in play in much of the discussion of modeling and simulating. That category is the direct experiment—that is, our everyday practice of producing effects in the lab as a means of characterizing the type of system that is being manipulated in the lab. We perform some direct experiment, perhaps on an instance of the type rat—perhaps we inject one with some drug—and then we have experimental knowledge about rats. It's not very much knowledge, because it is (among other things) statistically weak; but it is knowledge about rats because it's an experiment on rats. Accumulating more data on rats by experimenting on more rats will produce more and more robust knowledge of rats. The folk category of direct experiment has little in the way of clear application conditions—there are no tables of what counts as a direct experiment. Instead it is one of those categories that is known by its negative instances, the experiments that are not direct: analogical, thought, simulated, and so forth. These latter types of experiment constrain the class of direct experiment to whatever does not fall within their parameters. Whatever is not not direct is direct. But there is nothing left—all experimentation is indirect. That's its nature.

The last few chapters have shown that analogical experiments can provide experimental knowledge that is as robustly experimental as the knowledge produced by their "direct" experimental cousins. They have not, however, challenged the presupposition that there is a clear distinction to be drawn between analogical and direct experiments. In fact there is no such clear distinction to be drawn. In the final part of this chapter I will argue that the limit of analogical experiment, as the morphisms between the analogous subsystems approach more and more closely to the identity map, is not direct experiment, but isolated observation. That is, what we have been calling direct experiments are really just analogical experiments where the users of the experiments are content with the strength of the support for the basic inference and feel no special need to argue for it or to attempt to establish it. That conclusion will suggest two things at least. First, it will tell us that the difference between direct and analogical experimental systems is one not of warrant, but of comfort and background beliefs—thus the analogical may well be as robust as the direct experiment epistemologically, if not even more robust. Second, it tells us that the distinction between analogical experimentation and direct experimentation is one of entrenchment and picks out no natural kind distinctions.

What I am arguing for is a conceptual point. Even if my argument is right that there is no clear concept that is direct experiment, there will be many experiments that seem obviously and intuitively to count as direct, so the point may appear too narrow to do much work. But the work the point is supposed to be doing is not to categorize experiments into direct and indirect, but to show that the logic of experimental inference is obscured by thinking of the folk concept as signaling the presence of a distinct class of experiment with its own epistemological standards. Rather, it is to show that all experimental practice is conceptually on the same footing, and that the distinction being flagged by direct versus indirect is a quantitative rather than a qualitative difference in support for the basic experimental inference.

Consider some examples of apparently direct experiments, like the examples we teach to children to illustrate the methods of science. Members of a grade school classroom decide to see what the connection is between light and growth for plants. They choose lima beans as their test subjects. Some lima beans are sprouted, and of those some are put in the dark, and others receive varying amounts of light during the day, but otherwise the conditions of the plants are all the same. Once the students make and record their observations, they are in a position to say something about the effect of light on lima beans' growth because they have directly experimented on lima beans. And they are prepared to make some counterfactual inferences about the unplanted beans: were they to have been given such and such light, they would have grown according to such and such a schedule. For an even more direct example, consider repeated interventions into the behavior of a spring. Over many trials I note the extension of the spring produced by hanging different masses from it. I am now in a position to say something about the relation between the mass of the suspended object (and hence the force) and the extension of the spring. Hooke's famous discovery *ut tensio sic vis* says it very well: there is a linear relation between the extension and the force applied. I can make some counterfactual inferences of my own: were I to have put on such and such mass, the extension would have been thus and so. The students and I are also in a position to make some predictions, obvious subjunctive variations on our counterfactuals: were we to do x, then y would follow. Are we right about all of this? Is this how direct experiment really works? Yes, more or less, and within limits, as for all empirical inference.

How does this differ from the indirect cases? I observe the behavior of sound waves in a Bose–Einstein condensate and conclude that cosmological pair production is governed by some equation; I induce bladder cancer in rats using saccharin and institute a limited ban on saccharin; I imagine a solar-system-sized gravity wave detector and conclude that gravity must be quantum mechanical; I simulate many scenarios in a computer and conclude that there's a hurricane coming. These experiments, though valuable and interesting, simply do not have the robust epistemological support of the direct experiments, because the systems about which I am drawing conclusions are different from those on which I intervened. At least that's the standard assumption. But it is wrong.

26.2. The More Direct, the Less Direct

One might think that there are at least some clear cases of direct experimentation. For example, aren't experiments on the entire class of electrons themselves the result of observations of token instances of electrons, meaning that only the induction from a sample to the population of which it is a sample is necessary? And in the case of elementary particles like electrons (assuming they are elementary), the proximal system is not an analogue of the distal system but is, in fact, properly identical to it. However, in the case of electrons specifically (and, I think, more generally), the closer to identical the proximal systems are to their distal analogues, the further our observations are from touching those proximal systems themselves—and the more the observations that generate our experimental knowledge are of intermediary proximal systems. Consider measuring the charge on the electron as in Millikan's experiment. We are using proximal electrons to find out about distal electrons, so this should count intuitively as a direct experiment. But what does Millikan—what do we—observe in this case? We observe accelerations of oil drops in the atmosphere, *not* properties of electrons.[1] Those observations connect us to information-bearing signals from features of the electrons that are on those oil drops, and in turn to the general class of electrons. But they do so because structural features of the systems to which we have observational access are analogical to those of the systems on which we are experimenting: facts about the accelerations of charged oil drops bear information about the charges on their surfaces.

The empiricists of the seventeenth and eighteenth centuries got this much right: we do not engage observationally with types in the world. Instead we engage with individuals, and even then only over highly restricted temporal spans. It is only because these individuals bear information about other things of their same type that our practices of deriving type knowledge from token observations (and then further token knowledge about other members of the

---

1. Our sensory input, of course, is colored dots whose visual positions are correlated with the visual position of the hands (colored lines) of stopwatches or the numbers (arrangements of illuminated pixels) displayed on a digital watch. But that kind of phenomenalism is not really on point here. We can concede as much here as any realist would want.

type) works as well as it does. For things like tables and chairs and dogs and cats, we don't have to explicitly construct and analyze the infomorphisms that allow us to derive this token-to-type-to-token knowledge—we grow up applying them! But their presence is what makes those inferences go. Generally, the things we interact with and can observe are highly complex and in that sense unique. We are pretty good, however, at finding the ways in which they are analogous to other things and generating knowledge about those other things thereby. No matter how similar they are to other things, however, they are unique individuals the observation of which allows experimental knowledge of other things only because of how these things bear information about those, and only via the analogical connections between them that support that information flow. There is just no good sense in which normal experiments are direct and other sorts are merely analogical. It's all analogical.

Simulation Experimenting

27.1. Introduction: Projectable Predicates in Simulated Systems

To succeed in giving knowledge of their targets, simulational experiments must fulfill a number of desiderata, but they all amount to the fact that the simulans and the simulatum should be infomorphic: distinct tokens in (the classification in use for) the simulatum should be mapped to distinct tokens in (the classification in use for) the simulans; and there should be a well-defined time-evolution function, taking tokens in the simulation from one state to another. This time-evolution function should be such that a token in the simulans representing some token in the simulatum should be connected by that operator to the token in the simulans that represents the time-evolved version of the token in the simulatum; and the types in the simulans should be such that if two tokens in the simulatum are of the same type, then so too should the tokens they map to in the simulans.

The requirement that these features be in place will guide us in our selection of the appropriate classifications for the simulatum and the simulans, so that we can relate them infomorphically. It may be that we do not know the time-evolution function of the simulatum; we may have only some series of data about initial and final states for a range of tokens, and be trying to learn that function. More commonly, we may know the function well enough but be unable to solve for final state properties even given that function and initial state data, and might need the simulation to tell us about those final states.

For example, suppose I want to simulate the behavior of a gravitating system, say a system of planets about a sun. Then there should be a token in the simulans for each planet and for the sun. The salient properties of the planetary system relevant to its operation as a gravitating system are the masses of the bodies, their relative velocities, and their distances.

There is, in my view, no important conceptual distinction between experimenting with computer simulations and experimenting with other kinds of material systems. The distinction is one of practicality alone. I do have a general proposal, conceived in collaboration with Walter Warwick, about how to address persistent and apparently intractable difficulties of seeing how to make experimental use of certain kinds of computer simulation. Roughly, I propose to adopt a more explicitly empirical stance toward these simulations. In conjunction with the more common method of attempting to modify the simulation and its code to allow a classification that is congenial to whatever infomorphism is appropriate to the analogy we have in mind, Warwick and I suggested that taking the simulations as we find them, and attempting to determine what classifications they support qua going dynamical systems, is likely to yield useful analogies for a range of distal systems. I cannot report a groundswell of studies along those lines, but I remain committed to the principle of treating going simulations that don't do what we want as themselves objects of empirical study.

Warwick and I (Mattingly and Warwick 2009) analyzed simulations and the way they give experimental knowledge, and their connection with analogical systems and the way *they* give experimental knowledge. We concluded that they function the same way. I think the analysis is sound, and it offers an interesting way of conceptualizing the kinds of argument I have been making in this book. We began with the idea that in order for us to gain knowledge about the world from other systems in the world (even the imaginative world of the simulation), the systems we are looking at must have ontologies that are stable in the right way. Our thought was that the right way is the way Goodman ([1955] 1983) put it: the stabilities we are looking for are those that support counterfactual, causal reasoning with respect to the description of the system of which they are parts. So we can, if we like, think of various (temporally persistent yet evolving) arrangements of pixels on a video game screen as objects and reason about them successfully once we know the appropriate classification determined by the video game itself. The causes that are doing work in the system, despite being merely analogical or simulational, we called projectable predicates in Goodman's sense. The problem is that normal predicates may not apply in the context, or if they do they may not be useful for drawing conclusions about distal systems. But as we argued, if we can find a classification of the proximal system into objects with projectable predicates, then causal reasoning about them can go on as normal. We put it this way, speaking of natural versus analogical versus simulational laws (Mattingly and Warwick 2009, 477–78):

Our basic orientation is that it is fruitful to view computer simulation as a kind of analogue experiment. Returning to Goodman's distinction between projectible [sic] and non-projectible predicates, we can make the following observation. In general, in computer simulations, and in analogue simulation more generally, we do not know the natural kinds and very often we do not have a good sense for what predicates of the simulation are projectible. More seriously we do not know what the relation is between the projectible predicates of the one system and those of the other—especially since we are often trying to use the simulation to find those predicates that are the most fruitful for framing our hypotheses.

To get clear on the issues that face us, it will be worthwhile to introduce some categories that will allow us to relate the general aspects of analogue physical systems to those of computer simulations. What we are trying to do is relate one system to another by some mathematical relation between them. Generally we have a theory of some experimental setup, including a method for turning our finite data into mathematical models involving perhaps continuous values of various parameters. To investigate some system type by conducting an experiment we abstract away irrelevant features of the experimental setup using our theory of the experiment and construct this data model. We can then compare that model to the various mathematical models that arise in our theoretical descriptions of the target system.

In the more general case of analogue systems there is the added complication of restricting the parts of the experimental system we allow to be relevant despite being just as significant in magnitude and scope as other parts. For example in the Bose–Einstein condensate we pay attention only to certain vibrational modes even though other modes are present and impact the state of the system—we do not identify those modes with particle anti-particle pairs, we do identify these modes with them.

Even with these complications however the basic framework is the same. We have two systems, the target and the experiment. To bring the results of our experiment to bear on the target we need a more or less explicit theory of that experiment—less in cases where the similarity between the experiment and the target is obvious, more in cases where the connection is more tenuous. An experiment in Newtonian particle mechanics using a billiard table needs a much less explicit theory of the experiment than an experiment in quantum cosmology using a Bose–Einstein condensate.

We have seen that even in the latter case it is possible to provide a very general analysis of the causal artifacts of the system and their complicated interactions with each other. We did not discuss the breakdown of the model, but it turns out that the system can be made stable and the analogy itself can be made stable across a wide range of parameter values—and most importantly, that range is known.

Our theory of the experiment in this case functions as it should despite the fact that the underlying features of the experimental system are so far removed from the salient, artifactual, causal agents of the experiment as a model of the target system.

This indicates, I think, that the work of this chapter has already been done implicitly in my treatment of analogical experiments. However, I do think it is worthwhile to treat simulations separately as experimental systems on their own terms.

27.2. Characterizing Simulations

There are a number of competing definitions of simulation in the literature. What is at stake between these definitions seems to be captured by two basic distinctions. On the one hand is whether or not (and if so, how) simulations must be understood as modeling the processes governing the systems they are simulating, and on the other hand is whether or not the intentions of agents are relevant to characterizing simulation. Regarding the first, I will argue that we are best served by seeing simulations as coming in a range of cases, some of which mimic the underlying processes that are being simulated more tightly than do others. However, I will also argue that a key feature of a simulation is that the transition between its individual states must be generated (in a sense to be made clear) from its prior states. Failure to adhere to this requirement will result in at most a record of a system or process, not a simulation of that system or process.[1] Regarding the second, I will argue that the intentions of agents are entirely misplaced as an evaluation of whether or not some system is simulating some other system. I will attempt to make clear that the apparent virtues of appealing to intentions serve rather to obscure than to illuminate the nature of simulation. Principally, this is because there is little to be gained from answering the question of what simulations are. Instead we should focus on how simulations give us knowledge of the systems they are simulating. They give that knowledge, however, not in virtue of any particular ontology but rather by regulating information flow between the simulating (proximal) system and the target (distal) system. In this respect—the only respect that matters for understanding how simulations give experimental knowledge—we will see that they do so just like any other experimental systems.

Rather than agents' intentions, we should (as usual) focus on the state of agents' knowledge, and that focus should be reserved for discussions about

---

1. Thanks to Richard Fry for urging me to be clear about this point.

the role simulations play in generating knowledge of the systems they are simulating.

We would like to be able to say that the process X simulates the process Y in respect φ when each of the states x_i of x is φ-related to a single state y_k of y such that if $i < j$ and x_i corresponds to y_k and x_j corresponds to y_l then $k < l$. The φ relation could be, for example, something like the relation between a Ping-Pong ball and a well-defined set of pixels on a screen in the game Pong that represents the location of the ball in the game. We would also want the process X to be causally independent of the process Y while the individual states of X are functionally related to the states of Y. This would be something to the effect that there is a function f on the earlier states of X that outputs the current state, as well as a function g *between* the states of X and Y such that if x_N corresponds to y_n then $f(x_0, \ldots, x_{N-1}) = N$ iff $g(y_n) = N$. This captures the idea that the X process is generated functionally by features of the Y process, but we don't demand, say, that these functions are invertible. Typically we will want to be able to read off behaviors of the Y process from features of the X process. That's the point to having the simulation—but in my view that should not be built in to what it means to simulate. To get knowledge from the simulation is a separate task.

This characterization is permissive in the sense that it allows any thing/process to simulate any other thing/process. The point to the definition is not, as is usual, to exclude some things. Rather, it is to make manifest the way one system simulating another may be used to gain knowledge of the latter from knowledge of the former. What is added in that case is that the states of the simulating system bear information about the states of the simulated system, as proximal and distal systems.

One restriction this characterization does make is that the process that generates the X-states be independent of the outputs of the Y process. This is reflected in the causal independence requirement. Thus, to be a simulation will require more than mere correlation (as Bunge has it) and more than mere representation of Y-states by X-states (as Parker has it); it will require that an independent process generates the X-states from the prior states in the sequence. Other kinds of process can model the states of some given process—e.g., a sequence of still photos of the original process, a movie, a recorded voice reading a description of the process. But a simulation should require that the states do more than mirror each other (indeed, a mirror image would not count as a simulation of the thing imaged because the processes are not causally independent); instead the one should be capable of standing in for the other. I have tried to capture the idea of "standing in" with the characterization I just described.

But the above characterization is lacking in at least one respect. It only captures the idea of simulation when there is no input possible. That can be fixed by supposing that if there is some input to the simulated system, there will be a simulation of that input to the simulating system. That said, how does this characterization do on an intuitively clear case such as automatic cameras simulating human retinas? If we let the flow of light into the camera count as the simulation of the flow of light into the eye, then it seems to do OK—each state of the camera is generated by a combination function of the input light signal and the prior state of the camera system, and given the state of light impinging on the retina, the state of the camera film under that impingement is also given.

27.3. Computation and Representation

Winsberg tells us that "while having a mimetic quality is not in itself what gives a simulation interesting methodological and epistemological features, it is an important sign of other features which do. The extensive use of realistic images in simulation is a stepping stone that simulationists use in order to make inferences from their data. It is also a tool they use in order to draw comparisons between simulation results and real systems; a move that is part of the process of sanctioning their results. It is the drawing of inferences and sanctioning of results that give rise to interesting philosophical connections between simulation and experimental practice" (2003, 113). This sounds largely correct, but I don't think we should focus as Winsberg does on the imagistic sense of mimesis. What makes simulations sources of experimental knowledge is precisely their mimesis—that they are systems with classifications that evolve under a given time-development operation, which classifications are infomorphic to distal systems of interest.

When they are working properly, simulations are proximal experimental systems, and they are mimicking the distal system of interest. That mimicking is through the lens, of course, of the salient infomorphism. So only some parts are mimetic: the projectable parts that are related to the distal system through that infomorphism. When it is not true that they are mimicking in that way, they are not working properly: we may still call them simulations, but they are not simulating any distal system according to any infomorphism we can use to gain experimental knowledge.

Winsberg (2010) makes an important point about the direction of flow of the algorithm for constructing and employing simulations. He notes that a typical feature of model building is that features of one simulation—or attempt at a simulation—are used as inputs to the construction process. So the models informing the treatment are sometimes built out of wrong physics in the hope that the simulated physics that emerges from the total simulation will itself be correct, or will exhibit features closer to those of the target system than would a total simulation whose underlying model of the physics might more accurately represent the underlying physics of the target system.

In this connection Winsberg outlines the important notion of *scientific fiction*. These are bits of modeling practice that are not even meant to be reflections of the system being modeled: they are neither approximations nor limiting cases nor idealizations. Instead these items are straightforward attempts to control some behavior of the model that is essential for prediction but which we haven't any ability to model in a physically correct way. His central example is the "silogen" atoms that provide the bridge between scale domains in Abraham's model of the propagation of cracks in solids—a model that marries Continuum Mechanics to Molecular Dynamics to Quantum Mechanics in the various regions of the crack. But it does this by allowing each theory to govern its own domain and to "communicate" across domain boundaries through a kluge whose purpose is only to transfer data about certain variables of interest from one part of the description to another. By introducing the fake "silogen" atoms, for example, the model can capture features of the energy modeled in one domain and encode them in a way that is accessible to the model of its neighboring domain. But from my perspective this sounds remarkably like what is needed to conclude that simulations are experiments. We do not, on my proposal, need all aspects of some idealized experimental concept to be unambiguously realized in order to be engaged in the generation of experimental knowledge. Indeed, if we did, we would never have any experimental knowledge. In general, the underlying features of a worldly system or a simulated system that allow them to function as they do to bear information need not be accessible to the experimenter who is taking advantage of those features. We would always like to know more, but for the experiment at hand, the classification schemes and their infomorphism suffice. The scientific fictions allow for the isolation and characterization of the local logic that maps to the logic of real solids.

I don't suggest that Winsberg is unaware of the following point, but it does need to be emphasized that the tinkering feature, and the practice of using the behavior of the device to inform corrections to the further construction and elaboration of the device, is present in the case of actual experiments as well. One difference there, perhaps, is that such tinkering comes along with tinkering on the "theory of the experiment" (or, better, theory of the device) in a way that is absent in the simulation case where the model rather than the theory is the target of the tinkering. That is to say, sometimes observation of the behavior of an experimental system results in a recharacterization of the device, and other times it results in attempts to restructure the device to preserve the prior characterization. I do not see that Winsberg has success-fully argued that we are more likely to do the latter in the case of simula-tions and the former in the case of standard experiments—but in any case the larger point remains that there is two-way flow even in the case of standard experiments.

27.4. Materiality: Interventions and Ontology

There has been much interesting work lately on the connection between sim-ulation and experiment. Some of that work has illuminated what it is about various activities that is uniquely experimental, and in particular how it is that observations of one system that is radically distinct physically and onto-logically from another can yet give experimental knowledge about that other (Parker 2009, Morgan 2005, Winsberg 2010, Guala 2002).

Many believe that simulations cannot count as robust experimental prac-tices, especially in the sense of being apt to generate experimental knowledge of worldly systems. Their (alleged) failure is often grounded in their lack of materiality, in the sense that the systems generated by our simulating equip-ment, the systems that we can observe[2] thereby, are not material real-world systems. One point of disagreement has been precisely whether to focus on the similarity of effective dynamics or on material similarity. The import of this lack of materiality is taken to be twofold: simulations lack the kind of ontic similarity to their targets (the distal systems of interest) that is found in standard laboratory experiments; and simulations cannot be intervened on, as is required for real experiments. Recent work has shown, I think, that shared ontology is simply not doing the work it is taken to be doing—what does that work is also present in the case of simulations, after all—and that, in fact, simulations can be intervened on in the appropriate sense.

While there is some convergence of views about how to draw lines between simulation and experiment, significant disagreement remains. There is little uniformity to the accounts of what criterial features separate actual experiment from other sorts of experiment. This lack of uniformity has led to a signifi-cant amount of wrangling back and forth over the ontology of experiments. I believe that this lack of uniformity obscures what is in some sense implicitly

---

2. This observation is not solely visual, as in images on a computer monitor (though that as-pect is important). It also includes other sensory modalities, as well as (crucially) data readouts of the state of the simulated system.

acknowledged by all: experimental practice is a method or cluster of methods for extracting information about some segment of the world by controlled manipulation and/or observation of some distinct segment of the world.[3]

I have been speaking as though simulation concerns only computational simulation. Computational simulations are but one (albeit important) class of simulations. Indeed, I think that the systems involved in analogical experimentation are themselves simulations. But computer simulation is peculiar in that the analogy is even more indirect, because the physical system of the computer is not strictly the proximal system of the experiment. Instead the proximal system is the funny quasi system that we find out about by taking status readouts from the computer and by seeing what is happening in a visual representation, as for instance provided by the computer's monitor. In any case, the prevalence of computational simulations in the literature tends to allow their features to be definitive of what counts as simulation. Some draw a distinction between "digital" and "analog" simulation, but that does not appear to capture the key feature that distinguishes computational simulations from other kinds of simulation. Rather, the digital/analog distinction points to one of the key questions one can ask about a simulation: is it a dynamical simulation of a dynamical system (or its features), a kinematic simulation of a dynamical system (or its features), a dynamical simulation of a kinematic system (or its features), or a kinematic simulation of a kinematic system (or its features)? And within all of these various options there is the further question of how the kinematics and dynamics are related to each other. Is the simulation in virtue of the similarity between the equations of motion governing the two systems, or in virtue of the similar time evolutions of the systems (or their features) that may arise even for systems governed by very different forces, constraints, etc.? For example, it is an interesting feature of Hooke's law systems that, in certain circumstances, their behaviors are the same as the behaviors of systems governed by inverse square law forces. In this respect a mass on a spring that is free to rotate about its attachment point on a frictionless surface can be used to simulate the orbital path of a gravitationally bound system, at least for regimes where the spring remains governed by Hooke's law across the full range of the simulated orbit's major and minor axes. This would count, I take it, as a simulation of an orbital mechanical system. I believe what

3. I intend this to apply to even the degenerate case of gaining experimental knowledge of a single device. In that case, the full time specification of the device (or perhaps its future development) is the distal experimental system, while the apparatus at those individual times when it has been subjected to treatment is the proximal.

I say below applies, *mutatis mutandis*, to all manner of simulations, but I will explicitly focus only on computational simulations.

27.5. Ontology in Computer Simulations

Morgan is concerned to understand what some call a key feature of experimentation: that experiments can produce new "experimentally isolated regularities [which] are *noteworthy* precisely because they don't fit into current theoretical accounts (they neither confirm nor deny the theory)" (2005, 318). Drawing a contrast between an experiment which "recreates a part of the real world inside the artificial environment of the laboratory" and a model which is "an 'artificial' world—artificial because made out of mathematics, diagrams, or alternative physical domain materials" (2005, 320), she focuses on the material distinctions between various systems and the other systems they are representing. This materiality consideration is important to Morgan, but it seems like a red herring. The real contrast is probably between recreating part of the actual world and creating an artificial world. But even this is not entirely clear. The difficulty, of course, is that once we have a physical model up and running, it's real and a part of the actual world, and without the right map between an experiment and the rest of the world, the recreated part is as isolated from that world as the model is. So the real question, as always, concerns the status of the mapping.

In Morgan's previous article, "Experiments without Material Intervention," vicarious experiments are the main concern. Here an important issue is her idea that what demarcates experimental from other practices is the intervention. Her effort is to expand the scope of things that count as interventions to include modeling and simulating. Principally, the idea is to put the thought experimentation of economists on the spectrum (at the far end, to be sure, and as a limiting case) of things that count as interventions. While I am not committed to the interventionist standard, this does strike me as a worthwhile move. In service to that idea, Morgan focuses on differences between lab experiments and economic modeling: the achievement of experimental control (2003, 219); and the "production of results" where instead of intervening as in the lab, we use math and logic to generate our results, so what we have is basically the "contrast between experimental demonstration and mathematical demonstration" (2003, 220).

She gives an interesting account of and contrast between two experiments on similar but importantly distinct systems: very thin slices of a bone from a cow that reveal enough of the structure of the bone that it can be digitized and then examined minutely in the computer; and a computer-generated "stylized" bone image. All of the manipulations of the things being examined (the computer images) are of the same sort. She calls this a "mixed method of demonstration" and points out that here the mathematical model is itself the tool of the experimentalist— the model is not exactly being experimented on in this case (2003, 222–23).

What Morgan is asking about in the later part of this paper is how we determine whether and to what extent some one system is similar enough to others to derive appropriate results. This can be parsed as the question: are "representatives of" something apt as "representatives for" those other some- things? She gives no settled answer that I could make out, but appeals gener- ally to materiality and particularity as indicative of the strength of the rela- tion. Basically, though, she rejects "representative of" and "representative for" as useful categories in the mathematical mode. I find myself unconvinced here. First she argues that there are things said in the case of mathemati- cal modeling that are not aptly said in the case of physical representatives: for example, while mathematical models can denote situations in the mate- rial world, a mouse cannot denote a human (2003, 229). But I don't have any problem with seeing denotation in the case of the representation relations for analogical systems. Clearly, in the water/electricity model I considered earlier, pressure denotes voltage, etc. It seems to me more a delicacy of ex- pression than a conceptual distinction that makes us reluctant to say of living things that they denote—but in their role as model organisms, that does seem to be how mice are being used. I am not entirely committed to this, however, and there is an easier way to see through the issue. In the same way that, as Patrick Suppes showed us a long time ago, experiments never come directly into contact with the theories they bear on, but rather interact through many mediating models (of data, of the experiment itself, etc.), so too experimental systems do not come directly into contact with the systems they bear on. In- stead their contact is by means of the infomorphisms between the classifica- tions by which we capture the components of interest to us and the behavior of those components. The elements of those classifications are the means by which the physical systems can correctly be said to represent each other. They represent each other *by means of* the mathematical representations.

But the question of representation here is not the main one. Instead, mate- riality is at issue. But there is a lack of clarity about how more straightforward cases of physical systems giving us the kind of information they do about other physical systems obscures what is going on in the case of simulations. It is not the shared ontology that is doing the work; rather, the shared ontology is related to our long-standing familiarity with the strength of analogical dy- namics and so forth in various cases. That familiarity, that knowledge about when some facts are information-bearing and when they are not, is doing all the work. The way that work is done, however, is (even if implicitly) through our recognition of which tokens are normal and which are not. Easy familiarity with a classification and its apt usage, caused by a similar ontology, should not be confused with the similar ontology itself supporting that apt usage.

Replying to Morgan and Guala, Winsberg also dismisses the distinction between one class of experiment and another as grounded in materiality, and locates the difference in the attitude of the experimentalist. "In simulation, the argument that the object can be used to stand in for the target—that their behaviors can be counted on to be relevantly similar—is supported by, or grounded in, certain aspects of model-building practice. We now need to spell out what those are. What separates ordinary experiments from simulations are the answers to these questions: Why do the researchers believe the object can serve as a good stand-in for the target? What kind of background knowledge do they invoke, implicitly or explicitly—and does their audience need to accept—in order to argue persuasively that one can learn about the target by studying the object?" (2010, 64).

Winsberg's questions are on point, but his conclusion goes too far. It is clear that some systems' failure to share material constitutions does not tell us immediately that their basic, relevant dynamics are not the same. It is precisely the fact of shared dynamics that we do find in certain analogue systems, and there the shared dynamics provides an identity map between the logics in the distal and the proximal systems. As Parker points out, shared materiality, in addition to being unnecessary, is not sufficient. The failure of scale invariance in fluid dynamics, for example, blocks an identity map between the classifications of small-scale models of fluid flow and those of their larger cousins (though there will be, in general, maps weaker than identity that support experimentally based inferences despite changes of scale).

I'll return shortly to the real difficulty for Winsberg's view—that he takes justification to be a necessary feature of epistemology, and moreover thinks that distinctions between things that offer justification can somehow ground a difference in activity. But the real issue is over what allows information to flow.

Parker (2009) takes Guala (2002) and Morgan (2003, 2005) to task for relying too much on the notion of shared materiality. Guala says that "in a genuine experiment the same 'material' causes as those in the target system are at work; in a simulation they are not, and the correspondence relation (of similarity or analogy) is purely formal in character" (2002, 9). Parker's (2009, 495) conclusion is that

> it is the observed behavior of a material/physical system—the programmed digital computer—that constitutes the immediate results of a computer

experiment, and accounts of the epistemology of computer simulation should reflect this.

It is not the case that traditional experiments have greater potential to make strong inferences back to the world than computer experiments do; for both types of experiment, the justification of inferences about target systems requires that scientists have good reason to think that the relevant similarities—whether material, formal or some combination of the two—obtain between the experimental and target systems, where relevance is a function of the particular questions being asked about the target system. Since relevant similarity is what ultimately matters when it comes to justifying particular inferences about target systems, the intense focus on materiality here is somewhat misplaced.

Parker also worries that Morgan's focus on materiality is misplaced, even if correct under some appropriate *ceteris paribus* clauses (2009, 492). But despite this worry, I think we can see that Morgan is right-headed in her view that "ontological equivalence provides epistemological power" (2005, 326), but the view needs unpacking. What is doing the work here is the fact that true ontological equivalence produces dynamical equivalence. The knowledge that two systems are dynamically equivalent supports an identity map between their separate classifications. In the sense outlined here, Guala and Morgan are correct, but while shared materiality (Guala) and ontological equivalence (Morgan) do in fact support these identity maps, they are not, as Parker correctly notes, necessary for them. Whenever we see that the proper classification of the proximal dynamics does in fact support a logic that has an identity map (or even something weaker) to the logic of the proper classification of the distal dynamics, we can see that observational information about the one system can generate experimental knowledge of the other. While Guala's concern overall seems to be with the formal character of the relations in the case of simulating, that concern may best be characterized as a concern with grounds for various inferences in the different cases. If it is our built-in representation scheme, the activity is different than when it is one discovered in nature. I don't think this is quite right myself, but it is not exactly the kind of mistake his talk of materiality would indicate that he is likely to make. Still, Guala's identity between "material causes" is probably best seen as a species of dynamical equivalence, and this, as we have seen, is epistemologically easiest to ground when the systems of interest are ontologically equivalent, although ontological equivalence is not necessary for that grounding.[4]

---

4. I think the prospects for unpacking the notion of ontological equivalence are rather better than Winsberg makes out. As a first pass we might suggest equivalence in the case that we have reason to think that any scientific classification we adopt of the one system as a whole can be

27.6. Interventions on Computer Simulations

Recall that for Parker, "an experiment can be characterized as an investigative activity that involves intervening on a system in order to see how properties of interest of the system change, if at all, in light of that intervention" (2009, 487). While intervention is not necessary for experimentation, still it can be interesting to see whether we can secure the pragmatic benefits of intervening in the case of simulations. Parker continues on, showing clearly that the physical objects the computer comprises are themselves, in some respects, objects of intervention. That is correct, of course, and when we set the simulation going we do so by interacting with those material components. I think the sense of intervention there is weak, however, because it isn't obvious that we are intervening on the simulated system in a way comparable to how we intervene with other systems *just because* we set it up in that way. And moreover, saying that materiality is not the issue with the reliability of simulations because they are materially grounded doesn't quite get the correct sense of experimental intervention.

Morrison (2015) provides an interesting take on the nature of simulations that situates them squarely (in some contexts) within experimental practice. She tells us that the line between experiments and simulations is difficult to draw sharply—in part because of the way simulations can generate observations (even though that is mostly in conjunction with more traditional experimental methods), and in part because simulation is often an intervention into a system we're trying to find out about. Parker (2009) does draw that line sharply: computer simulations will not qualify as experiments because they are not "activities undertaken by inquiring agents," while "computer simulation studies" will because they involve intervening on a system (the digital computer) to see how properties of interest change as a result of the intervention (2009, 498). Here Parker is using a definition of computer simulation that I rejected in my account of simulations. Her use of that definition is related to her own rejection of computer simulation studies as experiments: as she states things, the subsequent states of the computer simulation are not dynamically related to each other, and so they are static. It follows from her view that we do not observe any system by means of the computer simulation; rather, we display a collection of the records of an observation, or of state descriptions. But this collection cannot bear us information, because as the observation of a (mere) time development it gives no support to any

---

applied to the other as a whole. However, the issue isn't really important for getting clear on the experimental status of simulations.

counterfactual claims. Absent the dynamical connection between the states, we have no clear connection to any further system, which connection is the fundamental defining feature of experimentation. I do not agree with Parker on this point, because the states of a computer simulation are related to each other quasi-causally: the state of the simulation is generated by the state of the computer as it executes the program that makes it transition from one state to another. Moreover, the state of the simulated system and its simulated dynamics, when the simulation is doing its work, counterfactually support inferences about its subsequent state. We'll come back to this.

I do agree with Morrison that the distinction between simulation (study) and experiment is blurred, but not for the reasons she states. This is because intervention, as I have shown, is a conceptual red herring. But the connection to intervention is not idle here; what we see, when we intervene into simulations, is that these interventions can produce information-bearing signals in the well-calibrated channels the simulation comprises, signals that can clearly generate experimental knowledge. Indeed, the case for simulations as experiments is even better than Morrison imagines *because* intervention is not relevant. All that is relevant is whether or not the proximal system (the simulation) stands in the right kind of infomorphic relation to the distal system of interest. Given that it does, an intervention is a convenient way of generating signals that bear the information supported by that relation. So while intervention is not a necessary feature of experimental systems, its presence is a good indicator of their power, and in fact we do have the capacity to intervene on computer simulations.

27.7. Simulation and Calibration

Given that materiality does not provide salient criteria for judging the experimental status of simulation, it will be more fruitful to focus on shared dynamics, and in particular on maps between the dynamics of salient classifications of the systems being compared. As we have seen, simulations are also analogical systems as I have used that term in the preceding several chapters. The behavior of the one stands in for the behavior of the other. It is simply that we map parts of the classification of the one to those of the other with functions that are less than identity. And indeed, that failure of identity can be of a radical sort, as we saw. In the type of analogical system that is theatrical acting, the classifications are very tightly related, where the hand of a human (actor), say, bears information about the hand of a human (character). But in the hydrological models of electric circuits I considered before, there is a clear simulation even though the classifications have radically distinct elements: charge is represented by fluid, and the flow of liquid bears information about the electric current. I believe that computer simulations are themselves just another kind of analogical model, and experiments by their means are just another kind of experiment. I think that the main difficulty of seeing them this way is in the ambiguity between finding the analogy in the case of more standard worldly models, and creating the simulation in accordance with the analogy in the case of computer simulation. But this same ambiguity is present already in the case of non–computer-based simulations. Things are perhaps exacerbated in the case of computer simulation by the nonfundamentality of the physics of electronic gate control—the foundation of computer simulation—with respect to the *manifest physical law* of the simulation itself. In the case of computer simulation we don't have as straightforward an insight into that manifest physics, and so lack the right classification grounding the analogical relation as we do in the case of other, more worldly seeming analogues.

There may well be something quite distinct about computer simulations, but I don't see it. They present novel complications in finding infomorphisms— but that is a novel practical difference, not a novel experimental epistemological difference. As a conceptual matter, there is no good way of drawing the distinction to which I am appealing intuitively above. The nature of the experimental knowledge generated in the general class of analogical experiments (which as we saw covers all manner of what normally count as experiments) is no different from that generated with computer simulations. What throws even the clearest thinkers off the track here, I believe, is a continued reliance on a justificatory conception of knowledge generation. Thus, for example, Eric Winsberg (2010, 66–67) tries to draw the conceptual distinction between experimentation and simulation this way:

> The conceptual distinction between experiment and simulation is now clear: When an investigation fundamentally requires, by way of relevant background knowledge, possession of principles deemed reliable for building models of the target systems; and the purported reliability of those principles, such as it is, is used to justify using the object to stand in for the target; and when a belief in the adequacy of those principles is used to sanction the external validity of the study, then the activity in question is a simulation. Otherwise, it is an experiment.

This will not work for me, though, as it continues to confuse the notion of experimental knowledge with other things. What do we call activities where the researcher is concerned equally about the connection between the dynamics of the proximal system and that of the distal system, and about the nature of the dynamics of the former? Winsberg overlooks that the entire structure of calibration is about providing the kind of sanctions he is talking about here. My activity does not become simulation when I am calibrating my device.

A much better way to understand things here is that sometimes I am calibrating my devices and sometimes I am using them to generate experimental knowledge. As we saw from part I, I can be doing one at the same time that you are doing another with respect to the same device. Like any other proximal system in the lab or in the wild, analogical or "direct," simulations must be calibrated in order to generate experimental knowledge. When I am using my observations to find out about the information-bearing properties of the simulation system, then I am calibrating; when I am using them to gain new knowledge of a distal system based on the information they bear about it, then I am experimenting and getting experimental knowledge. Here as always, my intentions and concerns are less relevant than the use to which I am putting my observations, the role they play in my knowledge economy. Even so, whether I am calibrating my simulation or experimenting with it, I am still simulating. Whether some activity counts as simulating is simply orthogonal to whether or not it counts as experimenting.

27.8. Simulations and Knowledge

What we should have learned from the preceding sections is that little of substance hangs on precisely how one defines "simulation." The important issue is the role simulations play in generating knowledge. Using simulations to generate knowledge functions in the same way as experimentation generally, but the points of possible information flow failure differ.

It is true that everything is similar to everything else in some respects, and thus that every system simulates every other in some sense. But this is not very helpful in understanding what allows simulations to generate knowledge of other systems. That understanding can best be gained by characterizing the simulation relation as an information-bearing relation. For while it is true that every system can bear information about every other, in our sense of bearing information we already have a well-worked-out theory of how they do so, and what knowledge can be gained about some from the information carried by another.

Having identified the infomorphism that displays the correct similarity conditions between the proximal and distal systems, we observe the proximal system and establish that it comprises tokens of certain types that are of interest. Which types are of interest to the simulator will be driven by the character of the infomorphism between the two systems. That infomorphism, finally, will be used to support inferences about the distal system that are based on the proximal system being of the types that it is. When we find, by this process, that the distal system is of some given type, we then know that experimentally.

One thing that is perhaps troubling about simulations is that they contain those scientific fictions. Such fictions are apparently sharp dis-analogies in the generation of the (simulated) observational data between the target system and the way we attempt to represent that system in the code-generating results within the computer itself. While the fictions fail to faithfully represent the process under consideration, they do capture the important transitional features the model needs in order to simulate the systems in question. The equations employed in their analysis to generate the output of the simulation have almost nothing to do with the true account of the transitional dynamics the fictions are representing. However, they do generate an output that bears sufficient similarity to the target system that a human user can draw appropriate inferences about the behavior of the latter. We are in a peculiar situation in these instances because we have good reason to think that the simulations are telling us something correct, yet we do not have a good theory of how they are doing so.

So how can we say that these systems are giving us experimental knowledge of their targets? The answer is that they can't be giving us anything else. Consider two cases. In one we have a simulation program with a number of fictions perhaps that cannot be used to generate reliable predictions about its target. In the other, things are similar, except that it is a reliable predictor. It may well be the case, as a matter of logic, that with the first simulation we have just been unlucky and that soon it will begin generating very reliable data. Similarly, it may well be that we've just been lucky with the second and that soon it will stop generating reliable data. However, that is more a skeptical worry than a question about what is happening when we do have knowledge generated by simulations. We are now attempting to understand how it is that we have knowledge of the target system when we do have it. And as we saw before, the only way to gain knowledge about some system is for information to flow. In the first case we have, by supposition, found that we cannot generate knowledge of the target by performing the simulation in question, and in the second we have found that we can. So the inferences that we draw from the sequence of states in the first simulation, when used to generate inferences about the sequence of states in the target system, produce false empirical claims. But in the second those inferences are used to generate inferences that produce true empirical claims. Let's put this into the language of information flow.

The underlying logic of the computer itself is in the very complicated and humanly unknowable processes governing the transitions in computer memory, processes that with contemporary computers can comprise upward of trillions of operations a second. However, the images that are produced by the computer, and the data generated alongside those images, have their own logic. The logic of the simulation itself is unknown, but the logic of the output states, regarded as its own kind of dynamical system, can be known. Generally this logic will be (quasi-)nondeterministic, but that is no obstacle to generating knowledge of the distal system. The key is that the logic governing the simulated normal tokens, output by the computer, is known—in these cases perhaps by inspection. We then assert that the transition rules between token states of the world are those governing these simulated tokens. When we *can* gain knowledge, as in the second case, it is because the connection between the logics of the simulated output and that of the world itself does bear information. Notice that logic here is not the same as dynamics. The point to the scientific fictions is that we can't get the right dynamics programmed into the computer in a straightforward way. Instead we might say that here we have an infomorphism between the kinematics of the simulation and that of the world itself. Here "kinematics" is being used rather loosely to indicate the time varying features of the system in the absence of understanding of the laws governing those features. In the first case the problem is that we are unable to generate an infomorphism. For whatever reason, the logic of the output from the simulation is simply too different from the logic of the world to allow information to flow from the distal system that is the world, through the proximal system of the simulation, into us. In the second case, however, we can generate the necessary infomorphism.

But wait. Have I really answered the question here? How does this work? What gives the users of the simulation the knowledge that there is really an infomorphism between their simulation and the world? That question can only be answered in specific cases by appeal to the specific circumstances of the case. Each modeling practice will have its own tools and techniques first for generating the simulation and then for appraising its accuracy and reliability. But that is not specific to simulation studies; it is common to all experimental practice. How do we know that a given experimental arrangement is capable of generating knowledge of its target system? By showing that there is an infomorphism of the appropriate type and strength to generate sufficient information flow between it and the researcher. How do we show that? By digging into the specifics of the case in the way we have done throughout this book. In the case of simulations we must first check to see that the tokens of the output are normal, and that we are adept at determining their logic. Then we have the task of performing many simulations and checking the tokens so generated against our token observations of the weather, or materials of interest, or what have you.

Let's return to the scientific fictions mentioned above. Far from making the simulations less like scientific experiments, these fictions serve to assure us that the appropriate logic associated to the simulation (the proximal system) really is infomorphic to the logic of the distal system. The simulation's guts—the instructions we pass to the computer, and the kludges that tie together and smooth out the differences between the bits of the overlapping domains of these instructions—are not the simulation and do not directly enter into an infomorphic relation with the distal system. The relations between the observed properties of the simulated systems themselves compose the local logic of the proximal system. The novel work of the simulation, and what separates simulation practice from other experimental practices, is (as Winsberg seems to suggest) tightly connected with the direct control simulationists exert over the properties of the underlying systems that generate their simulations. Simulationists have their own advantages in trying to produce systems that are appropriately infomorphic with systems of interest. Very often they have more direct control over the environmental features of the "laboratory" (that is, the computer in which the simulations are being produced). Their corresponding disadvantage is in having to exercise constant control in order to keep the non–environmentally influenced features of things in appropriate infomorphic relation to the distal systems. But as in other types of experiment, once we have produced the simulations and found appropriate infomorphisms it is a straightforward matter to draw experimental conclusions from them about the distal systems they simulate.

I want to reiterate that the problem of getting simulations that are appropriately infomorphic to systems of interest is a practical one, and that any conceptual problems with doing so plague all versions of experimentation, from laboratory RCTs to natural RCTs to prospective studies to the classes of experiment we just looked at as examples of analogical experimentation. They are *all* analogical, and they all require swapping out identity relations for something weaker, something merely analogical. For some experiments the problem of finding the analogy is easily solved; for some experiments our conviction that the analogy holds is easily gained, and thus so is our acceptance of the results. The ease or difficulty of such things does not matter, though, for the question of whether or not information flows in the experiment. If it does, and that causes our belief, then experimental knowledge has been generated.

In the same way that diagrams ground inferences that provide us with knowledge of their represented systems by bearing information in the way they do, experiments generate knowledge when their bearing information grounds appropriate inferences. I have been trying to consider the generation of experimental knowledge, and to leave aside questions of the ontology of experiment and what should and should not count as an experiment. Instead, I have considered the question of whether we can gain information about one system by considering features of some other system; then I have considered the further question of whether the kind of information we have about the latter system is the result of experimental intervention. Because the answer to both of these questions is "yes," it makes sense to consider the information to be what has provided experimental knowledge of the former system.

But the experimentalist as such (whether experimenting on a simulation or not) needn't care how this is all done.[5] What she needs to care about is whether information flows between the simulans and the simulatum. To calibrate her device she may need to check all of these things, but that is a different task altogether.

---

5. Compare users of very complicated devices (Western blot machines in biology labs, for example) that turn preparations into observable traces. Or a more pedestrian example: I can still get knowledge about what a person at the other end of the telephone line is saying even if my own theory about how my telephone generates its output is based on hopelessly incorrect physics.

28

Background for Thought Experiments

I am going to argue that thought experimenting is simply another form of experimentation in science. These experiments are not arguments, not a priori platonic insights, not mental models—they're experiments. Here, again and schematically, is what happens in standard cases where we get scientific knowledge from experimentation. First a proximal system of some sort is prepared or found. Then observations are made of that system, perhaps as the result of our interventions on it. Finally conclusions are drawn about the features of some distal system (or systems) based on the properties of the proximal system. Those conclusions, when they amount to scientific knowledge, must reflect the flow of information from the distal system through the observed properties of the proximal system, into the heads of researchers.

On the surface, thought experiments seem very different from any other type of experiment.[1] The key difference is that, in the case of thought experiments, there appears to be no proximal system to observe in the first place. That seems to be enough to disqualify them as real experiments; but the trouble with such an easy dismissal is that they do seem, in some cases, to give us empirical knowledge. So how can they do this?

A number of people have considered some version of the question, "Are thought experiments really experiments?" Their answers are varied, ranging

1. My discussion of thought experiments will mimic in some respects my discussion of simulations. The view expressed there is sufficiently controversial that, in conjunction with how controversial the rest of the book must seem, I will not appeal to it here. Rather, I will treat of thought experiments independently from simulations. Those who were convinced by my account of simulations will find more here to agree with; those who were not will, perhaps, be convinced here and so reconsider.

from *yes, but they're nonempirical experiments dealing with the realm of the forms* (Brown), to *yes, and sometimes they're no different really from other kinds of experiment* (Gooding, Gendler), to *heck no, they're just fancied up arguments, and probably using classical logic at that* (Norton)—to paraphrase a bit. What all of these responses have in common is that they agree that knowledge of nature can, in some sense, be generated by thought experiments. I am going to focus on two of these views: (1) they are experiments, but they're not empirical and the proximal system is basically the world of Plato's forms and the truths that reside there; and (2) they aren't experiments, but they are empirically grounded arguments, and when they give knowledge it's because the argument is sound. I will reject both of these views, but I will keep the experimental part of (1) and the empirical part of (2). That is to say, in my view there really is a proximal system to observe, and it's empirically accessible, though not exactly worldly—it is a thought system.

The remainder of this chapter will be devoted primarily to undermining the two views just mentioned; I will present, explain, and defend my own in chapter 29. I will start with a brief overview of some of the discussion of the nature of thought experimentation that has taken place over the last few decades, starting with Roy Sorensen's idea that all thought experiments are one of two types of paradox. Next I will very briefly present and critique James Brown's view that thought experiments are nonempirical methods for gaining access to laws of nature. Then I will explain why the dominant view of thought experiments in the literature, John Norton's argument view, is so far from being a viable account that it is in fact an obvious nonstarter.

28.1. Thought Experiments as Paradoxes

Sorensen's interesting view assimilates thought experiments to experiments by considering the idea of a "discounted performance" (1992, 192) somewhat akin to a future discounting of an investment. The basic idea is that performing an experiment is not, as such, a necessary feature of the experiment. We can, says Sorensen, inch ever closer to the kind of idealized situation encountered in thought experiments by seeing them as a limiting case of an experiment that is further and further delayed. In this case the delay is due not to any contingent features like lack of funding, but to the experiment's impossibility—for example, because we would need infinite refinements of materials to realize it.

Thought experiments, then, are a special kind of nonperformed experiment. As Sorensen says, "A thought experiment is an experiment that purports to achieve its aim without the benefit of execution. The aim of any experiment is to answer or raise its question rationally" (1992, 205), while an "experiment is a procedure for answering or raising a question about the relationship between variables by varying one (or more) of them and tracking any response by the other or others" (1992, 186). He tells us that an experiment must have a cognitive aim—that is, it must be for the purpose of learning something. I'm not entirely convinced about that, but since I don't really have a *substantive* definition of experiment, I won't worry about it. More pressing to me is the question of whether thought experiments can give experimental knowledge. If the answer is yes, then I think we should call them experiments. So in outline his idea seems fine to me: experimental knowledge is the kind of cognitive outcome we normally want from experiments.

How, though, are thought experiments supposed to achieve their aims without being executed? As I understand it, Sorensen's account makes thought experimentation happen all at once, so to speak. We describe a scenario, and we say what would be the result of such a scenario were it actually possible (though we know at the outset that it is not). Merely to understand the description of the setup is supposedly sufficient to give knowledge of what is in doubt. Thus to hear a description of Galileo's experiment of tying a musket ball to a cannon ball to test the Aristotelean conception of the celerity of fall as proportional to weight is to learn the intended lesson of the thought experiment. So . . . if we tie the two masses together, we have a larger mass with more celerity than either alone, but we also have a smaller mass with less celerity dragging back the larger mass with antecedently more celerity . . . and now everyone should understand the point, and the experiment at the heart of this thought experiment need not be performed. Were the situation realized, there would be a contradiction.

There is a lot more to say about Sorensen's interesting view, but this is enough to communicate the idea that, for him, thought experiments are deployments of a paradox. In this case Aristotle's account is itself paradoxical, and that paradox is made manifest by describing the dual aspect scenario it would entail were it a correct account of our or any world. But as we will see, I don't believe thought experiments function quite like this. Nor do I think (as Sorenson does) that thought experiments are principally in the business of modal appraisal. Later I will recharacterize the Galileo refutation of Aristotle in a slightly different way, a way that I think gives experimental knowledge of his result, whereas I do *not* think that the discussion above is really of something that gives experimental knowledge.

28.2. Encounters with Abstracta

Brown (1992; [1991] 2011) outlines a conception of thought experiments that involves, essentially, seeing the truths of certain things (laws of nature, for example) in a way that cannot be grounded in empiricism. I very much appreciate the basic outline of his account of thought experiments—that they are a kind of sensing, and give us knowledge in the way that sensing does. But he missteps radically when he thinks that he has reason to reject empiricism based on the results of thought experiments. He gives three reasons to think that Galileo's thought experiment gives us nonnecessary but a priori knowledge (that is, a priori knowledge about the contingent features of our universe). The first argument is a simple one, but it is profoundly mistaken. He says that the thought experiment gives knowledge without new data ([1991] 2011, 99):

> *There have been no new empirical data.* I suppose this is almost true by definition; being a thought experiment rules out new empirical input. I think everyone will agree with this; certainly Kuhn (1964) and Norton (1991) do. It's not that there are no empirical data involved in the thought experiment. The emphasis here should be on *new* sensory input; it is this that is lacking in the thought experiment. What we are trying to explain is the *transition* from the old to the new theory and that is not readily explained in terms of empirical input unless there is new empirical input.

What Brown does best is illustrate the kind of experience one is having when conducting a thought experiment. But his commitment to a priori synthetic knowledge makes the view impossible, and in fact conflicts with his apparent commitment to the perceptual character of thought experimentation.

Note the clear and unwarranted equivocation between new data and new sensory input. I find out about myself all the time, for example, by thinking "What do I want for dinner? Was that really a good movie, or was I simply swept up in my childish attraction to Tom Cruise?" and so on. The results of these wonderings are new data, but they are not sensory in the way Brown is denying thought experiments can be—because they come into my thoughts not from the surface of my body, but rather by way of other thoughts. But the second and more worrying point—worrying for anyone, not just those who have my view of thought experimentation—is that learning simply does not require novelty. My memory of an event, as I reflect about it, can give me new knowledge that I didn't have before, and yet it all comes from the physical encounter with the world that I am recalling. Indeed, there is a new quasi experience in such cases. While there are no new *sensory* inputs, if by that we refer to body-surface inputs, there are new *perceptual* inputs. My imaginative faculty functions in a perceptual way, and can well generate novel perceptions.

28.3. Just Arguments

Norton has articulated and defended what has come to seem the common-sense view of thought experiments. It is a sober, empirically grounded conception. Thought experiments, according to Norton, are just arguments. When they're good arguments they give knowledge; when they aren't they don't. Yes, there is a lot of talk of imagination and seeing things and all that, and such arguments certainly seem fancy. But when all is said and done, they are just arguments like any other.

Norton's influential view that thought experiments are arguments follows easily from a simple mistake: the belief that thought experiments are persuasive discourse. But they are not. Thought experiments are ways to generate observational data for those who perform the experiments. (In this sense Mach was right all along about the nature of thought experiments.) These data can then be used to derive experimental knowledge, in the same way that data from other observations are used. This claim commits me to providing an analysis of thought experimental error and of how to protect against it. I propose some first steps in that direction in chapter 29 when I lay out my own view. Here, however, I will examine Norton's account of why thought experiments can be trusted. It will become evident how his equivocation between thought experiments as persuasive discourse and the results of thought experiments being used in persuasive discourse underwrites his theory, and why that's a mistake. In short, everything Norton says about thought experiments holds for other sorts of experiments. If the former are just arguments, so too are the latter. But the latter aren't, so the former aren't. Now *that* (unlike a thought experiment) was an argument.

28.4. Experimental Controls for Thought Experiments

Everyone seems to agree that at least sometimes thought experiments can be trusted, but Norton, in his examination of *why* we should trust them, attempts to get insight into their nature by sharpening the question this way: given that not all thought experiments succeed in generating knowledge, what separates those that do from those that don't? His answer is that there is a mark that attaches to those that succeed that is absent from those that fail. The mark, he says, is that successful thought experiments are good arguments. I will appraise this response negatively and offer an alternative that is more in keeping with viewing thought experiments as experiments. My proposal is less clean than Norton's, but then experiment is a messy business.

I now consider Norton's view that thought experiments can be trusted, when they can be trusted, because they're valid arguments. I then show that everyday laboratory experiments share the key features Norton uses to conclude that thought experiments are arguments: they admit of experiment/antiexperiment pairs; and they are sometimes governed by faulty logic. I conclude that the best way to understand how thought experiments can be trusted is by treating them as experiments.

Pace Brown, Norton must be right that there is nothing about thought experimentation that conflicts with empiricism. That is just because empiricism is true, and so if a view is predicated on the failure of empiricism, that view must be wrong. I disagree, however, about what follows from this commitment to empiricism. Norton argues that because thought experimentation is not magic and cannot be experimental, it must be argumentation.

One could instead allow that thought experimentation must be compatible with empiricism, and claim that it is, nevertheless, a form of experimentation. When we reject (as we should) the idea that we can have insight into any nonempirically accessible domain, we do not thereby necessarily reject the view that thought experimenting is experimenting. Rather, we reject the view that our ground for drawing the experimental conclusions that we do draw can be anything other than an empirical ground. This is a far cry from the conclusion that thought experimentation is argumentation, even argumentation in disguise. We can and should take it as a basic presupposition of our scientific practices that the source of all information about the natural world is the natural world itself. Once we are convinced that thought experimentation can help to establish empirical conclusions, then it is our job to determine how they do so.

Norton (2004) poses a challenge to those who think there is more to thought experiments than argumentation in disguise. He develops the notion of thought experiment/anti–thought experiment pairs. These pairs are such that one member of the pair appears to establish some proposition, while the other appears to establish its denial. These experiments cannot both succeed. So a successful account of thought experimentation will have to show why at least one of the pair fails. Norton says that taking them as arguments clearly satisfies his requirement. And having established in this manner that they must be arguments, Norton proceeds to explain what general logical features they must have in order to give us empirical knowledge.

Norton tells us that there must be some "mark that identifies successful thought experiments, that is, those that succeed in justifying their outcomes" (2004, 53). His argument for his conclusion is straightforward: (1) the mark must be internal to the thought experiment, because a "thought experiment is portable and moves wherever its written account goes" (2004, 54); and (2) the mark cannot be brute but must be a structural property shared by all thought experiments that succeed. But he's already told us what such structural features amount to: argument schemas (2004, 53). Therefore the mark is that thought experiments are governed by logic (2004, 54).

Norton is not quite right about what follows from the portability of thought experiments, though he is right that they are portable. They do not, however, travel wherever their written accounts go. To argue that is to assume from the outset that the thought experiment is fully captured by its written account. But that assumption is false. Once all knowing subjects are dead and gone, there may well be written accounts of thought experiments still lying about. There will not, however, be any thought experiments, because there will not be any thoughts. But thought experiments are portable even so—they travel with those capable of performing them. It may well be that I require a written (or some other form of) account in order to perform a thought experiment. But the key thing is that the subject who performs the thought experiment be able to generate *in thought* the thought experimental system. So even were there to be a mark, there is little reason to think that portability requires that the mark be structural.

On the other hand, why think that there is such a mark at all? Norton's answer is this: "Without it, we have no way of determining whether some new thought experiment will succeed in justifying its result; and no way to check that a claim of successful justification is properly made" (2004, 53). First note the return to an epistemological standard grounded in justification; I won't worry about that here. But notice also the return to a kind of verificationist standard. It is generally a mistake to try to ground ontology in verifiability requirements, and Norton's mark requirement is no exception. I won't worry much about that here either—for there is a more pressing issue. The real trouble with Norton's story is a persistent equivocation between "thought experimentation being governed by a logic, possibly of very generalized form" (2004, 45) and thought experiments being arguments. And indeed Norton probably does succeed to some extent in showing that thought experimentation must be governed by logic. But so what? Everything is governed by *some* logic or other.

To think that this establishes that thought experiments are arguments is to miss the obvious analogy between thought experiments and other experiments in this regard. In order to function in the way that they do, experiments in the lab must be governed by a logic, possibly of very generalized form. That doesn't prevent those experiments from involving essentially observations of situations in the world. Similarly, that thought experiments are governed by logic does not tell us whether or not they involve essentially observations of situations constructed in thought. It's just a non sequitur. Norton's entire story about the distinction between thought experiments and other kinds of experiment arises from a very obvious confusion between being governed by logic and being a logical argument. It would be easy enough for Norton to reply that the kind of logic operating in the case of physical causation—Schrödinger equation time development, the logic that connects the past, present, and future of physical systems—is very different from the kind operating in thought experiments. But that wouldn't work. His claim about thought experiments being arguments simply does not follow from the fact that they are governed by logic. The only recourse would be to show that they are *in fact* arguments. Norton doesn't, and can't, do this because they aren't. Their *results* are used in arguments, as are the results of all kinds of other experiments; but, like them, they're just experiments.

28.4.1. WHY TRUST OTHER FORMS OF EXPERIMENT?

The requirement that there be some mark that separates successful from unsuccessful thought experiments fails to take into account what it is that separates successful from unsuccessful experiments of other types. For example, what is the mark by which successful experiments in the real world are known? Is there even a univocal sense of success for experiments? Sometimes an experiment succeeds in persuading us that some object in the world has specific properties, so it is successful in that sense. But maybe we're after success in the sense of establishing that some totally other system has some particular features—for instance, that all electrons have a specific charge based on measuring the properties of a finite number of floating oil droplets. And maybe we're after neither because we're using the experiment to calibrate some other device. Suppose we treat success as giving experimental knowledge. Then what tells us the method by which we can *determine* whether it will "succeed in justifying its result"? How can we "check that a claim of successful justification is properly made"?

In the same way that we can wonder about the mark that attaches to successful thought experiments, we can wonder about the mark that attaches to successful non–thought experiments—that is, the mark that attaches to successful real-world experiments. And there just doesn't seem to be one. But according to Norton's reasoning there should be one, and as with thought experiments it should be internal. Notice that experiments are also portable in the relevant sense. It is not correct to say that they travel wherever their *specifications* do—but we use those specifications to construct the experiment wherever we find the materials to do so. As with thought experiments, the mark for worldly experiments cannot be brute and must attach to every successful experiment.

Even were he correct to conclude that the thought experiment must be expressible as a generalized argument rather than merely be governed by some logic, Norton would not have identified a mark. Rather, it is a fact about the thought experiment; but not all facts are marks. And the fact he uncovers would not aid one in determining whether the thought experiment succeeds.

Let's consider the case of material experiments that do not succeed. Why do we not believe in the phenomenon of cold fusion? Why do we no longer believe that saccharin causes bladder cancer in humans? In both cases the answer involves the attitude we take, as a community, to the status of the experiments that are supposed to have provided key observations in establishing the experimental knowledge claims in question.

28.4.2. COLD FUSION,
AN EXPERIMENT/ANTIEXPERIMENT PAIR

We do not believe that Pons and Fleischmann (and subsequent investigators) have produced sufficiently persuasive experimental evidence in favor of cold fusion. We do not trust their key observations, because their experimental protocols were far enough removed from what we desire that we cannot accept their theory of the experiment, and also because attempts to replicate their observations failed. We may well believe that the observation statements they make concerning their experimental setup are correct, or that their experimental setup was as described, or that their analysis does reveal the connection between their setup as described and the way the world would have to be in order for their results and experiment to be as described. However, we do not believe all of these together. There have in fact been fascinating analyses of precisely where things go wrong in their experiments, but I cannot go into them here.[2]

Naturally Pons and Fleischmann will have their champions. But for most of us their case is an object lesson in just how difficult it is to really perform an experiment at all. Certainly Pons and Fleischmann had electric current sources in their lab; certainly they had some palladium samples with deuterium compressed in; certainly they recorded some numbers in their notebooks based on readings that appeared on their instruments. Even so, most of us will deny that they performed an experiment that established any knowledge at all about cold fusion. There was a situation in the lab. There were actions

2. See Gieryn (1992), for example.

performed that appear to correspond to actions that, performed in other contexts, would constitute experimental interventions. But there is a clear sense in which there was no generation of experimental knowledge.

The principal issue from our perspective seems to be that their experimental results were impossible to replicate, despite others performing what are nearly indistinguishable interventions on nearly indistinguishable setups. In other words, we have here examples of experiment/antiexperiment pairs. Look to any experimenter who took Pons and Fleischmann's report of their experimental arrangement and constructed that arrangement in the lab. Failing to detect cold fusion there is consistent with a successful experiment.[3] Pons and Fleischmann's efforts are constitutive of an unsuccessful experiment. How do we determine which succeeds and which fails? Given competing experiments, we appeal not to internal but to external factors: How many researchers can, and how many cannot replicate it? Which researchers do we trust? Does one result violate some strongly held physical principles? By answering these kinds of question we come to have confidence in one or the other of the pair being unsuccessful. But what matters for experimental knowledge is which, if any, of these situations can support information flow and thereby generate knowledge.

This case is consistent with believing that sometimes thought experiments produce novel experimental observational facts and sometimes they do not. If we think that thought experiments are like other forms of experiment, then there is no mystery about thought experiment/anti–thought experiment pairs, for such pairs are found in all branches of experimental science. Thought experimentation and other forms of experimentation are, on this point, comparable.

28.4.3. TAB COLA, VICTIM OF AN EXPERIMENT GOVERNED BY A FAULTY LOGIC

Note that the logic that is faulty here is really just the logic relating normal tokens of the rat systems to normal tokens of the human systems.

Regarding the alleged carcinogenicity of saccharin, we are pretty clear that what we do not trust is the connection that was supposed to obtain between the observed properties of some peculiar population of rats and those of humans at large on being dosed with comparable amounts of saccharin. It just

---

3. Compare in this context replications of Joe Weber's experimental arrangements for measuring gravitational waves.

turns out that these observations do not suffice to underwrite conclusions about nonrat populations.

The document "Evidence of Carcinogenicity of Sodium Saccharin" produced by the Reproductive and Cancer Hazard Assessment Section of the Office of Environmental Health Hazard Assessment of the California Environmental Protection Agency (Krowech et al. 2003) has a useful discussion of the various studies that are meant to show the carcinogenicity of saccharin. A striking feature of these data is summarized this way: "The notable differences in tumorigenicity in studies with different exposure periods supports the hypothesis that critical cancer-related effects of sodium saccharin occur during periods of organ growth, namely cell proliferation or cell division within the urinary bladder, which primarily occurs post-gestationally, during the first three weeks of life in the rat (Cohen and Ellwein, 1990)" (2003, 22). Another feature is noticeable in the last reported rat study of Fukushima et al. (1983), which used four different strains of rat and detected bladder cancer in only one, the ACI strain (1983, 21). The long and short of most documents addressing the issue today is "that the formation of a crystal deposit in rat urine is associated with the later development of bladder cancer in rats. . . . In humans or other species (mouse, monkey) all of the necessary conditions leading to formation of the urinary deposit in rats do not occur" (Health Canada 2007, 2).

In other words, rat metabolism of saccharin is not sufficiently similar to that process in humans to support inferences about humans based on the observation of rats. Let me spell out the connection with Norton's view: Here there is clearly a logic governing the experimentation, where one could reconstruct premises to the effect that if p is a property of rat metabolism of saccharin then p is also a property of human metabolism of saccharin. The falsity of that premise explains the failure of this experiment to generate experimental knowledge regarding human bladder cancer. However, it certainly does not show that this experiment fails because it is a bad argument. Instead it shows that these perfectly good experiments (and their valid observation results) do not figure into a successful *but separate* argument about human bladder cancer.

In both of these cases we are confronted with difficulties attendant to the process of establishing experimental knowledge. In these cases and many others (Hertz versus Thomson and Perrin, Weber's flawed gravitational wave experiments, the many unsuccessful results in the social sciences, etc.) we see how difficult it is to come to know the world on the basis of acute observation of complicated and contrived situations in the material world. Or rather, we see how difficult it is for the vast majority of us to come to know the world on the basis of descriptions of such situations. And yet we do know about the world in this way. The fact that sometimes things go wrong does not prompt us to reject this as a valuable and even reliable mode of knowledge production, nor to call it something other than experiment. Instead, we attempt to develop and display methods that help us to safely adjudicate questions of experimental fact production, theories of experiments, and the relation between experimental/observational facts and the knowledge we hope to derive from them. We do not judge experiments as having succeeded or failed based on whether they possess some internal mark that is common to all good experiments; instead we use whatever tools expedience dictates for the given situation.

28.5. Trusting Thought Experiments because They're Experiments

Why are we reluctant to count thought experiments as providing us with experimental knowledge? I think it's because we're asking the question the wrong way. When we ask, "What kind of thing is a thought experiment?" and we answer that it is a kind of argument, or it is a sequence of mental models, or it is a direct encounter with the realm of the forms, we are proposing all of these answers because we see thought experiments as in some measure lesser than or defective versions of more standard kinds of experiment. Even Sorenson, who comes closest of all in saying that thought experiments are the limits of real-world experiments, cannot quite find his way to the conviction that thought experiments give positive knowledge about the natural world, rather than merely whatever follows from their refutation of possibilities or their refutation of a claim of impossibility. But that is in fact what thought experiments do: they give knowledge of nature in an observational way. We should be asking instead, "What do thought experiments produce?" And what they produce, at least sometimes, are novel observations (see Gooding 2002 and Gendler 2004).

The reason we distrust thought experiments is that we have a wrong view of experiment: that is, we think of experiments as interventions into material systems, with the observations that result giving experimental knowledge. Consequently thought experiments are thought not to be relevantly interventional, and to give no relevant novel observations, and so are not thought able to give experimental knowledge. One must be clear on the model of experiment which is operating in the background and which allows us to see them that way. And as we have seen, that model is something like the following: we intervene on some system in the world; we observe its behavior; we have experimental knowledge. Thought experiments do not involve any relevant intervention on worldly systems; they do not afford relevant observations of worldly systems; so we don't have experimental knowledge. I endorse Gooding's claim that thought experiments are much more similar to our experimental practice than is normally supposed, but not his view that experiments have much to do with embodiment. Experimental knowledge is distinct from the experimental activities with which it is normally associated. The reason for this is that virtually all experimental knowledge arises far (temporally, materially, conceptually, etc.) from the place of experimental activity. Experimental activity is, in my view, the storehouse of the observational basis of experimental knowledge. Experimental practice, and the kind of craft knowledge associated with it, is conceptually far removed from the kind of knowledge of nature we have, since the scientific revolution, taken experimentation to provide. And perhaps that distance is why it has proven so difficult to understand experimental knowledge when we cannot detach experimental practice from the knowledge it produces.

Norton has given important clarifications about the working of thought experiments. He has made clear that an important role for thought experiments is to establish empirical knowledge claims. Where he moves too quickly in his analysis is at those points where the connection is drawn between the *use* of thought experiments and the *nature* of thought experiments. When Norton contends that thought experiments are arguments, he focuses only on the uses to which the results of thought experiments are put and the constraints of practice on that use. But he does not adequately consider similar claims in the context of actual experiments. Norton's position seems to be that because thought experiments are appealed to in the context of persuasive discourse, then they are themselves a kind of persuasive discourse. But that is not right. For precisely the same arguments that he uses to establish that thought experiments are arguments can be redone, *mutatis mutandis*, in the case of material experiments. They are used in persuasive discourse; they may appear in countervailing pairs; they are governed by a generalized logic. Indeed we could redo these arguments with all forms of experimentation: standard lab experiments; natural experiments; simulations; analogical experiments; and made up stories even. All of these are appealed to by those who would persuade us of some conclusion, and yet none of them are, themselves, persuasive discourse. What is distracting in the case of thought experiments is, perhaps, that they are not material and so appear to be much more like discourse than these other things are. But that is an illusion, and a very obvious one. For the thought experiment is not the chat about the thought experiment that I appeal to when I attempt to persuade you of some novel empirical claim. Rather, that chat is a *report* on what I claim I found when I performed the thought experiment and what I claim you would find were you to perform the experiment yourself. Similarly, the chat about some experiment performed in the lab is not the experiment performed, but a report on what I claim was found when I or others performed that experiment and what I claim you would find were you to perform it yourself.

It is true that the observation reports from thought experiments, even when they are not what we expected going in, must still figure into some kind of good argument in order to generate experimental knowledge. That, however, is a feature shared by all forms of experiment, and so does not single out thought experiments. Nor, consequently, can it suffice to establish that thought experiments are no more than arguments without doing the same for other forms of experiments. Again, if all Norton really means is that some logic or other characterizes the connection between initial and final state in the thought experiment, then so be it. But the same is true for the initial and final states of any other experiment. Generally the logic is that of causation, or the Schrödinger equation. But that doesn't make *those* experiments arguments.

How would one fare in an attempt to argue that all actual experiments are themselves arguments? Not well, I think. Experiments are just a different kind of thing. They are spatially and temporally extended material practices that help to move information from one part of the world to some other part. They do this by generating observational knowledge of some system, and then we appeal to our other background data and theory to establish experimental knowledge of other systems. Experimental activity is, on this view, the source of the observational basis of experimental knowledge.

So if we are not to trust thought experiments because they are arguments, then I propose that we trust them because they are experiments. And to do that we will need to treat them as experiments. But should we trust thought experiments to generate experimental knowledge? Can merely thinking about some situation increase my knowledge about how things are in the world outside the mind?

28.6. Standards on Thought Experiments

Thought experiments can certainly fail, as can other experiments. The intended situation may not be constructible in thought, as we will see with Eppley and Hannah's experiment (and yet even in that case one can get empirical knowledge from recognizing precisely how it is that the situation is unconstructible). One's judgment about how things will play out for situations that are constructible may be incorrect. One's conclusion that these observations support inferences about material systems of interest may not be validly drawn.

These are serious issues, and one might well despair of the possibility of securing experimental knowledge at all with such an unsuitable tool. But note, as I have repeatedly emphasized, that material experiments suffer from these same challenges. While we may have in front of ourselves some material system and a method of intervening on that system, that method may well not function as we think; the material system in front of us may not be of the sort that we think it is, and importantly may not be of the sort that it must be in order that our interventions produce the observations we need in order to draw further conclusions (even if it has passed all the tests we know how to apply to it to be sure that it is the right sort of system). Or the results of the intervention may not turn out as our data indicate—that is, our records may well not correspond to the actual events we are attempting to measure and record; and our conclusion that the observation supports inferences about material systems of interest may not be validly drawn. Let me be clear at this point. I am not arguing that there is no difference between thought experiments and other experiments. Rather, I claim that grounds for suspicion or caution about thought experiments of the sorts I identified above are not, as a conceptual matter, the appropriate way to understand their differences.

If thought experiments really are experiments, why are there so many bad ones? It's all very well to point to a few favorites and claim that they can generate experimental knowledge. But unlike the case with material systems, it's a trivial matter to simply suppose the outcome one wants and say it results from a thought experiment. So the question gets at an important issue, but I think it is not framed exactly correctly. While there may be some thought experiments that turn out to yield false or ambiguous results, many of what count as bad thought experiments simply fail to qualify as experiments at all—thought or otherwise. The difference here between *purported* thought experiments and other kinds of experiment is in the standards we apply to each before we endorse them as experiments. We have no good method in place for evaluating whether some description of a situation counts as a thought experiment. In short, in the case of thought experimentation, we lack a good notion of experimental controls. Thought experiments are easy to pose and difficult to evaluate. One might well take this situation as an indication that thought experimentation is markedly different from other forms. As I have argued, that would be too quick. Instead I think the right approach is to call for higher standards and to develop methods of assessment that can successfully distinguish between what does and does not count as a thought experiment. We should not uncritically accept as thought experiment everything advertised as such.

Still, a significant obstacle to efforts to understand the nature of thought experiments and their role in producing experimental knowledge is that we have very little to go on when we attempt to assess their constructibility. But without some method for assessing that, we cannot easily determine whether we are dealing with a thought experiment rather than some vague imagining, or some other kind of thought situation that does not bear at all on empirical matters. An obvious suggestion for judging whether something or other should count as a thought experiment is that the situation imagined should be constructible in thought. What is missing, then, is something like the protocols established by the Royal Society for vetting and endorsing matters of scientific fact, for conferring the status of "observation," but for the case of thought observation. The problem with thought experiments is not that they are not really performed, but that we lack appropriate standards for appraising whether they have been performed properly. It is not possible to be sure whether the observation reports from someone else's thought experiment are genuine or simply made up. While we would like a process of appraising proposed thought experiments, so that some proposals are rejected and others are accepted as legitimate experiments, we don't yet have anything that works very well. And if nothing else, this discussion will, I hope, direct the attention of philosophers toward the task of developing such standards.

Even before we have such a process in place, though, we should be able to recognize the parity between bad or defective thought experiments and bad or defective other types of experiment and derive some guidance therefrom. How do we judge the success of thought experiments in adding to our observational basis? By paying close attention to their articulation in order to judge whether the situation is really constructible in thought, by attending to the appropriate time-evolution operator governing the thought situation, by having others attempt to replicate the experiment, and by varying the parameters in order to see how sensitive the results are to the particularities of the initial state. In short, we judge thought experiments by the standards that govern the rest of our experimental practice.

28.7. Two Ways to Learn

In this final section, I will address one possible confusion that is sometimes used as a reason to reject the experimental status of thought experiments. The idea is that because in some cases I could have found out exactly what I found out when I performed the thought experiment by *instead* performing a logical derivation, then there is no more to thought experiments than logical derivation. This path is related to the argument conclusion discussed above. So consider the following point.

In electrodynamics we have a situation where the magnetic field and the electric field have been found to be dual aspects of one and the same electromagnetic field. Which of these features is observable in any given situation is a feature of the situation. Sometimes I can measure a nonzero electric field and a vanishing magnetic field. That's my situation. But if you are traveling along past me with a constant velocity, you will generally measure a different value for these fields. I cannot *observe* what you will observe, but I can find out about it by measuring what I do observe and using the values I find along with the Lorentz transformations to figure out yours. I might then send you an email with my results for your field strength values, and you might find them out that way. On the other hand, you might measure them and find them out that way. I would think that which you choose determines whether or not you have observational knowledge of your field strength values. In the same way, I think that there are many avenues of finding out about the same empirical things. I do not, however, believe that *because* you could have found out something by observation, you did so. The possibility of one mode of activity does not, generally, rule out the possibility of some other mode.

Norton's mistaken view that thought experiments are merely arguments follows in part from his mistaken idea that thought experiments are fundamentally persuasive discourse. They aren't, though their results, like the results from other kinds of experiment, are *used in* persuasive discourse. This is related to another mistake: thinking that because I could have made an argument, I was making an argument. That's as silly as saying that because Hooke could have been told that the force goes as the tension and so could have learned it from testimony, in fact he did not find out about it experimentally. The one has little to do with the other. The fact that some clever person with better facility with the calculus than I have could have solved some problem with a deductively valid derivation in the first-order predicate calculus has little bearing on how I solved the problem. Why do we think geniuses like Gauss are so interesting? In part because of the *way* in which they solve problems. Gauss *could* have added up all the numbers from 1 to 100 as his classmates were allegedly trying to do. Instead he added 1 to 100 and multiplied by 50. The semantic entailments are the same, but the actual thing done is very different.

Here is one more example. Einstein suggested a version of the following thought experiment in an attempt to show the equivalence between gravitational fields and accelerated frames of reference.[4] (I change it slightly here to make more clear its observational import.) Imagine someone trapped in an

4. Not everyone thinks he succeeded in this, but he did succeed in displaying some observational similarities between accelerated and gravitational systems.

elevator, unable to see the outside and trying to determine whether the elevator is accelerating uniformly upward or is suspended in a gravitational field. At first there doesn't seem to be anything to distinguish between them. Stepping on a scale will have the same effect, playing handball will go exactly the same in either situation, etc. The features by which we normally experience the gravitational field of earth are indirect—it is by the earth deflecting us from our otherwise unaccelerated state that we "feel" the gravitational field. But there is *one* apparent difference. Recalling from special relativity that light is indifferent to the state of motion of its source, we might wonder what will happen if we shine a laser across the elevator. Propagating the uniformly accelerated elevator upward with the laser shining horizontally will result in a beam with a curved path. That's a difference between the two systems, you might think. In fact, though, this is a way to see that light travels on curved paths in gravitational fields, so rather than finding a difference between the situations, one can learn about the gravitational field of general relativity this way. One could very well have discovered this by calculation, but calculation is not necessary. And calculation is, in fact, not how I first understood what was going on in the example. In my case at least it was the imaginative propagation of the light beam that informed me about what was happening.

When I perform a thought experiment, when I construct a system in thought and allow it to evolve according to a given rule of time development, I am doing one thing. If someone else wants to find out the same thing by performing an experiment on a worldly approximation to my system, or by making some kind of an argument, so be it. Let's let a thousand flowers bloom, and let us not assume that because one flower is blooming over there, there isn't one blooming over here.

29

Thought Experimental Knowledge

In chapter 28 I showed that objections to thought experiments being experiments were misguided. In this chapter I will give a positive account of thought experimental knowledge *as* experimental. That will complete my discussion of thought experiments as well as my discussion of experimental knowledge generally. In a way it makes sense to read this account of thought experiments as the point to which the book has been directed. I have moved by stages to the position that even thought experimentation, the furthest activity from the standard account of laboratory experiments, will be seen to give knowledge that is every bit as experimental as any other kind of experimentation. The previous parts of the book showed how experiment works: in the first case by looking at how replication, intervention, and calibration figure into experimentation; in the second by illustrating experimentation as a regulation of information flow between the researcher and the rest of the world. This part has been illustrating that activities further and further removed from the RCT of science class exemplify that the way one system's features bear on another, experimentally, is always analogical. All of that has taken place by reference to systems that are physically instantiated: even simulations are observed with physical senses. With all of that in place, there is no reason to disregard thought experimentation. First, there is a system on which we can, if we like, intervene. Second, data are taken regarding the behavior of that system under time development. Third, there is an appropriate infomorphism between that system and the distal system of interest. Objections to each of these three claims are easy to meet.

   In addition to arguing directly that thought experiments give experimental knowledge, I will outline a criterion of constructibility for what should count as thought experimentation. As I will make clear, not everything that is called thought experimentation merits the name, just as not everything that is called experimentation does. The divide that separates thought experimentation from other forms is narrow. Indeed, these practices differ epistemically at only one point—the validation of some particular observation statement (or class of statements) at the heart of the experimental knowledge claim. This point looms large in nearly every discussion of scientific thought experimentation. Both sorts of observation, however, share the feature that they play their role in establishing experimental knowledge not in isolation, but against a vast network of such claims and habits of inference and material technology and other background views. Even so, the observations in thought experiments seem less secure than in other kinds of experiment. The thought experimentalist makes experimental knowledge claims based on the expectation that we will endorse her judgment that such and such would be observed were we ourselves to construct the situation she describes (either materially, or in thought in the case of idealizable but not realizable situations). Someone else who has built, calibrated, and deployed some device can, by contrast, say, "This is what I did observe when I realized the situation I described." Yet this is not all there is to their experimental knowledge claim. Moreover, the security of the observational result may be no greater in their case than in hers.

29.1. Just Experiments Plus Standards

My claim here is simple: thought experimentation just is experimentation; it can give experimental knowledge. There are many arguments against this view, but it seems to me that all but one are bad, and the better of those bad arguments I looked at in chapter 28. The good argument doesn't work either, but it does need to be taken very seriously. That argument is pretty simple and it goes like this. Thought experiments do not generate novel data; experiments require novel data: therefore thought experiments are not experiments. The reason this argument doesn't work is that both premises are false, though they are at first blush extremely plausible. We have already seen, in the case of replication without new data, that it is possible to perform an experiment without ever generating novel data. In that case the novel calibration provided by new protocols sufficed to generate different observations than those in the original experiment even though there were no new data. I anticipate, however, that not everyone will be satisfied with this claim. First, not everyone will believe it. I do think the argument I made for that conclusion in part I of this book is pretty compelling, but it's sufficiently odd that some will strongly resist it. On the other hand, even if it's granted as a general matter that experiments don't *always* need new data, we still have the problem that if thought experiments *never* generate novel data, then there aren't any thought experimental data at all to start with. So the conclusion that we could replicate thought experiments without generating new data would be pretty cold comfort. I do agree with this myself, in fact. Thought experimentation is going to have to generate novel data at some point in order to count as experimentation.

I'm going to argue that thought experimentation *does* generate novel data, but I will do so more toward the end of this chapter. First I need to be clear about what it is that we are doing when we perform a thought experiment. Once I have a relatively explicit model of thought experimentation to work with, I will show that these experiments do produce novel data, and thus are capable of generating experimental knowledge. On that basis I will conclude that they are, after all, experiments.

29.2. What Is Thought Experimentation?

We can begin here with Elke Brendel's (2004) useful categorization. She focuses on the reconstructibility of thought experiments as logical arguments providing an essential goodness requirement, and as following from the demands of naturalism. Argumentative reconstruction is what provides an answer to what she calls here the "informativeness problem." How do we get information? We know it is *possible* to get this information when we display the thought experiment reconstructed as an argument. Some such response is necessary to avoid the strength of Brown's alternative suggestion that thought experiments are informative because of their *non*natural character.

Brendel also claims that thought experiments rely ineluctably on intuitions. She gives a naturalistic account of those, claiming that they are "best regarded as mental propositional attitudes which are accompanied by a strong feeling of certainty" (2004, 96). In this way she argues that her account makes clear how thought experiments can fail; intuitions themselves suffer from "fallibility, relative instability and fragility" in certain domains (2004, 90). Because very often these intuitions are providing important background conditions for the thought experiment, when they go wrong so too will the experiment.

Brendel's account of what makes thought experiments *experiments* is that their proper functioning constrains them to reveal the "functional dependency of variables by planned and controlled data change" (2004, 91). She also (confusingly, given Hacking's establishment of what we might call the independence of experiment and theory) claims that they are like material experiments because both rely on "background assumptions or background theories" (2004, 91). But either this claim is trivial, in the sense that everything we do relies on whatever it is that constitutes the background for our investigation, or it takes us back to a conception of experiment that relies too much on theory for its constitution. Experiments do not *rely on* background theories for their proper functioning, although it make take a lot of theoretical knowledge to get an experimental apparatus up and running properly. What requires background theories is not the experiment as such, but rather the drawing of correct inferences from the experiment—the generation of experimental knowledge. And that feature material experiments and thought experiments do have in common.

Brendel's overall view of thought experiments, though, is in many respects congenial to mine. She takes it that the "main difference between thought experiments and real [*sic*] experiments lies in the fact that according to the intentions of the 'thought experimenter,' the aims of the thought experiment can be achieved without needing to perform a real experiment" (2004, 91), echoing Sorensen's definition. As stated, though, this is hopelessly circular. But presumably what she has in mind is that some interventionist experiments can be performed without that intervention being on a material system. She also, correctly in my view, outlines a productive role for thought experiments, whereby they give support to theories rather than playing only a destructive role (2004, 92). Brendel analyzes Galileo's falling body thought experiment and notes that we cannot, as Brown seems to think we can, derive the contradiction without appeal to further features of the system. For example, she reminds us that in fact bodies do not all fall at the same rate in any given medium; bodies' weights are irrelevant to their rate of fall only when there is no resistance in the medium. This is a well-known point, but Brendel's observation helps us to focus on the necessity of giving a complete enough description of the initial situation, as well as of the presumed local law of time evolution of the system.

While she accepts Norton's conclusion that thought experiments can always be reconstructed as arguments, and indeed that this provides the foundation for how we extract information from them (2004, 95), Brendel is reluctant to concede that thought experimentation just *is* argumentation. There is, she claims, something special in the way thought experiments are used to restructure prior empirical data in order to provide this information (2004, 95, 96).

Her view, then, can be summarized this way: thought experiments are methods of extracting information from empirical facts and theories that we already know about, and they function by the exercise of a capacity that we do seem to have: we can see by the inspection of our intuitions that these facts and theories have certain properties, properties that *could* be revealed by a logical analysis but often are not. The modal operator here is crucial for underwriting the informativeness of these experiments.

It is here that I reluctantly part company with Brendel. First, thought experiments are not epistemically good because they can be shown to have good logical structure; rather, the reason they can be shown to have that structure (when they can) is because they're good. Second, like Norton and virtually every other analyst of thought experiments, Brendel has failed to take seriously the structural identity between thought experiments and material experiments. That identity shows that if what gives thought experiments their validity is logical reconstructibility, then that is also what gives validity to material experiments. Since the latter is false, so too is the former.

The reason it is so hard to see that thought experiments are structurally identical to material experiments is that material experimenting is much easier to see as separate from the uses to which it is put. That is to say, doing things in the lab is easily distinguished from telling someone what we were doing in the lab and from attempting to persuade someone that those doings generate experimental knowledge. With thought experimentation, however, when I tell you what I did and attempt to convince you that it gives knowledge, it can sound like the telling and the experimenting are more or less the same thing. Moreover, it can sound like the *persuading* is somehow part of the thought experiment itself. And that is just not right. It is no more right than suggesting that material experiments, because they are used in persuasive discourse, are themselves persuasive discourse. In any case, experiments themselves are not used in persuasive discourse—their results are. Similarly, thought experimental results are used in persuasive discourse; the thought experiments themselves are not. And the thought experiments themselves are absolutely not persuasive discourse. That is just a category mistake.

What then is thought experimentation? In my view it is just like other experimentation. But instead of using occurrent data generated from some situation in a lab or in the wild (as in natural experimentation), it uses nonoccurrent data generated in the mind to support the inference that some distal token D is of type Δ. It proceeds, as experiments usually do, by attributing the type Π to a proximal token P on the basis of observations of that token, and then using that as a signal that bears the information that D is Δ. Generally, for the proximal attribution of Π to P one must appeal to prior empirical data (and in fact that is always the case, since no facts about the world follow from pure logic other than, at most, trivial implications like "if the world exists, then something exists," etc.), but sometimes the bit of data that goes beyond what other data we already have is a logico-mathematical fact. For example, perhaps Benford's law allows a thought experiment showing that someone was cheating on a tax return. . . .

What are the appropriate infomorphisms in the case of thought experimentation? As always, the plan is to use them to connect up the local logic of the proximal system with that of some distal system. So we first specify the distal system of interest. Once that is done, we consider the proximal system and its associated classification structure. What do we know about its local logic? Generally, in the case of thought experiments we will proceed backward. In material experiments we begin with some initial state of the system, and once we have seen how it behaves under various circumstances (typically under interventions and controls), we are able to infer its local logic, the logic that reveals the causal relations among tokens of various types. This is how it works at least in cases where we can unproblematically relate the system to itself at various times. In the case of thought experimentation we very often appeal to background knowledge to generate a new state from an initial state; rather than using worldly causation, we use the causation inherent in the situation as constructed in thought. We should not make either of the complementary mistakes of thinking that this causation is the same as material causation, or thinking that it is unconstrained and can be whatever we want. The specification of the (thought)-temporal flow in the experiment must be part of the specification of the experiment itself, just as much as the (thought)-situation itself. Without these in place, we cannot say with any clarity how to classify the initial system, nor how to evolve that classification in (thought)-time.

As is true for material experiments, thought experiments always involve multiple classes of induction. So even when we are able to generate the right local classification structure and to use it to generate new data from old, we must still be careful that the map we construct, connecting the proximal to the distal system, really will allow information to flow: in other words, that it really is an infomorphism. Only in the case that we do have an infomorphism will we be able to generate experimental knowledge using our thought experimental apparatus.

The intuitions Brendel consigns to the background of thought experimentation thus are actually front and center. But the propositional attitude is not the key thing. Instead the key is the target of those propositional attitudes. Once a system is constructed in thought and its initial state is specified, that system must be evolved in (thought)-time; and for that, some time-evolution operator must operate. Sometimes thought experimenters will suppress discussion of the operator, assuming that we will all, in conducting the experiment, apply the same, correct operator to the initial state of our own construction of the thought system. When that is not true, though, we can easily run into difficulties for replication.

29.3. Thought Experiments Defined

A thought experiment should be a mental representation of a situation that could be constructed in the sense that the specific initial state that would result in that situation is possible (even using some laws of nature different from those of this world) and the time evolution of that initial state into the situation of interest is possible. And the time evolution governing that state should show the system so modeled to be of some type. That is the thought analogue of intervention on a proximal system establishing that it is of some type. We gain experimental knowledge of some distal system when the proximal system being of that type bears the information that the distal system is itself of some other type. This is just as I have proposed and argued for as what it is to have experimental knowledge generally.

Thought experimenting, then, is a kind of experimenting.

29.4. True Experiments

The glaring difference between thought experiments and other kinds of experiment is that there is no true proximal system in the world that is the subject of the experiment. Of course one might say that there is such a system: the brain of the thought experimenter. But to learn of a particular human that her brain state was thus at the start and so at the finish of a thought experiment tells us little about the system in the world she was representing with those brain states. The fact that her imaginative state transitioned into another in accordance with a given rule, though, may well bear us information about that worldly system.[1]

Surely it is right that the mere fact that one has two, or several, or even a continuous series of images is not sufficient to establish any significant empirical facts. We all ought to be Humean enough to recognize that an image in and of itself has no connection to any others in a given series of images. And noting that I am the kind of person who tends to associate such images is also not enough (*pace* Hume) to derive empirical import. What is necessary

---

1. I do not mean to rule out the very unlikely possibility that in the future we may be able to read mental content from brain state. But presently we can't get thought-experimental knowledge that way.

is a reason to think that this sequence of images counts as a proximal system with certain properties, and that the observation of it bears the information that some distal system has a given property. Whether or not the things we have said so far in this chapter suffice for establishing that one can get empirical knowledge out of thought experiments, we are still confronted with the problem that we are lacking a proximal experimental system. Cannot all of the things we've said so far be fit into, say, Norton's conception of thought experiments as fancied up arguments?

I want to challenge the idea that there is no proximal experimental system in the case of thought experiments, and I want to do so much more strongly that does Parker in her suggestion that computer simulations are experimental systems. I will claim that thought experiments are themselves full-fledged experiments. Recall the discussion of replication from part I. We showed that one could replicate experiments without ever producing new observational data. That is, we showed that reconsidering the data generated in some sequence of interventions could count as performing a new experiment. Here I will claim that we can generate novel experimental observations imaginatively, and that those count as observations that a proximal system has some particular properties. Moreover, that the proximal system has those properties can be used to support the inference that some distal system has the properties that it does. How?

Our body of background knowledge is such that in certain cases we can use that knowledge first to construct imaginary situations that we know are realizable; then to propagate those situations forward in time using known laws to construct imaginary situations that, on the basis of that construction, we come to know are realizable; then finally to note of those new situations that they have the properties they do. And then we are back in familiar territory where we use the tools we've discussed at length already to infer the properties of actual systems in the world. This does not count merely as noting the properties of some agent imagining some situation. When the thought experiment is conducted properly, we have observed the properties of a realizable though not realized system.

Here is an example. I recall one day when we were celebrating the national holidays in Chile. My father-in-law wanted some champagne chilled, but we were short on time. We were going to put the bottle in the freezer, but we thought that would take too long to chill it. So we put it in a bucket of ice water. When my sister-in-law saw what we were doing, she said that we should add salt to the water. I had never heard of this before, but very quickly came to believe it would work. Why? I first constructed the imaginary situation of the water, the ice, the salt, and the champagne. Using my background knowledge about salt water (that it melts at a lower temperature than unsalted water), I was able to propagate the system forward in time—using the fact that the ice cubes in a salt bath are further above their melting point than those in a nonsalt bath, I could know that they would melt faster as well and in melting faster would absorb heat from their surroundings (including the champagne bottle) faster. On the basis of imaginatively propagating the systems forward in time using known physics, I generated novel observational (albeit imaginative) data. These data could then be used to support the inference that the worldly system would itself behave in this same manner.

One might at this point trot out the Norton objection that I merely made an argument. But that's simply not what happened. I did use a little argument *just now* in explaining my choice of a method for propagating the thought system forward in time, and why it worked to infer on that basis how the real-world system would behave. But what I did, in conducting the thought experiment, is construct a realizable though nonactual system and allow it to evolve to a novel state. Norton's story is attractive, but wrong. We cannot be said to have been making an argument in the carrying out of some activity simply because we can *later* justify our account of that activity using argument. As we saw above, precisely the same reasoning, were it correct, would show that throughout history there never has been an experiment, only a series of arguments.

We tend to discount the pronouncements of thought experimentation, officially, because they offer no new observations. That's fair in only this sense: it's true that they don't offer new observations in the strictest sense of that term. But in general, neither do other kinds of experiment. Instead they give us novel arrangements of the same old observations that we have grown up making. For example, Millikan's electron charge experiment consists principally in him checking the rate at which an oil drop moves back and forth across some small area under his microscope. He observes this by seeing a point of light cross his field of view. The novelty there is not in the content of the experience or the observation, but in its arrangement. Similarly, in thought experimentation our results are not novel observations but novel arrangements of observations. No, I am not going all the way down to Carnapian construction of the world from bare sensory data here. I simply want to point out that looking into a microscope, the source of the observations themselves, isn't the main event here. The main event is the sequences of initial states yielding to final states. The causes in the case of thought experiment are less direct than in other sorts: we propagate the thought system forward in time using our understanding of the laws of nature (or the stipulated laws we are testing), while in some other experiments the system is propagated forward in time by nature itself. This wrinkle demands extra caution, then, but it does not support a conclusion that thought experiments are different in kind from other forms of experiment.

Here is another way to see the same point. I will pose it as a puzzle that empiricists haven't addressed with sufficient clarity, and it is a puzzle that arises in quite this way only for empiricists. Just what is an observation? Of course many similar puzzles appear in its train: what is a *novel* observation? what are data? what counts as new data? Now for someone like Hume, I think, this is all pretty easy. Thinking is just a species of perceiving. Indeed the only real difference, at bottom, between sensing and thinking is that the former is somehow more urgent or more vivid or more lively than the latter.

What if we begin here? That is, suppose we acknowledge that for an empiricist there is no *specific* difference between thinking and feeling, that they are each instances of the species *perceiving*. When we do that it becomes not only impossible to draw a sharp distinction between observations of thought systems and observations of extramental systems as cognitive activities, but pointless as well. We can draw a distinction between the systems themselves, to be sure, along the lines of one being worldly and the other being imaginative. But that won't by itself tell us what information these systems bear about others. Only direct examination of the specific features of those individual systems will do that. Once we see that, however, it is easy to see that a distinction between thought and other experimentation that is predicated on the fact that in the latter we get new observational data and in the former we do not is one without a difference. Observations require not only perceptual promptings, but cognitive uptake as well.

To clarify this a little bit, it might be useful to consider some familiar activities: Looking at a picture and noting things about it that can be observed. Recalling those things in memory. Recalling the picture in memory and noting those things. Recalling the picture in memory and noting new things that were not noted at the time. Calling up a mental representation of a uniformly colored object and registering the color in awareness. Calling up a mental representation of a two-colored object and registering both colors in awareness. Calling up that same representation and determining which color is darker by being directly aware of it. Doing the same but by knowing facts about the ordinal arrangement of color from light to dark. In addition to direct experience with the world, the promptings I get from memories as well as from imaginative situations can be observational. The real issue then is with making sure that those observations are of systems that can bear information about other systems. For this we will need standards. If thought experimental systems are to play the role of proximal systems whose observable features can provide experimental knowledge, they need to be constructible.

29.5. Aside on Observation at the Beginnings of Science

I am going to consider, in this connection, a distinction between plain old sensing and observation that arises in Pomata's (2011) and Daston's (2011) trenchant discussions of the development of the scientific category of observation. One very important point is that observation is much more than sensing—it is the diligent, structured, prolonged engagement with the object of sense, repeatedly shining the beam of attention at that object to extract observational data. I claim that when we engage imaginatively with thought experimental scenarios, we are doing observation in just that way.

What is observation? In her analysis of the development of observation as what she calls an "epistemic genre," Pomata (2011) points out that observation as understood in the late Renaissance could be seen either as obedience to a rule, or as attentive watching, and hadn't yet entered into the canon of natural philosophy as the fundamental source of our knowledge. What she means by genre are "standardized textual formats—textual tools we may call them— handed down by tradition for the expression and communication of some kind of content. In the case of epistemic genres, this content is seen by authors and readers as primarily cognitive in character" (2011, 48). She offers a fascinating account of the development of observation from bit player to workhorse of the burgeoning scientific revolution.

As a genre, observation still required some further development. In the next essay in that same volume, Daston (2011) details the transition of observation from epistemic genre to epistemic category. Observations as genre are characteristically "singular events" involving an "effort to separate observation from conjecture" and the creation of "virtual communities of observers." According to Daston, the transition from observation as epistemic genre to observation as epistemic category—that is, "an object of reflection that had found its way into philosophical lexica and methodological treatises"—had taken place by about 1750 (2011, 81). What is the importance of this? For my purposes, the main significance is the development of a number of products. These products—reports—are the results of sustained attention from specially qualified people, and are distinguished in that way from other remarks made by the unqualified. Observations are thus not the sensing of things in the world so much as the results of that sensing—the data generated by it.

Here the key feature is the repeated attention paid to the objects of our awareness. Observation understood this way is not significantly bound to the novelty of the sensory experience itself. Rather, it is bound up with attentive re-experiencing. As Daston puts it: "So pencil-thin and intense was the beam of attention that it could hardly be sustained over long periods. Hence the observer must return over and over again to the same object, picking out different details, different aspects each time, and multiply confirming what had already been observed" (2011, 99–100).

I find this a remarkable development because it closes the circle on Pomata's view of the origins of observation. Or perhaps better, it is the reimposition of Pomata's first sense: obedience, where the rule is the rule of the disciplined naturalist.

Objections to the possibility of novel observations in the context of thought experiments arise in pursuit of a red herring—the idea that observation and sensing are broadly the same. Even empirically minded philosophers are not generally well acquainted with the deeply structured and nuanced origins of those categories, like observation, that we tend to use indiscriminately in discussing the nature of scientific knowledge. And that lack of familiarity makes it all too easy to collapse these categories into one. Observation is fundamental to the scientific enterprise, especially its experimental parts, but it is not simple, either in the sense of being easy to do or in the sense of being conceptually simple. Talk of data, of observation, etc., is always targeting deeply structured and complicated phenomena.

So are there novel observations in thought experiments? Of course there are. Sustained, repeated, focused attention to the objects the thought experiment comprises continue to yield observational data long after these objects have been constructed in thought. They yield them the same way that sifting data, comparing outcomes, looking from another point of view, and so on yield novel observations in more standard laboratory contexts.

29.6 A. Constructibility Standard for Thought Experiments

All that remains, then, is to sort out what classes of representations are capable of giving experimental knowledge.

There are many things that are said to be thought experiments, in philosophy, in science, and even in daily life. These things are given more or less detail about how they are to function, how they are to connect up with the external world, how their initial conditions make sense, and so forth. Standard accounts of thought experiments seem to include all of these things under the category of thought experiment, and so fail by allowing in too much. They seem to allow anything to count as long as its description doesn't contain an out-and-out contradiction. Indeed, some allow even that, as long as we have reason to think that these supposings give us insight into what we're wondering about.[2] As long as we preface some descriptive talk with "imagine that" or "suppose that," we have succeeded in producing a thought experiment. If the idea of thought experiment required no more than that they generate insight or understanding or some other state than knowledge, such views would probably be OK. But if they are to generate knowledge, and specifically experimental knowledge, they are going to have to be the kind of thing that one could observe. For that they are going to have to be the kind of thing that could be built in thought.

This amounts to a *constructibility in thought* criterion on thought experiments. My claim is that many things that share certain features with thought experiments fail to be thought experiments and are instead some other kind of imaginative exercise. Likewise, many things that share many features with experiments fail to be experiments. In the same way that filling a room with desks and chairs and notebooks and postdocs and electrical equipment doesn't by itself count as conducting an experiment on the foundations of electrodynamics, so too filling your head with images and other sorts of thoughts all having to do with electrical apparatus doesn't by itself count as conducting a thought experiment on the foundations of electrodynamics. We can do a lot of things in a lab that don't count as experimenting, and we can do a lot of things with our minds that don't count as thought experimenting. There are two important reasons some believe that thought experiments need not be constructible in thought in order to work: (1) they see a need for certain kinds of idealization that only thought can give in order to generate what appears to be reliable thought experimental knowledge; or (2) they think that sometimes the work of thought experiments is to display certain apparent possibilities as, in fact, impossible via a species of *reductio ad absurdum*—something that can only be done in thought. These are important uses of thought experimentation, and if they were to violate the constructibility standard that might be reason to abandon the standard. But they don't.

The best thought experiments have very little ambiguity in their descriptions, and consequently more clarity in their execution. Eppley and Hannah's (1977) thought experiment is a great example of this. They proposed that the possibility of a theory of gravity that is a quantum-classical hybrid—quantum on the matter side and classical general relativity on the gravitation side—could be experimentally tested. Their articulation of what would be involved in such a test is not at the level of an engineering specification, but it does give

2. Recall Sorensen's version of Galileo's experiment about free fall.

enough detail to allow a realistic construction of the initial situation and the development of its time evolution in thought; thus it promised to make possible experimental knowledge of the universe at large. Their argument about what to conclude was less successful, but their description of how to carry out the experiment was exemplary. The experiment was supposed to show that no theory of quantum mechanical matter could be coupled to a theory with a classical gravitational field.

A simple musing that would not rise to the level of a thought experiment might go like this: We can use a classical gravitational wave to violate the Heisenberg uncertainty principle. Since classical gravitational waves do not obey the quantum mechanical relation between momentum and frequency $p = h\nu$, we could find a very low momentum, very high frequency gravitational wave. This wave would of course have a very short wavelength. So we could localize some quantum mechanical object very precisely both in momentum space and in position space. This is not really a thought experiment, because it tells us little about how we could make the necessary measurement, and gives us no reason to think we could actually make the measurement in question—even in thought.

Eppley and Hannah do much more than this. They tell us how to imagine a device that will generate the appropriate beam of gravitational radiation, and how we might detect it and use it to find the precise momentum and position of some material object. And their description is no mere musing over the imagined consequence of a hybrid theory. While they do not provide actual engineering drawings specifying how to build the machine they are suggesting, they come close. And they give detailed specifications of the source of the probing beam, the collimator, the target, the detector apparatus, etc. Apparently all that is lacking is a funding agency with deep enough pockets to support the building of a device ten times the diameter of the solar system. Yes, it's big! But that's no problem for a thought experiment. Their only constraint, which they acknowledge, is that no more matter and time and space than are in the universe itself should be required. That is the kind of explicit standard that one can evaluate.

So what kind of knowledge does this experiment give us? This case is interesting, in part because the device does not function at all as Eppley and Hannah claim and in part because it still gives us experimental knowledge, but of a very different type than they thought. One can analyze the constructibility demands that they make to evaluate its (thought-)possibility. I did just that (Mattingly 2006) and found that the machine they propose was so massive that it would collapse into its own black hole before being built. It would also, if built, generate from its own mass far too much heat to be in the necessary quiescent state to carry out their proposed observations. There were many other problems with it as well, but the point to all these problems is that they followed explicitly from the construction details Eppley and Hannah gave. Without those details we might well imagine that we could do the exact experiment that they suggest. But with those details we can learn some things about gravitation, about the constraints that c, G, and $\hbar$ put on the kinds of matter that can exist in our universe. Further, we can perhaps learn experimentally—by way of our observation that any such system would collapse into its own black hole—that in our universe no device is capable of experimentally violating the Heisenberg uncertainty relations *even if* there are nonquantum radiation fields generated by gravitating matter. I don't think there are, but I do think we may have experimental knowledge that such devices are impossible. At least we have strong experimental constraints on what could work for the task.

29.7. Realizable, Idealizable, "Constructible"

A significant obstacle to efforts to understand the nature of thought experiments and their role in producing experimental knowledge is that we often have very little to go on when we attempt to assess their constructibility. But without some method for assessing that, we cannot easily determine whether we are dealing with a thought experiment rather than some vague imagining, or some other kind of thought situation that does not bear at all on empirical matters. My rather obvious suggestion for judging whether something or other should count as a thought experiment is that the situation imagined should be constructible, and that the only things preventing it are contingent facts about our material resources or present technological limitations (or perhaps the particular physics of our own world). Two obvious difficulties arise immediately. First, this doesn't really count as an operationalized assessment procedure, because we cannot in general determine unambiguously whether what prevents the construction of some situation is our present lack of technological prowess or some hidden impossibility implicit in the description of the situation. Second, even were we to have that, we might be ruling out too much. It has been pointed out by many authors that much of our thought experimental practice involves idealized situations, and as such there is no possible technological improvement that would allow them to be realized.

I have argued that Eppley and Hannah's thought experiment fails because their device cannot be built, and that even were it built it would not function as advertised, not least because the device itself would give off an unacceptable amount of heat and so would not allow the measurements their experimental design would require. One might think that with sufficient ingenuity the excess heat could be dissipated. But that is not so. The heat at the surface of the detector is a space-time phenomenon, characteristic of the surface gravity of the device. As such it is an ineliminable feature of the experimental apparatus and cannot be mitigated. The measurements required to detect any alleged violation of the Heisenberg uncertainty principle simply cannot be made.

Shumelda (2013) demurs on some of my claims, and I think incorrectly. He argues that thought experiments may be useful even when they cannot be constructed. I think there are things we imagine and muse about that can be of use even when they do not rise to the level of thought experimentation. But without the ability to construct something in thought, and to take observations and learn thereby about distal systems, we can derive from them no experimental knowledge. I prefer not to call such things thought experiments, but I won't worry so much about the term. However, Shumelda resists strongly the idea that thought experiments should be constructible, even in thought. His worry is that my constructibility demand is too exacting. The problem, he says, is that nothing in the world is ideal, but we commonly use idealizations in thought experiments. For example, frictionless surfaces do not exist in the world, perfectly spherical objects do not exist in the world, and so forth.

Shumelda appeals to just this point in his discussion of Eppley and Hannah's thought experiment. Idealizations can be used fruitfully and properly in thought experiments, though, while apparently there is no possible procedure to construct them explicitly in thought—they are possessed of stipulated but not justified properties. Does this not make trouble for any attempt to develop evaluation procedures for purported thought experimental constructions? Perhaps, but the situation may not be as easily assessed as it seems. Some situations that are commonly taken to be unrealizable may not be. Frictionless surfaces, for example, may indeed be realized by exploiting the Meisner effect, a quantum effect that results in superconductors repelling magnetic fields and thus generating magnetic levitation. Even should there be damping of the motion of magnetic bodies on superconducting surfaces (due, say, to the drag of the magnetic field), the friction, the result of the now nonexistent normal force between surface and object, has been eliminated. Such objections miss the true point to the constructibility demand.

Shumelda takes it that my view requires constructibility while no ideal physical systems can be constructed. But he misses a key feature of my requirement. The constructibility I require is not a *physical* requirement; rather, it is a requirement that the thought system be constructible on its own terms. What does this mean? Simply that if a demand arising from the constraints of the experimental description requires that the thought system have a property,

then that property must be realizable in thought *along with* any other properties that arise from the constraints of the description.

For just this reason Shumelda makes an important mistake in his reading of my analysis of Eppley and Hannah's experiment. He argues from the view that constructibility according to the laws of physics is part of my general story about thought experiments. It is not, even though the constructibility demand itself is. The Eppley and Hannah experiment is specifically framed in order to refute one particular possibility for how to reconcile quantum mechanics and general relativity: the semiclassical approach. Their idea is to construct what Sorensen calls an "alethic refuter" by showing that the laws of physics allow the construction of the machine, and then allow a specific series of observations to be made with the machine. It is thus *their own constraints* on the experiment to which I appeal in my analysis, not a generalized demand that thought experiments be constructible according to the laws of physics. In fact I have very little confidence that we know what the laws of physics are, though we do probably have very good grounds for constraining their phenomenology in a number of important domains. This is, I take it, more or less our situation with respect to all the laws of nature. We have many going theories of various domains, good reason to think that some of these will be overthrown in the course of our further investigations, and pretty good reason to think that a large number of experimental/empirical findings in those domains will stand the test of time. My constructibility demand, then, has little to do with actual laws of nature. Instead the demand is that the thought experimenter be answerable to our probing questions in the same way that an experimenter looking for funding to build a device would have to be, both before funding is granted and afterward during an audit to check whether the device really does what it is supposed to. Moreover, the experimental protocol must be rendered precisely enough that anyone (subject to background knowledge constraints) could perform the thought experiment if asked. That in no way rules out idealization.

When a thought experiment involves a frictionless plane, the constructibility demand is not that such a plane be realizable in the world, but rather that the law of time development of the thought system involve no forces arising from friction. In this sense it is not a demand that something extra be generated as a component of the system, but rather that something not be part of the logic of the classification. Some constructibility demands are of this negative sort. Others are positive demands on the rule of time development of the system. For example, if mass is included in the system, and gravitational forces are as well, then the time-development operator must include these things, and consistently so, on pain of incoherence. Thus if it is a key part of my experimental apparatus that matter fields obey the linearized semiclassical Einstein equation, then a constraint arising from that requirement is that they do so! Thus when I critique Eppley and Hannah's experiment for failing to be constructible using our current physical models, it is because they themselves adopt that as their constructibility standard. I require in general only that a thought experiment be constructible in its own terms—these terms both set the standard and detail the classifications that will be available to us in our attempts to generate experimental knowledge. Eppley and Hannah were testing a particular theory as a possible account of our universe, and thus those empirical standards had to apply.

Let me distinguish two cases: (1) a mere wondering involving something presented as if it were a thought experiment; and (2) a fully articulated description of some proposed situation using either only background theories/ beliefs or consequences of the theory that is under test. In the first case there is not enough detail or background structure to say we have even a candidate for "thought experiment"; in the second it is clear that we do have such a candidate. But it is not yet clear which will turn out to be genuine. Some examples will clarify the situation.

For case (1) consider testing economic theories by imagining what would have happened if your favored candidate had won election rather than the other person. One sometimes hears things like "well, if we had done the right thing and elected my candidate we would finally see that supply-side economics is [choose one: insane; the only real way to run an economy]." Clearly this doesn't count as a thought experiment at all, much less one that could tell us anything about economics.

For case (2) consider again the classic case of Galileo's proposed test of the Aristotelian theory of free fall, which apparently says that heavier objects fall faster than lighter objects. One can, clearly, envision joining two different masses together—so perhaps one is a stone box that the other, also a smaller, lighter stone, just fits inside. Here we take advantage of well-supported background knowledge to convince ourselves that the initial conditions can be realized. We then use the theory under test to try to propagate the thought system forward in time. Let's imagine that we turn the box over and let the stone box/stone system fall. What we find, after puzzling over it for a bit, is that the theory tells us something impossible. We already know, from all of our background beliefs, that the system will fall together as one, and in any case the only way for them to come apart is by the lighter one moving ahead of the heavier. But the Aristotelian theory tells us that the smaller stone must slow down the large stone, and so the pair must fall more slowly than the large stone alone. But how tightly must the smaller stone fit inside the larger before we consider it to be one system, and so to count as one system?

The problem here is not the construction in thought of the situation itself; the problem is evolving it in time. This *fails* in my view as a thought experiment because it cannot be evolved in time according to a realizable-in-thought time-evolution operator. That is to say, Aristotle's conception of free fall demands two distinct time-evolution operators: the one where heavy bodies are slowed by having to pull along lighter bodies, and the one where there is only one, now very heavy, body. Nothing tells us how to choose. We cannot make the trial at all in this case. Observing that there are implicitly two incompatible time-evolution operators is what generates experimental knowledge. Observational facts about a proximal system (that it requires *each* of these time evolutions according to the theory) gives knowledge about distal systems. On the other hand, we can if we like look at this a different way. We can suppose that the two time-evolution operators must yield the same result, if they each have the correct functional dependence between mass and acceleration under free fall. We then find that that function must be a constant. Perhaps a thought experiment along those lines could give knowledge. Still, there is no way to do the original experiment, because there is a fundamental ambiguity in the specification of a (thought-)time-evolution operator.

Now let's dig a little deeper and say a little more about about case (2), to give ourselves some insight into possible experimental controls for thought experiments. In this case we can draw a few more distinctions: between (2i) a situation that can be neither realized nor idealized, even though the conditions on its idealization/realization are clearly allowed by the theory in question; (2ii) a situation that may perhaps be idealized; and (2iii) a situation that may perhaps be realized. (2i) is what we might call a negative instance of a thought experiment: a proposed situation that can be seen to be impossible because the conditions necessary for its construction are impossible under our background knowledge, even though the theory in question allows both the situation and its construction. This is what Sorenson calls an alethic refuter. If we were to come to see, on the basis of observing various situations, *that* the description given cannot be constructed even in thought, then I suppose it would make sense to call our knowledge of that fact experimental knowledge. Cases (2ii) and (2iii) are, I would say, well characterized as thought experiments, but even in those cases there will be situations that can be seen to be impossible to bring to bear on the proposed test question. These would still appear to be thought experiments, but not very informative ones. I will say that the experimental situation is constructible in these cases.

A hard test case in this domain comes from Norton's dome example. It should be worth analyzing. A frictionless ball, rolled with the right speed toward a frictionless dome with shape given by $h = (2/3\,g)\, r^{3/2}$, will roll to the top of the dome and stop. That doesn't seem so unusual at first glance, because we can see doing that with our normal balls and normal domes that are subject to friction. So the thought doesn't seem impossible to realize with some special class of idealized thought materials. But this cannot be a physical dome, even of a very idealized sort, because the existence of the shape is incompatible with determinism. (Perhaps rather than saying that it is unphysical, we should say that any space-time with a Cauchy surface has no domes of that shape.) It is hard to believe that determinism itself constrains the geometry of matter fields, but it does seem to. Here it is not easy to say what we should conclude for the analysis of thought experimentation. But we should say at least this: any metaphysician who uncritically thinks that certain kinds of material constituents are metaphysically possible based on their apparent imaginability is mistaken. This in particular makes it clear that no thought experiment involving so-called gunk has any significance at all for metaphysics. Absent an account of the physics of gunky matter, we simply cannot evaluate its possibility.[3] The constraints on the nature of matter, and indeed on the *possible* nature of matter, are too complicated to yield to bare imaginative faculties. In order to evaluate a given thought experiment, we must show its constructibility in thought. Or we need to show that any such construction is not possible. Absent one or the other of these two methods, we cannot say that a thought experiment has been performed. The realms of thought and of matter are the same in this respect: if you want to demonstrate something with an experiment, you must show it.

How would we apply this principle in the case of Norton's dome? We could, for example, say that for a deterministic universe—or at least a normal Newtonian universe that fails of determinism in normal ways—the Lifschitz conditions on differential equations obtain. Finding out that a situation requires violating those conditions tells us that the thought experiment will not have an infomorphism of an appropriate sort to say much about whether our world is deterministic (or fails of it in only normal ways). Universes without Lifschitz conditions don't bear such information about universes that do have them. This kind of situation is nothing special in the case of thought experiments. All that is going on here is that we found that universes with Lifschitz

---

3. This should be enough to rule out all conversations involving metaphysical gunk, but it probably will not, more's the pity.

conditions are a special case; experimental results that apply to one domain but not another abound in all branches of experimental practice.

The constructible experiments then are capable of yielding new experimental knowledge. But how can we tell in any given instance which kind of imaginative situation we are dealing with? Perhaps we cannot always tell. We may think we have experimental knowledge as the result of a thought experiment, and yet not have it. That's OK—it's something else that happens in all branches of experimental practice. Still, we want to know how to avoid such situations going forward. The parlor economics test above is pretty easy to detect as a bogus thought experiment. But what procedures can be put in place to detect less obviously bogus thought experiments? First we should try all the normal procedures for testing the legitimacy of standard experimental reports.

Generally one should ask as many detailed questions about the thought experimental situation as one would ask about an experiment outside of thought: "Where does this wire go?" "How do we know that the core of the system reaches a quiescent state?" "How do we read out the data at the end?" and so on. Generally, thought experiments are given leeway to be vague when corresponding material experiments are not, and that is just the wrong way to go about things. What is often overlooked is that, in some respects, thought experimentation is *harder* than material experimentation, not easier. Once my physical machine is up and running in the lab, at least I know for sure that there is a device there that I can observe and that is producing data. But with thought experiments a great deal of the work is involved *precisely* in being sure that we have properly generated and are in possession of an observable situation.

This discussion has only scratched the surface of how to go about evaluating thought experiments generally, but it has made clear that they can work just like other kinds. Full discussion of experimental controls will have to await another occasion.

Conclusion to Part III, and the Book

30.1. Thought Experiments as the General Form of Empirical Claim

My final outrageous suggestion will also be my final positive argument. It begins with this question: what is the semantic content of remarks like "we know from experiment that ...."? The claim does not seem to be that we have constructed this precise situation in the lab and have seen what happens. In the first place, many such claims are completely general. As such they can never have direct experimental situations that correspond to them. In the second place, it seems to be pretty common to take experiments of a quite restricted sort as bearing on situations of a quite different sort. For example, I think we know from experiment that spontaneous generation of cellular organisms does not take place. But it does not follow directly from experiment. Rather, the claim seems (when used properly) to have the content that we are confident in the observational status of some properties of a proximal system, or perhaps many such systems, and we are confident in the strength of analogy between some distal system and this proximal system. In other words, we are confident that the systems are appropriately infomorphic to allow knowledge of the properties of the distal system based on the properties of the proximal system.

And what gives us this confidence in the observational status of the proximal system's properties? Is it direct observation? Not generally; the observations are most often some sort of proxy. Consider what happens when we observe the properties of subatomic particles in accelerator detector chambers. What we observe when we see indicator panels far from the scene of the action (or computer readouts) is the heat dumped into a calorimeter, or the number of charges that interact with an array of charge-coupled devices, or the curvature of some path. We simply cannot observe directly the mass, say, of some elementary particle. Similarly, much of what goes on in molecular biology is an inference to possessed properties rather than direct observation—chemical byproducts tell the story of what processes are taking place in the interior of cells, for instance. Perhaps our tools of observation will improve so dramatically over the coming decades that we will be able to make such observations directly. But for now, we cannot.

What is the appropriate reaction at this point? How should one respond to the charge that observation in science is not precisely direct observation of the proximal systems of interest? I think the right answer is "so what?" It is a mistake to think that the reach of our scientific knowledge, indeed our observational knowledge, is no greater than the reach of our human sensory apparatus. Indeed, recognizing that this *is* a mistake was at the heart of the scientific revolution of the seventeenth century. Direct observation is an almost vanishingly small portion of our observational knowledge.

Is the method of extending our observational powers using indirect methods infallible? By no means. We can certainly, even when we take great care, assign properties to proximal systems that they simply do not possess. And we can do so even though we perform these activities in the laboratory with controlled conditions, calibrated instruments, good theories of the devices we are using, etc. We are fallible, our instruments are fallible, science is fallible. But all of these things are correctable as well.

How does this bear on the status of thought experiments? Thought experiments share the feature that they do not involve direct observations of proximal systems in the laboratory. Instead they provide observations of systems constructed in thought. And about this I think the correct response is, also, "so what?" What is it about being a thought construction that bears on the question of whether that system is observable? whether observations can give knowledge of the system's features? whether that knowledge bears information about other systems (either thought constructs or worldly objects)? In my opinion, it is nothing beyond what bears on these same questions in the case of other experiments. (Can we build it/find it/etc.? I don't know, let's find out. Can we observe it? Let's see. And so on.) We need to answer those questions, of course. But they should be answered by considering the kinds of thing we really can find out about such systems, not by appealing to some vague notion about the lack of reality that would underwrite a wholesale rejection of the status of such systems as potential proximal experimental systems.

Thought experiments are in principle legitimate sources of experimental knowledge (and empirical knowledge generally); and I further argue that essentially every empirical claim that appeals to experimental evidence is thought experimental. The conclusion follows readily from considering the basic structure of empirical claims. One wants to be able to say of some system that, were some intervention to happen, then some conclusion would follow, and that this is a known property of the system. Or one wants to be able to say that having prepared some system in accordance with some procedure, the system is now known to have some property. So generally one says that "the data show this or that" or "experiment has established thus and so." But the support we offer is generally not direct experimental support of the type "this system is identical to that system, and that system has property π so this system has property π." Rather, the support we offer is generally indirect: "this system is sufficiently similar to systems of that sort; that sort of system has property ξ; the appropriate property of that sort in this system is π; so this system has property π." While that is all pretty abstract, the point is easily concretized. Here are some examples: taking this penicillin for ten days will cure your tuberculosis; cooking this piece of pork above 144°C will destroy any trichinosis larvae in it; drilling a shaft in this cliff face will not lead to a collapse; and so on.

When we know the theory, and we have constructed the initial setup (your lungs, the pork, the cliff, etc.) in thought, then developing it in thought-time according to the theory, noting its new properties, and inferring new properties about the worldly system on that basis seems best characterized as performing a thought experiment. Instead of performing a thought experiment, could I instead construct an argument of some sort? Perhaps. But I do not think that is what we generally do when making such predictions. Instead I think we evolve the thought system forward in time and apply a near-identity transformation to read off the further properties of the material system from our observations of the thought system. Indeed, explaining any experimental result will usually involve the implicit appeal to maps between the proximal system whose behavior is being explained and a range of distal systems. These maps will generally go both ways: if the distal system is one about which we have less experimental knowledge, then the maps will serve to establish inferences about their behavior, resulting in experimental knowledge of those systems; and maps from distal systems we already know a great deal about will give even further knowledge of the proximal system by showing that these new experimental results support inferences with respect to this system following from experimental results in that other system. Other experiments give us the correct time-development operator for the thought system and help to keep us empirically grounded, and the thought experiment provides the specific new data we need for this particular prediction. Seen in this way, thought experiments abound and in fact provide the bulk of the knowledge we say we have from experiment.

30.2. Final Word

Thought experiments seem different from other kinds of experiment, in part because they seem to do this: find out the properties of some system given the evolution law of the thought experiment. Other kinds of experiment seem to do this: given the properties of some system, find the evolution law. At first blush thought experiments appear to have all the information about the time evolution of the thought system built in, and so cannot be generating information for the thought experimenter *about* the time evolution of the thought system. That is not entirely correct, though, for two reasons. First, while the thought experimental system and its construction, as well as the time evolution of the system, are under my control, I may still notice things— get signals—that I hadn't noticed before when I examine the properties of the thought system after that time evolution. Second, other experiments *also* have all the information about their time evolution built in, but by the laws of nature themselves.

Thought experiments also seem different because they appear to concern themselves only with the local system and showing that it would have such and such properties were it built this way and were the laws that way, while other experiments appear to be vehicles for learning about distal systems. This also is not quite right. When I find properties of some system in my lab and say that systems elsewhere will behave in the same way, I have two assumptions looming large: the lab system and systems outside of the lab are relevantly similar; and the law in the lab is the same as the law outside of the lab. Thought experiments are no different in that respect. I construct the thought experimental system and its rule of evolution *so that* its properties can reveal those of distal systems. When I succeed in this, I have experimental knowledge.

Acknowledgments

Profound thanks are due to many. Here I will single out just a few whose contributions simply cannot go without acknowledgment. The Tryst writers' workshop, that is, Bryce and Burstein, for getting me to clarify where this project was heading. The Flynn and Eric, directors of the Matabor Artist Residences, for a crucial summer of lodging and support in the early days of the project. Members of graduate seminars over the years who read and commented helpfully on scattered bits and pieces before there was a book. Richard Fry, obviously, for many pushes, prods, and helpful exchanges over the entire life of the project. Ruth Espinosa, for arranging many venues over the years for the presentation of various parts of the book. At the very beginning, Jon Barwise for turning me on to the profound explanatory power of information flow. At the very end, the Hammerschlag–Bronstein pomodoro club.

Bibliography

Adams, Fred. 2005. "Tracking Theories of Knowledge." *Veritas: Revista de filosofia da PUCRS* 50 (4): 11–35.

Affentranger, Beat. 2000. *The Spectacle of the Growth of Knowledge and Swift's Satires on Science.* Parkland, FL: Dissertation.com.

Angrist, Joshua, and Alan Krueger. 1991. "Does Compulsory School Attendance Affect Schooling and Earnings?" *Quarterly Journal of Economics* 106 (4): 979–1014.

Atmanspacher, Harald, and Sabine Maasen, eds. 2016. *Reproducibility: Principles, Problems, Practices, and Prospects.* Hoboken, NJ: John Wiley & Sons.

Bacon, Francis. 1620. *New Organon.* Early Modern Texts. https://www.earlymoderntexts.com /assets/pdfs/bacon1620.pdf.

Barceló, Carlos, Stefano Liberati, and Matt Visser. 2000. "Analog Gravity from Bose-Einstein Condensates." arXiv:gr-qc/0011026v1.

Barceló, Carlos, Stefano Liberati, and Matt Visser. 2003. "Probing Semiclassical Analogue Gravity in Bose-Einstein Condensates with Widely Tunable Interactions." arXiv:cond-mat /0307491v2.

Barceló, Carlos, Stefano Liberati, and Matt Visser. 2011. "Analogue Gravity." *Living Reviews in Relativity* 14 (2011): 3.

Barwise, Jon, and Gerry Seligman. 1997. *Information Flow: The Logic of Distributed Systems.* Cambridge: Cambridge University Press.

Baumann, P. 2006. "Information, Closure, and Knowledge: On Jäger's Objection to Dretske" *Erkenntnis* 64 (3): 403–8.

Beisbart, Claus, and John Norton. 2012. "Why Monte Carlo Simulations Are Inferences and Not Experiments." *International Studies in the Philosophy of Science* 26 (4): 403–22.

Bertazzi, Pier, Angela Pesatori, Dario Consonni, Adriana Tironi, Maria Teresa Landi, and Carlo Zocchetti. "Cancer Incidence in a Population Accidentally Exposed to 2,3,7,8-Tetrachlorodibenzo -para-dioxin." *Epidemiology* 4 (5): 398–406.

Bhatt, Arun. 2010. "Evolution of Clinical Research: A History before and beyond James Lind." *Perspectives in Clinical Research* 1 (1): 6–10.

Bogen, James, and James Woodward. 1988. "Saving the Phenomena." *Philosophical Review* 97 (3): 303–52.

Boyle, Robert. 1669. *A Continuation of New Experiments Physico-mechanical, Touching the Weight and Spring of the Air, and Their Effects.* Oxford: Printed by Henry Hall for Richard Davis.

Brendel, Elke. 2004. "Intuition Pumps and the Proper Use of Thought Experiments." *Dialectica* 58 (1): 89–108.

Brody, H., P. Vinten-Johansen, N. Paneth, and M. R. Rip. 1999. "John Snow Revisited: Getting a Handle on the Broad Street Pump." *Pharos of Alpha Omega Alpha Honor Medical Society* 62 (1): 2–8.

Brown, James Robert. 1992. "Why Empiricism Won't Work." *Philosophy of Science* 1992 (2): 271–79.

Brown, James Robert. (1991) 2011. *The Laboratory of the Mind.* 2nd ed. New York: Routledge.

Buchwald, Jed. 1995. "Why Hertz Was Right about Cathode Rays." In *Scientific Practice: Theories and Stories of Doing Physics*, edited by J. Z. Buchwald, 151–69. Chicago: University of Chicago Press.

Cauwels, Anje, Benjamin Vandendriessche, and Peter Brouckaert. 2013. "Of Mice, Men, and Inflammation." *PNAS* 110 (34): E3150.

Chabris, Christopher, and Daniel Simons. 2010. *The Invisible Gorilla and Other Ways Our Intuitions Deceive Us.* New York: Broadway.

Chen, Ruey-Lin. 2007. "The Structure of Experimentation and the Replication Degree— Reconsidering the Replication of Hertz's Cathode Ray Experiment." In *Naturalized Epistemology and Philosophy of Science*, edited by Chienkuo Mi and Ruey-lin Chen. Amsterdam: Rodopi.

Coase, R. H. 1960. "The Problem of Social Cost." *Journal of Law & Economics* 3:1–44.

Collins, H. M. 1985. *Changing Order: Replication and Induction in Scientific Practice.* London: Sage.

Collins, H. M. 1991. "The Meaning of Replication and the Science of Economics." *History of Political Economy* 23 (1): 123–42.

Danshita, Ippei, Masanori Hanada, and Masaki Tezuka. 2017. "Creating and Probing the Sachdev-Ye-Kitaev Model with Ultracold Gases: Towards Experimental Studies of Quantum Gravity." arXiv:1606.02454v2.

Dardashti, Radin, Karim Thébault, and Eric Winsberg. 2017. "Confirmation via Analogue Simulation: What Dumb Holes Could Tell Us about Gravity." *British Journal for Philosophy of Science* 68 (1): 1–35.

Daston, Lorraine. 2011. "The Empire of Observation, 1600–1800." In *Histories of Scientific Observation*, edited by Lorraine Daston and Elizabeth Lumbeck. Chicago: University of Chicago Press.

Davis, M. 2008. "A Prescription for Human Immunology." *Immunity* 29 (6): 835–38.

Dear, Peter. 2014. *Discipline and Experience: The Mathematical Way in the Scientific Revolution.* Chicago: University of Chicago Press.

Douglas, Heather. 2000. "Inductive Risk and Values in Science." *Philosophy of Science* 67 (4): 559–79.

Dretske, Fred. 1981. *Knowledge and the Flow of Information.* Cambridge, MA: MIT Press.

Dretske, Fred. 2006. "Information and Closure." *Erkenntnis* 64 (3): 409–13.

Eells, Ellery. 1983. "Objective Probability Theory Theory." *Synthese* 57 (3): 387–442.

Eppley, K., and E. Hannah. 1977. "The Necessity of Quantizing the Gravitational Field." *Foundations of Physics* 7:51–58.

Feest, Uljana. 2016. "The Experimenters' Regress Reconsidered: Replication, Tacit Knowledge, and the Dynamics of Knowledge Generation." *Studies in History and Philosophy of Science* 58:34–45.

Fidler, Fiona, and John Wilcox. 2018. "Reproducibility of Scientific Results." In *The Stanford Encyclopedia of Philosophy*, Winter 2018 ed., edited by Edward N. Zalta. https://plato.stanford.edu/archives/win2018/entries/scientific-reproducibility.

Fingerhut, Marilyn A., William E. Halperin, David A. Marlow, Laurie A. Piacitelli, Patricia A. Honchar, Marie H. Sweeney, Alice L. Greife, Patricia A. Dill, Kyle Steenland, and Anthony J. Suruda. 1991. "Cancer Mortality in Workers Exposed to 2,3,7,8-Tetrachlorodibenzo-*p*-Dioxin." *New England Journal of Medicine* 324 (4): 212–17.

Fisher, Ronald. (1935) 1974. *The Design of Experiments.* New York: Hafner/Macmillan.

Foley, Richard. 1987. "Dretske's 'Information-Theoretic' Account of Knowledge." *Synthese* 70 (2): 159–84.

Franklin, Allan. 1986. *The Neglect of Experiment.* Cambridge: Cambridge University Press.

Franklin, Allan. 1997. "Calibration." *Perspectives on Science* 5 (1): 31–80.

Franklin, Allan, and H. M. Collins. 2016. "Two Kinds of Case Study and a New Agreement." In *The Philosophy of Historical Case Studies*, edited by Tilman Sauer and Raphael Scholl, 95–122. Cham, Switzerland: Springer.

Fukushima, S., Y. Kurata, M. Shibata, E. Ikawa, and N. Ito. 1983. "Promoting Effect of Sodium o-Phenylphenate and o-Phenylphenol on Two-Stage Urinary Bladder Carcinogenesis in Rats." *GANN Japanese Journal of Cancer Research* 74 (5): 635–32.

Gallison, Peter. 1987. *How Experiments End.* Chicago: University of Chicago Press.

Garwin, Richard L. 1974. "Detection of Gravity Waves Challenged." *Physics Today* 27 (12): 10–12.

Gendler, Tamar Szabó. 2004. "Thought Experiments Rethought—and Reperceived." *Philosophy of Science* 71 (5): 1152–63.

Geroch, Robert. 1985. *Mathematical Physics.* Chicago: University of Chicago Press.

Gettier, Edmund. 1963. "Is Justified True Belief Knowledge?" *Analysis* 23 (6): 21–23.

Gieryn, T. F. 1992. "The Ballad of Pons and Fleischmann: Experiment and Narrative in the (Un)making of Cold Fusion." In *The Social Dimensions of Science*, edited by E. McMullin, 217–42. Notre Dame, IN: University of Notre Dame Press.

Gooding, David C. 1992. "What Is Experimental about Thought Experiments?" *PSA: Proceedings of the Biennial Meeting of the Philosophy of Science Association* 1992 (2): 280–90.

Goodman, Nelson. (1955) 1983. *Fact, Fiction, and Forecast.* 4th ed. Cambridge, MA: Harvard University Press.

Gordon, W. 1923. "Zur Lichtfortpflanzung nach der Relativitätstheorie." *Annalen der Physik* 377 (22): 421–56.

Guala, Francesco. 2002. "Models, Simulations, and Experiments." In *Model-Based Reasoning: Science, Technology, Values*, edited by L. Magnani and N. J. Nersessian, 59–74. New York: Kluwer.

Guerrini, Anita. 2003. *Experimenting with Humans and Animals: From Galen to Animal Rights.* Baltimore: Johns Hopkins University Press.

Hacking, Ian. 1983. *Representing and Intervening: Introductory Topics in the Philosophy of Natural Science.* Cambridge: Cambridge University Press.

Hájek, Alan. 2003. "What Conditional Probability Could Not Be." *Synthese* 137 (3): 273–323.

Hájek, Alan. 2019. "Interpretations of Probability." In *The Stanford Encyclopedia of Philosophy*, Fall 2019 ed., edited by Edward N. Zalta. https://plato.stanford.edu/archives/fall2019/entries/probability-interpret/.

Hammack, Judd, and Gardner Mallard Brown Jr. 1974. *Waterfowl and Wetlands: Toward Bio-Economic Analysis.* Baltimore: Johns Hopkins University Press.

Harder, Douglas. n.d. "Water Analogy to Circuits." University of Waterloo, https://ece.uwater loo.ca/~dwharder/Analogy.

Hartley, R. V. L. 1928. "Transmission of Information." *Bell Systems Technical Journal* 7 (3): 535–63.

Hayday, A., and M. Peakman. 2008. "The Habitual, Diverse and Surmountable Obstacles to Human Immunology Research." *Nature Immunology* 9 (6): 575–80.

Health Canada. 2007. "Information Document on the Proposal to Reinstate Saccharin for Use as a Sweetener in Foods in Canada." https://www.canada.ca/content/dam/hc-sc/migration /hc-sc/fn-an/alt_formats/hpfb-dgpsa/pdf/securit/saccharin_prop-eng.pdf.

Herschel, J. F. W. 1831. *A Preliminary Discourse on the Study of Natural Philosophy*. London: Longman, Rees, Orme, Brown, Green, and Taylor.

Hertz, H. 1898. "Experiments on the Cathode Discharge." In *Miscellaneous Papers*, translated by D. E. Jones, and G. A. Schott, 224–54. Leipzig: J. A. Barth.

Humphreys, P. 1985. "Why Propensities Cannot Be Probabilities." *Philosophical Review* 94 (4): 557–70.

Ioannidis, John. 2005. "Why Most Published Research Findings Are False." *PLOS Medicine* 2 (8): 696–701.

Jacobson, Ted. 2000. "Trans-Planckian Redshifts and the Substance of the Space-Time River." arXiv:hep-th/0001085v2.

Jäger, Christoph. 2006. "Skepticism, Information, and Closure: Dretske's Theory of Knowledge." *Erkenntnis* 61 (2): 187–201.

Junod, Suzanne White. 2013. "FDA and Clinical Drug Trials: A Short History." Washington, DC: US Food and Drug Administration. https://www.fda.gov/downloads/AboutFDA/History /ProductRegulation/UCM593494.pdf.

Kahneman, Daniel. 2012. "A Proposal to Deal with Questions about Priming Effects." Open letter to *Nature*. https://www.nature.com/news/polopoly_fs/7.6716.1349271308!/suppinfoFile /Kahneman%20Letter.pdf.

Kahneman, Daniel, Jack Knetsch, and Richard Thaler. 1991. "Experimental Tests of the Endowment Effect and the Coase Theorem." *Journal of Political Economy* 98 (6): 1325–48.

Kahneman, Daniel, Jack Knetsch, and Richard Thaler. 2008. "The Endowment Effect: Evidence of Losses Valued More than Gains." In *Handbook of Experimental Economics Results*, edited by Charles Plott and Vernon Smith, 939–48. Amsterdam: North-Holland.

Kempthorne, Oscar. 1992. "Intervention Experiments, Randomization, and Inference." In *Current Issues in Statistical Inference: Essays in Honor of D. Basu*, edited by Malay Ghosh and Pramod K. Pathak. Institute of Mathematical Statistics Lecture Notes Monograph Series 17:13–31. Chennai: Institute of Mathematical Statistics.

Kipper, Jens. 2016. "Safety, Closure, and the Flow of Information." *Erkenntnis* 81 (5): 1109–26.

Kociba, R. J., et al. 1978. "Results of a Two-Year Chronic Toxicity and Oncogenicity Study of 2, 3,7,8-Tetrachlorodibenzo-*p*-dioxin in Rats." *Toxicology and Applied Pharmacology* 46 (2): 279–303.

Krowech, Gail, John B. Faust, Brian Endlich, and Martha S. Sandy. 2003. "Evidence on the Carcinogenicity of Sodium Saccharin." Reproductive and Cancer Hazard Assessment Section Office of Environmental Health Hazard Assessment California Environmental Protection Agency. https://oehha.ca.gov/media/downloads/crnr/sodiumsaccharinfinalhid.pdf.

Kuval, Praveen, and Vernon Smith. 2008. "The Endowment Effect." In *Handbook of Experimental Economics Results*, edited by Charles Plott and Vernon Smith, 949–55. Amsterdam: North-Holland.

Latour, Bruno, and Steve Woolgar. (1979) 1986. *Laboratory Life: The Construction of Scientific Facts*. Princeton: Princeton University Press.

Liberati, Stefano. 2017. "Analogue Gravity Models of Emergent Gravity: Lessons and Pitfalls." *Journal of Physics Conference Series* 880:012009.

Loewer, Barry. 1983. "Information and Belief." *Behavioral and Brain Sciences* 6 (1): 75–76.

Magnani, L., and N. J. Nersessian, eds. 2002. *Model-Based Reasoning: Science, Technology, Values*. New York: Kluwer.

Maher, P. 2010. "Explication of Inductive Probability." *Journal of Philosophical Logic*, 39:593–616.

Mattingly, James, and Walter Warwick. 2009. "Projectible Predicates in Analogue and Simulated Systems." *Synthese* 169 (3): 465–82.

Mattingly, James. 2001. "The Replication of Hertz's Cathode Ray Experiments." *Studies in the History and Philosophy of Modern Physics* 32 (1): 53–75, 81–82.

Mattingly, James. 2006. "Why Eppley and Hannah's Thought Experiment Fails." *Physical Review D* 73 (6): 1–8.

Mattingly, James. 2021. "The Information-Theoretic Account of Knowledge, Closure, and the KK Thesis." Unpublished manuscript.

Morgan, Mary. 2003. "Experiments without Material Intervention: Model Experiments, Virtual Experiments, and Virtually Experiments." In *The Philosophy of Scientific Experimentation*, edited by Hans Radder. Pittsburgh: University of Pittsburgh Press.

Morgan, Mary. 2005. "Experiments versus Models: New Phenomena, Inference and Surprise." *Journal of Economic Methodology* 12 (2): 317–29.

Morrison, Margaret. 2015. *Reconstructing Reality: Models, Mathematics, and Simulations*. New York: Oxford University Press.

Norton, John D. 2004. "Why Thought Experiments Do Not Transcend Empiricism." In *Contemporary Debates in Philosophy of Science*, edited by Christopher Hitchcock, 44–66. Malden, MA: Blackwell.

Olson, Harry, et al. 2000. "Concordance of the Toxicity of Pharmaceuticals in Humans and in Animals." *Regulatory Toxicology and Pharmacology* 32 (1): 56–67.

Osterburg, Andrew, Philip Hexley, Dorothy Supp, Chad Robinson, Greg Noel, Cora Ogle, Steven Boyce, Bruce Aronow, and George Babcock. 2013. "Concerns over Interspecies Transcriptional Comparisons in Mice and Humans after Trauma." *PNAS* 110 (36): E3370.

Ostrand-Rosenberg, S. 2004. "Animal Models of Tumor Immunity, Immunotherapy and Cancer Vaccines." *Current Opinion in Immunology* 16 (2): 143–50.

Parker, Wendy. 2009. "Does Matter Really Matter? Computer Simulations, Experiments, and Materiality." *Synthese* 169 (3): 465, 483–96.

Perrin, J. 1895. "Nouvelles propriétés des rayons cathodiques." *Comptes Rendus* 121:1130. Translated and reprinted in *Source Book in Physics*, edited by W. F. Maggie, 580–83. Cambridge, MA: Harvard University Press, 1965.

Plott, Charles, and Vernon Smith, eds. 2008. *Handbook of Experimental Economics Results*. Amsterdam: North-Holland.

Pohjanvirta, R., and J. Tuomisto. 1994. "Short-Term Toxicity Of 2,3,7,8-Tetrachlorodibenzo-*p*-dioxin in Laboratory Animals: Effects, Mechanisms, and Animal Models." *Pharmacological Review* 46 (4): 483–549.

Pomata, Gianna. 2011. "Observation Rising: Birth of an Epistemic Genre, 1500–1650." In *Histories of Scientific Observation*, edited by Lorraine Daston and Elizabeth Lumbeck, 45–80. Chicago: University of Chicago Press.

Popper, Karl. (1959) 2005. *The Logic of Scientific Discovery*. New York: Routledge.

Pritchard, Duncan. 2005. *Epistemic Luck*. Oxford: Oxford University Press.

Radder, Hans. 1992. "Experimental Reproducibility and the Experimenters' Regress." *Proceedings of the Philosophy of Science Association* 1:63–73.

Radder, Hans. 1995. "Experimenting in the Natural Sciences: A Philosophical Approach." In *Scientific Practice: Theories and Stories of Doing Physics*, edited by J. Z. Buchwald, 56–86. Chicago: University of Chicago Press.

Ragland, Evan. 2017. " 'Making Trials' in Sixteenth- and Early Seventeenth-Century European Academic Medicine." *Isis* 108 (3): 503–28.

Roediger, Henry. 2012. "Psychology's Woes and a Partial Cure: The Value of Replication." *APS Observer* 25 (2). https://www.psychologicalscience.org/observer/psychologys-woes-and-a-partial-cure-the-value-of-replication.

Saltelli, Andrea, and Silvio Funtowicz. 2017. "What Is Science's Crisis Really About?" *Futures* 91:5–11.

Sargent, Rose-Mary. 1989. "Scientific Experiment and Legal Expertise: The Way of Experience in Seventeenth-Century England." *Studies in History and Philosophy of Science* 20 (1): 19–45.

Sargent, Rose-Mary. 1995. *The Diffident Naturalist: Robert Boyle and the Philosophy of Experiment*. Chicago: University of Chicago Press.

Schnabel, J. 2008. "Neuroscience: Standard Model." *Nature* 454 (7205): 682–85.

Schoenig, G. P., E. I. Goldenthal, R. G. Geil, C. H. Frith, W. R. Richter, and F. W. Carlborg. 1985. "Evaluation of the Dose Response and In Utero Exposure to Saccharin in the Rat." *Food Chemical Toxicology* 23 (4/5): 475–90.

Seok, Junhee, et al. 2013a. "Genomic Responses in Mouse Models Poorly Mimic Human Inflammatory Diseases." *PNAS* 110 (9): 3507–12.

Seok, Junhee, et al. 2013b. "Genomic Responses in Mouse Models Poorly Mimic Human Inflammatory Diseases." *PNAS* 110 (34): E3151.

Shackel, Nicholas. 2006. "Shutting Dretske's Door." *Erkenntnis* 64 (3): 393–401.

Shanks, Niall, Ray Greek, and Jean Greek. 2009. "Are Animal Models Predictive for Humans?" *Philosophy, Ethics, and Humanities in Medicine* 4:2.

Shannon, Claude, and Warren Weaver. (1949) 1963. *The Mathematical Theory of Communication*. Chicago: University of Illinois Press.

Shapin, Steven, and Simon Shaffer. 1985. *Leviathan and the Air-Pump*. Princeton: Princeton University Press.

Shapin, Steven. 1988. "The House of Experiment in Seventeenth-Century England." *Isis* 79 (3): 373–404.

Shapley, H. 1958. *Of Stars and Men*. Boston: Beacon Press.

Shumelda, Mark. 2013. "At the Limits of Possibility: Thought Experiments in Quantum Gravity." In *Thought Experiments in Science, Philosophy, and the Arts*, edited by Melanie Frappier, Letitia Meynell, and James Robert Brown, 141–63. New York: Routledge.

Simons, Daniel. 2014. "The Value of Direct Replication." *Perspectives on Psychological Science* 9 (1): 76–80.

Smith, Vernon. 1980. "Relevance of Laboratory Experiments to Testing Resource." In *Evaluation of Econometric Models*, edited by Jan Kmenta and James Ramsey, 345–77. New York: Academic Press.

Smith, Vernon. 1982. "Microeconomic Systems as an Experimental Science." *American Economic Review* 72 (5): 923–55.

Snow, John. 1849. *On the Mode of Communication of Cholera*. London: John Churchill.

Sorenson, Roy. 1992. *Thought Experiments*. Oxford: Oxford University Press.

Sprat, Thomas. 1667. *The History of the Royal Society of London*. London: T. R.

Steinle, Friedrich. 2016. "Stability and Replication of Experimental Results: A Historical Perspective." In *Reproducibility: Principles, Problems, Practices, and Prospects*, edited by Harald Atmanspacher and Sabine Maasen, 39–64. Hoboken, NJ: John Wiley & Sons.

Sterrett, Susan. 2017. "Experimentation on Analogue Models." In *Springer Handbook of Model-Based Science*, edited by Lorenzo Magnani and Tommaso Bertolotti, 857–78. Dordrecht: Springer.

Stolberg, M. 2006. "Inventing the Randomized Double-Blind Trial: The Nuremberg Salt Test of 1835." *Journal of the Royal Society of Medicine* 99 (12): 642–43.

Takao, Keizo, and Tsuyoshi Miyakawa. 2015. "Genomic Responses in Mouse Models Greatly Mimic Human Inflammatory Diseases." *PNAS* 112 (4): 1167–72.

Tetens, Holm. 2016. "Reproducibility, Objectivity, Invariance." In *Reproducibility: Principles, Problems, Practices, and Prospects*, edited by Harald Atmanspacher and Sabine Maasen, 13–20. Hoboken, NJ: John Wiley & Sons.

Thaler, Richard. 1980. "Toward a Positive Theory of Consumer Choice." *Journal of Economic Behavior and Organization* 1:39–60.

Thomson, J. J. 1897. "Cathode Rays." *Philosophical Magazine*, 5th ser., 44:293. Reprinted in *Source Book in Physics*, edited by W. F. Maggie, 583–97. Cambridge, MA: Harvard University Press, 1965.

Tiles, J. E. 1993. "Experiment as Intervention." *British Journal for the Philosophy of Science* 44 (3): 463–75.

Todhunter, Isaac. 1873. *The Conflict of Studies*. Cambridge: Macmillan.

Tomer, Christinger. 1992. Review of *Replication Research in the Social Sciences*, edited by James W. Neuliep. *Library Quarterly* 62 (4): 468.

Unruh, William. 1995. "Sonic Analogue of Black Holes and the Effects of High Frequencies on Black Hole Evaporation." *Physical Review D* 51 (6): 2827–38.

Unruh, William. 2008. "Dumb Holes: Analogues for Black Holes." *Philosophical Transactions of the Royal Society A* 366 (1877): 2905–13.

Visser, Matt, and Silke Weinfurtner. 2007. "Analogue Spacetimes: Toy Models for 'Quantum Gravity.'" arXiv:gr-qc/0712.0427v1.

von Herrath, M., and G. Nepom. 2005. "Lost in Translation: Barriers to Implementing Clinical Immunotherapeutics for Autoimmunity." *Journal of Experimental Medicine* 202 (9): 1159–62.

Weber, J. 1967. "Gravitational Radiation." *Physical Review Letters* 18 (13): 498.

Weber, J. 1968. "Gravitational-Wave-Detector Events." *Physical Review Letters* 20 (23): 1307.

Weber, J. 1969. "Evidence for Discovery of Gravitational Radiation." *Physical Review Letters* 22 (24): 1320.

Weber, J. 1970. "Gravitational Radiation Experiments." *Physical Review Letters* 24 (6):276.

Weber, Joseph. 1974. Reply to "Detection of Gravity Waves Challenged." *Physics Today* 27 (12): 12–13.

Wilde, Louis. 1981. "On the Use of Laboratory Experiments in Economics." In *Philosophy in Economics*, edited by J. C. Pitt, 137–48. Dordrecht: D. Reidel.

Winsberg, Eric. 2003. "Simulated Experiments: Methodology for a Virtual World." *Philosophy of Science* 70 (1): 105–25.

Winsberg, Eric. 2010. *Science in the Age of Computer Simulation*. Chicago: University of Chicago Press.

Woodcock, J., and R. Woosley. 2008. "The FDA Critical Path Initiative and Its Influence on New Drug Development." *Annual Review of Medicine* 59:1–12.

Woodward, James. 2003. *Making Things Happen: A Theory of Causal Explanation.* New York: Oxford University Press.

Woodward, James. 2010. "Data, Phenomena, Signal, and Noise." *Philosophy of Science* 77 (5): 792–803.

Wootton, David. 2016. *The Invention of Science: A New History of the Scientific Revolution.* New York: Harper Perennial.

Yong, Ed. 2012. "Replication Studies: Bad Copy." *Nature* 485 (7398): 298–300.

Zagzebski, Linda. 1994. "The Inescapability of Gettier Problems." *Philosophical Quarterly* 44 (174): 65–73.

Index